Machine Learning: Foundations, Methodologies, and Applications

Books published in this series focus on the theory and computational foundations, advanced methodologies and practical applications of machine learning, ideally combining mathematically rigorous treatments of a contemporary topics in machine learning with specific illustrations in relevant algorithm designs and demonstrations in real-world applications. The intended readership includes research students and researchers in computer science, computer engineering, electrical engineering, data science, and related areas seeking a convenient medium to track the progresses made in the foundations, methodologies, and applications of machine learning.

Topics considered include all areas of machine learning, including but not limited to:

- Decision tree
- Artificial neural networks
- Kernel learning
- Bayesian learning
- Ensemble methods
- Dimension reduction and metric learning
- Reinforcement learning
- Meta learning and learning to learn
- Imitation learning
- Computational learning theory
- Probabilistic graphical models
- Transfer learning
- Multi-view and multi-task learning
- Graph neural networks
- Generative adversarial networks
- Federated learning

This series includes monographs, introductory and advanced textbooks, and state-of-the-art collections. Furthermore, it supports Open Access publication mode.

Fengxiang He · Dacheng Tao

Foundations of Deep Learning

Springer

Fengxiang He
School of Informatics
University of Edinburgh
Edinburgh, United Kingdom

Dacheng Tao
College of Computing and Data Science
Nanyang Technological University
Singapore, Singapore

ISSN 2730-9908 ISSN 2730-9916 (electronic)
Machine Learning: Foundations, Methodologies, and Applications
ISBN 978-981-16-8232-2 ISBN 978-981-16-8233-9 (eBook)
https://doi.org/10.1007/978-981-16-8233-9

This Springer imprint is published by the registered company Springer Nature Singapore Pte Ltd.
The registered company address is: 152 Beach Road, #21-01/04 Gateway East, Singapore 189721, Singapore

Preface

In the ever-evolving research field of artificial intelligence, deep learning reigns supreme in the recent over 10 years, casting a spell of awe and wonder with its remarkable feats of intelligence. Imagine a world where machines can perceive, learn, reason, and behave like never before, unraveling the mysteries of data (especially broad data) with an insatiable appetite for knowledge. This is the promise of deep learning—a journey that has captivated the imagination of researchers and practitioners alike, propelling us into a future where the boundaries of possibility are constantly redrawn.

The concept of deep learning was first introduced by Dechter in 1986 and later expanded to brain-inspired algorithms by Aizenberg et al. in 2000. However, the story of deep learning originates from the humble beginnings of neural networks and boasts a rich, multifaceted history spanning several decades. It has gained significant traction in contemporary times, thanks to the visionary individuals who dared to dream of machines capable of learning from experience. Its origins can be traced back to the early 1940s when Warren McCulloch and Walter Pitts proposed a computational model of neural networks, laying the groundwork for what would eventually be recognized as artificial neural networks.

During the 1940s–1960s, the first wave of interest in deep learning, often referred to as "cybernetics" emerged. This period was characterized by pioneering efforts to understand the computational principles underlying biological neural networks. Notably, Frank Rosenblatt's development of the perceptron in 1957 marked a significant milestone, representing the earliest attempt to create a machine capable of learning from its environment.

However, the initial enthusiasm for neural networks waned in the late 1960s. Specifically, Minsky and Papert published the influential work "Perceptrons: An Introduction to Computational Geometry" in 1969, which had a significant impact and argued that Rosenblatt's single-layer neural network could not solve fundamental logic problems such as the "exclusive or" (XOR) problem. This marked the beginning of a period of decline in interest and research activity in the field, often referred to as the "AI winter". Technically, we can also attribute Rosenblatt's failure to the limitations in computing power and the inability to train deeper architectures effectively.

The resurgence of interest in neural networks came about in the 1980s, marking the second wave of deep learning research. This period, characterized as "connectionist", saw the development of more sophisticated neural network architectures and learning algorithms. Notably, the introduction of the backpropagation algorithm by Rumelhart, Hinton, and Williams in 1986 revolutionized the field by enabling efficient training of deep neural networks. More importantly, the significance of this paper lies not only in the proposal of an algorithm but also in a major turning point where neural networks moved from psychology and physiology to the field of machine learning. Despite significant advancements, the limitations of computational resources and the lack of large-scale datasets led to another decline in interest in neural networks during the 1990s, marking the end of the second wave.

The third wave of deep learning, which dawned around 2010, emerged from obscurity with a surge of game-changing advances, shaking the very foundations of artificial intelligence. Leading this charge was Hinton's inspiring research on deep belief networks and unsupervised pretraining in 2006, revolutionizing the landscape of deep neural network training. This momentum culminated in the remarkable emergence of AlexNet, a successor to LeNet, which swept the field with its triumphant victory in the 2012 ImageNet challenge, demonstrating the unparalleled capabilities of deep learning in computer vision. The pioneering efforts by luminaries sparked a revolution, unleashing a torrent of innovation that transcended disciplinary boundaries and reshaped societal norms. The pinnacle of recognition arrived with the 2018 ACM Turing Award, honoring Yoshua Bengio, Geoffrey Hinton, and Yann LeCun for their monumental contributions, firmly establishing deep learning as the vanguard of artificial intelligence innovation.

The success of deep learning is not only attributed to advancements in algorithms but also relies significantly on two critical pillars: the availability of vast amounts of data and the remarkable computational power of GPUs. Deep learning models thrive on large datasets, which provide the diverse and abundant examples necessary for training robust models. Additionally, the emergence of GPUs, particularly those developed by NVIDIA, has revolutionized the field by enabling accelerated training and inference processes. These GPUs are optimized for parallel processing, allowing deep learning algorithms to harness their immense computational capabilities for training complex models with unprecedented speed and efficiency. As a result, researchers and practitioners can explore more sophisticated architectures and tackle increasingly complex problems across various domains, from computer vision to natural language processing.

The impact of deep learning has been nothing short of transformative, permeating every facet of our lives with its boundless potential. From personalized recommendations and intelligent assistants to breakthroughs in healthcare and autonomous vehicles, deep learning has become the driving force behind some of the most awe-inspiring advancements of our time. Yet, amid the dazzling array of practical successes lies a profound challenge—the quest to unravel the mysteries of deep learning's inner working mechanisms.

Therefore, many researchers, drawing parallels to the ancient pursuit of alchemy, view deep learning with skepticism and fascination alike. Like alchemists seeking to transform base metals into gold or discover the elixir of life, practitioners of deep learning aim to discover efficient deep learning models by extensively tuning hyper-parameters. This comparison highlights the mystical nature of deep learning research, which, much like alchemy, involves a blend of experimentation, intuition, and a quest for transformative breakthroughs.

While the practical triumphs of deep learning have dazzled the world, the theoretical foundations upon which they stand remain shrouded in mystery. Many practitioners, enamored by the sheer power of deep learning algorithms, may overlook the critical importance of theoretical studies, dismissing them as mere abstractions divorced from reality. However, the truth is far more tantalizing—for it is through the lens of theory that we gain deeper insights into the workings of these enigmatic networks, unlocking new frontiers of understanding and paving the way for even greater innovations.

As the tide of deep learning continues to rise, fueled by the emergence of foundation models trained on broad data, the need for a solid theoretical foundation becomes more pressing than ever. Theoretical understandings to deep learning networks not only provide insights into the mechanisms governing deep learning networks but also ensure their safe and responsible deployment in real-world applications. Moreover, theoretical studies serve as the crucible in which new ideas are forged, leading to breakthroughs that push the boundaries of what is possible in artificial intelligence.

It is within this crucible that we present this monograph—a testament to the power of theory in unlocking the secrets of deep learning. Drawing upon a rich tapestry of research spanning our relevant journal papers, conference papers, ArXiv technical reports, and Dr. Fengxiang He's doctoral thesis, this monograph endeavors to weave together a narrative that illuminates the theoretical landscape of deep learning. Our goal is to provide readers with a roadmap to understanding the intricate workings of deep learning networks, empowering them to harness the full potential of AI while mitigating the risks inherent in its deployment.

Join us on this exhilarating journey into deep learning theory, where mysteries await to be unraveled and discoveries lie just beyond the horizon. Whether you are a seasoned researcher seeking to deepen your theoretical understanding or a practitioner eager to harness the power of deep learning in your applications, we invite you to embark on this adventure with us. Together, let us chart a course towards a future where artificial intelligence knows no bounds, guided by the light of theory and fueled by the fires of imagination.

Before introducing the organizational structure of this monograph, it's necessary to touch upon the concept of explainable deep learning. This field represents a fusion of empirical strategies and theoretical rigor, fostering a space for enlightenment and discovery. While empirical methods have driven significant advancements in understanding and interpreting deep learning models, their focus lies predominantly on elucidating and dissecting the performance of the output and intermediate layers, and their correlations with input data. Techniques for studying explainable deep learning typically encompass feature importance analysis and visualization methodologies, aiming to achieve both local and global interpretability. By integrating these approaches to offer multi-layered explanations, users gain a deeper understanding of the behavior and predictive outcomes of deep learning models. This monograph primarily seeks to explore theoretical underpinnings and enhance interpretability within the realm of deep learning. Through the crucible of theory, we aim to provide readers with a thorough understanding of the theoretical underpinnings that drive the success of deep learning models.

Specifically, this monograph is a general introduction to deep learning theory that can serve as a textbook for students and researchers in machine learning, statistics, and other related areas. The reader is assumed to be familiar with basic concepts in linear algebra, probability, and analysis of algorithms. It covers fundamental modern topics in deep learning theory while providing the theoretical basis and conceptual tools needed for the discussion and justification of algorithms. It also describes several key aspects of the application of these algorithms.

The journey of this monograph begins with an introduction to the current status of deep learning and its theory in Chap. 1. Chapters 2–4, forming Part I, review the previous developments in statistical learning theory, which may have a significant discrepancy from the requirements of developing deep learning theory.

In Part II, Chaps. 5–10, we present recent advances in deep learning theory from six facets: (1) complexity and capacity-based approaches for analyzing the generalizability of deep learning; (2) stochastic differential equations and their dynamic systems for modeling stochastic gradient descent and its variants, which characterize the optimization and generalization of deep learning, partially inspired by Bayesian inference; (3) the geometrical structures of the loss landscape that drives the trajectories of the dynamic systems; (4) the geometrical structures of the input space, shaped by the nonlinear activations; (5) the roles of over-parameterization of deep neural networks from both positive and negative perspectives; and (6) theoretical foundations of several special structures in network architectures.

In Part III, Chaps. 11–13, we present some emerging directions beyond the generalizability which is focused on the previous chapters. We first explore trustworthy deep learning, encompassing the growing concerns in ethics and security, and their relationship to generalizability. Finally, the emerging supermodel paradigm or foundation model trained on broad data is discussed.

Throughout this monograph, we aim to provide readers with a comprehensive understanding of deep learning theory, from its foundational principles to its latest advancements and ethical considerations. We hope that this exploration will inspire further research and innovation in the field of deep learning, driving progress towards a future where intelligent systems enhance our lives in guaranteed ways.

While we have strived for accuracy, this monograph might still not be perfect. If you encounter any typos, factual errors, citation inconsistencies, and reference missings, please don't hesitate to bring them to our attention. Your feedback is important to ensure the accuracy and clarity of this work for future readers.

From Sydney, Beijing to Singapore and Edinburgh
2019–2024

Fengxiang He
Dacheng Tao

Contents

Part II Deep Learning Theory

Chapter 1
Deep Learning: A (Currently) Black-Box Model

Deep learning refers to a subset of machine learning techniques that employ artificial neural networks characterized by multiple layers of interconnected nodes (or neurons) followed by nonlinear activation functions. Such networks are of significant size. These deep neural networks are designed to process and learn from experience, extracting complex patterns and features through successive layers of computation. Such experience can be obtained either from human-annotated electronic records such as datasets or from the learner's own interactions with its perceived environment. By leveraging these layered architectures, deep learning models can autonomously discover and represent intricate relationships within the data, enabling them to make sophisticated predictions and informed decisions based on learned knowledge.

To date, however, the development of deep learning has relied heavily on experiments that have not thoroughly explored or sought an understanding of its theoretical foundations. The black-box nature of deep learning introduces unmanageable risk into its applications since we do not know why deep learning works, when it may fail, or how to prevent failures.

1.1 Definitions and Terminology

A prevalent application of deep learning is object recognition, which aims to categorize images based on their characteristics. This section takes object recognition as an example to introduce the definitions and terminology used in this book.

- *Example*: Let $\mathcal{X}$ and $\mathcal{Y}$ be the input and output space, respectively. A datum used in the process of training and evaluation/inference. An example is usually defined as a feature–label pair

$$(x, y) \in \mathcal{X} \times \mathcal{Y},$$

F. He and D. Tao, *Foundations of Deep Learning*, Machine Learning: Foundations, Methodologies, and Applications, https://doi.org/10.1007/978-981-16-8233-9_1

where x is the feature and y is the corresponding label. For object recognition, images and their representations are the features, and the labels are the concepts depicted such as 'tiger', 'horse', and 'cheetah'.

- *Feature*: A feature is a collection of attributes that represents an example and is typically structured as a vector or matrix denoted as $x \in \mathcal{X}$. Features are often obtained through sensor measurements, such as those from optical cameras or light detection and ranging (LiDAR) devices. Certain deep learning models are specifically engineered for feature extraction, as they excel at producing highly informative feature representations.
- *Label*: An annotation of an example. It can be a categorical value, a continuous real value, or a combined set of such values. Deep learning models learn mappings between features and labels. In this book, we denote a label by $y \in \mathcal{Y}$.
- *Training sample*: A set of examples used to train a deep learning model:

$$S = \{z_1, \ldots, z_m\},$$

where $z_i = (x_i, y_i) \in \mathcal{Z} := \mathcal{X} \times \mathcal{Y}$ and m is the training sample size. Usually, deep learning used large-scale training samples.
- *Validation sample*: A set of examples used to validate various hyperparameters and techniques when training a deep learning model.
- *Test sample*: A set of examples used to evaluate a deep learning model.
- *Data distribution*: In typical scenarios, we make the assumption that all examples within the training, validation, and test samples are independent and identically distributed (i.i.d.) random variables drawn from the same underlying data distribution, denoted as $\mathcal{D}$, i.e.,

$$Z \sim \mathcal{D}, \ S \sim \mathcal{D}^m.$$

For clarity, we use upper-case and lower-case letters to represent random variables and their corresponding observations, respectively.

Remark 1.1 It's crucial to acknowledge that the assumption of i.i.d. data, where each data point is assumed to be drawn from the same distribution and is unrelated to other data points, may not hold in many real-world scenarios. A notable counterexample is observed in time series data, such as financial data, where observations can exhibit significant temporal dependencies and variability over time. Despite this departure from the i.i.d. assumption, deep learning models can still effectively capture complex patterns and trends in such data due to their ability to learn from sequential information. Although the temporal variability in time series data challenges the i.i.d. assumption, it typically does not critically impede the performance of deep learning models. Therefore, in this book, we maintain the use of the i.i.d. assumption for simplicity and to establish the fundamental principles and theoretical foundations of deep learning, recognizing that the flexibility and adaptability of these models allow them to handle real-world data complexities effectively.

Deep learning applications can be broadly categorized as follows.

- *Supervised learning*: In supervised learning, every training example is annotated or labelled. The form of annotation varies in different scenarios. In image classification, the annotations are image categories; in object detection, the annotations are the bounding boxes around different objects; and in semantic segmentation, all pixels are individually annotated.
- *Unsupervised learning*: In contrast, the training data remain unannotated in unsupervised learning, and the corresponding deep learning methods are designed for clustering, feature extraction, and reconstruction. Self-supervised learning is a special form of unsupervised learning in which knowledge, or, more specifically, label concepts, is learned through self-training without external interference. This learning regime is becoming increasingly popular, especially for training large-scale deep networks in super deep learning, an advanced era of deep learning. Unsupervised learning is attractive because of the often high expense of data labelling.
- *Semisupervised learning* and *weakly supervised learning*: These two learning regimes represent intermediate approaches that bridge the gap between fully supervised and unsupervised learning paradigms. Semisupervised learning (SSL) combines both labeled and unlabeled data for training predictive models. Most SSL approaches are based upon one of the following several key assumptions: smoothness (similar inputs produce similar outputs), clustering (samples in the same cluster have the same label), and the manifold assumption (data points lie on a lower-dimensional manifold where neighboring points have similar labels). Popular SSL techniques include pseudo-labeling, self-training, co-training, and graph-based models. On the other hand, weakly supervised learning (WSL) trains models using limited yet imprecise labels, which helps reduce the cost of exact and extensive data labeling. WSL approaches can be categorized into three main types: inexact supervision (coarse-grained labels), incomplete supervision (only part of the data is labeled), and inaccurate supervision (labels may be incorrect). Classical learning techniques such as multi-instance learning, active learning, and transductive learning have been integrated with the above WSL types. These methods leverage available data to enhance model performance despite the weak supervision. Both learning techniques play crucial roles in practical machine learning applications, offering solutions for leveraging partially labeled or imperfectly annotated data to train effective models with reduced human annotation effort.

Remark 1.2 In either supervised or self-supervised learning, a deep learning method learns a mapping from features space $\mathcal{X}$ to output space $\mathcal{Y}$. This idea is inherited from conventional machine learning algorithms.

Remark 1.3 Deep learning has shown significant value in dealing with the challenges arising in conventional reinforcement learning (RL), which is an important machine learning paradigm. In RL, a learner determines what action an agent needs to take in each specific state of the environment. Deep RL exploits deep learning for efficient feature extraction to boost the policy approximation ability in RL, which is a considerable barrier for conventional methods.

Supervised learning plays a significant role in deep learning applications. It has shown promising performance in the following tasks.

- *Classification*: The goal of classification is to assign an unordered categorical value to each example; i.e., the labels can be represented by integers $\mathcal{Y} \subset \mathbb{N}$. All categories are equal in status, with no preference among them. Categories are commonly defined using integers for convenience. Sometimes, a single example is simultaneously associated with multiple categories; we refer to this learning regime *multilabel classification*. In this regime, $\mathcal{Y} \subset \mathbb{N}^l$, where l is the number of possible categories for each example.
- *Ranking*: The goal of ranking is to assign a specific ordered categorical value to each example, where $\mathcal{Y} \subset \mathbb{N}$ represents the set of possible ordered categories. In contrast to classification, which assigns discrete labels to instances, ranking involves organizing a sample into a partially ordered sequence or hierarchy based on their attributes or features. This distinction reflects a fundamental difference in the objectives of ranking versus classification tasks. In ranking tasks, the emphasis is on establishing a relative order or preference among examples rather than assigning definitive class labels. This approach is common in information retrieval systems, recommendation systems, and personalized search engines, where the goal is to present items or documents in an order that maximizes relevance or user satisfaction. By considering partial orderings, ranking methods allow for more nuanced and flexible representations of data compared to strict classification schemes.
- *Regression*: The goal of regression is to predict continuous values or vectors for a given example: $\mathcal{Y} \subset \mathbb{R}^l$, where l is the dimensionality of the labels. In most application scenarios, these continuous values or vectors are real. However, complex-valued data can also be encountered. While traditionally used for regression tasks, regression analysis has also played a significant role in classification problems. Logistic regression exemplifies this, by formulating classification as a process of fitting a logistic function, which is technically a regression problem. The logistic function, also known as the sigmoid function, is defined as follows:

  $$f(x) = \frac{L}{1 - e^k(x - x_0)},$$

  where $x \in \mathbb{R}$ is a real number, $L, k \in \mathbb{R}^+$ are two positive real constants, and $x_0 \in \mathbb{R}$ is a real constant. The high-dimensional version is termed the *softmax function*, which is defined as follows:

  $$f(\mathbf{b}) = \frac{e^{\mathbf{b}}}{\sum_{j=1}^{l} e^{b_j}},$$

where $\mathbf{b} = (b_1, \ldots, b_l) \in \mathbb{R}^l$ is a real vector and $l \in \mathbb{N}^+$ is the length of $\mathbf{b}$. Deep learning models for classification tasks usually adopt the softmax function for activation in the output layer.

Notably, deep learning breaks free of the framework of learning a mapping from features to labels.

Unsupervised learning is also of significant interest for deep learning theory. Prevalent scenarios are listed as follows.

- *Feature extraction*: The goal of feature extraction is to transform the original data into a more compact and informative feature space that captures essential characteristics for subsequent analysis or modeling tasks. One effective heuristic approach to feature extraction involves utilizing the outputs of intermediate layers within a neural network as extracted features. These intermediate outputs, often referred to as mid-products, represent the activations and transformations applied to the input data as it passes through the network. By extracting features from these intermediate representations, we can leverage the hierarchical and abstract information learned by the neural network during training. Using mid-products as extracted features offers several advantages. First, these features encapsulate complex patterns and relationships present in the data, learned through the network's layers. Second, they provide a more meaningful and compact representation compared to raw input data, which can be noisy or high-dimensional. Finally, extracted features can be used as inputs to downstream tasks such as classification, regression, or clustering, facilitating more effective and efficient model training and inference. By and large, feature extraction using mid-products from neural network layers is powerful to capture and leverage hierarchical information encoded in data, enabling improved performance and interpretability in machine learning applications.
- *Data generation*: It aims to produce new examples that share the same distribution learned from the training data. Recent advancements, such as Large language models (LLMs), diffusion models (DMs) and generative adversarial networks (GANs), have proven their superior capabilities to generate new data from the given training data. LLMs understand text rules and patterns by integrating transformer blocks and training on vast amounts of text data. As the training data and model parameters increase, the model's language capabilities continually improve, even exhibiting emergent capabilities. LLMs possess unique in-context learning capabilities and can handle novel tasks efficiently and cost-effectively through techniques like supervised fine-tuning and prompt engineering. GPT is a recently emerged and widely used type of LLM. GANs are constructed by a generator and a discriminator. The generator produces synthetic data by emulating the patterns it has learned from the training data, while the discriminator performs as an evaluator, validating the authenticity of the generated data by comparing with real data. The training process progressively teaches the generator to produce new data with distributions approach to that of the training data, while the discriminator continually improves its capability to separate synthetic data from real data. VAEs learn to encode and

decode training data. The encoder maps input data points to a latent space representation, while the decoder takes a point from the latent space and reconstructs the original input data point. Therefore, VAEs are capable to generate new data by decoding low-dimensional points efficiently sampled from the latent space into meaningful outputs. This method is widely useful to synthesise images and audio waves, as it allows for the generation of diverse and creative outputs by manipulating the latent representations. DMs have recently been widely used to generate new data by mimicking training data through two steps: forward diffusion and reverse diffusion. In forward diffusion, noise is progressively added to an image over several steps, gradually transforming it into a completely noise-like picture. By contrast, reverse diffusion learns to invert the forward diffusion process. Starting from the noisy image, the model iteratively denoises it, to progressively reconstruct the original image. The above data generation models effectively learn to capture and replicate the sophisticated patterns present in the distribution of the training data. These generators aim to generate data instances that are indistinguishable from real data. Although these models have found widespread applications in tasks such as image generation, question answering, or broadly the synthesis of realistic data for various purposes, including deep learning model training, they face many challenges, such as data bias, low reliability, poor quality, scalability, ethical and privacy concerns, as well as interpretability and transparency.

- *Self-supervised learning*: It aims to autonomously discover the underlying structures in data without relying on manually annotated labels. Contrastive learning is a typical self-supervised learning, which improves the feature representation capability of deep learning models by using large amounts of unlabelled data. By simultaneously maximizing intra-class similarity and minimizing inter-class dissimilarity, it improves the model's representation capability, ensuring that the feature representations of data in the same category are alike and those of data from different categories are as distinct as possible. In general, it does not primarily focus on the intricate details of the instances but instead aims to distinguish the data at an abstract semantic feature space level. Consequently, the model is simplified, its optimization is more efficient, and its generalization ability is improved. For example, in the context of self-supervised learning like RotNet, images are artificially augmented by rotating them with specific angles (e.g., 0 degrees, 90 degrees) to create pairs of rotated and original images. The rotation angles themselves are used as labels to train the model to predict the rotation applied to each image. While these rotation labels may appear arbitrary or devoid of semantic meaning, the process allows the model to learn rich and informative features that capture intrinsic properties of the data, such as object shape, orientation, and texture. Representative algorithms in contrastive learning include: SimCLR: Enhances feature representation through a simple contrastive learning framework by using data augmentation and large-scale batch training; MoCo: Improves the efficiency of contrastive learning by storing sample pairs in a dynamic dictionary through a momentum contrast mechanism; and BYOL: Enhances the stability of contrastive learning by introducing a self-supervised mechanism without the need for negative samples. In order to define supervisory signals without human intervention, self-supervised learning explores and exploits the inherent structure of the data. By generating labels from data transformations or relationships (such as image rotations in RotNet), self-supervised learning helps the model learn effective representations that generalize well to downstream tasks, even in the absence of explicit labels. This approach has been extensively applied to computer vision, natural language processing, speech recognition, and bioinformatics, demonstrating its versatility and scalability in learning from unlabeled data.

1.2 Advances in Deep Learning

Compared with deep neural networks, most conventional machine learning models share four major barriers to achieving comparably elegant deep learning performance.

- Conventional machine learning models/algorithms usually have significantly lower representation capacity. Their capacity is sufficient for modelling *simple* data, such as what is necessary for income modelling and credit assessment, but considerably limits their ability to fit *complex* data, such as images and videos. For example, the vanilla support vector machine (SVM) algorithm can be used to train only linear classifiers. In contrast, deep learning models have a *uniform approximation ability*: they can approximate any continuous function, time series, or distribution at any accuracy as long as the networks are sufficiently wide. In addition, *stochastic gradient*-based optimizers, including *stochastic gradient descent* and its variants, have excellent capabilities for optimizing neural networks. Moreover, representation learning proceeds in a data-driven manner, which significantly reduces the human labour burden incurred.
- Conventional machine learning algorithms usually suffer from the *curse of dimensionality*: to maintain the robustness of a learning model/algorithm, the required training sample size increases rapidly as the dimensionality of the input data increases. In contrast, existing empirical studies have not yet observed the curse of dimensionality in deep learning. In fact, increasing the input dimensionality has been shown to boost performance. For example, some papers have reported that training on higher-resolution images can lead to considerably higher performance, even as the input dimensionality significantly increases.
- Conventional machine learning algorithms usually rely on restrictive assumptions, considerably limiting their potential application domains. For example, linear classifiers assume that the data are linearly separable, hidden Markov models assume that the state sequence is a Markov chain, and kernel methods assume that the data lie in the corresponding reproducing kernel Hilbert space. In contrast, deep learning breaks free of most of these assumptions.
- Some conventional machine learning algorithms have limited capabilities in dealing with large-scale data. A prime example is Markov chain Monte Carlo, a prevalent Bayesian inference method that faces significant difficulty in scaling to the processing of big data.

- Deep learning algorithms usually operate in an end-to-end manner. Compared with their multistage counterparts, end-to-end methods have two major advantages: (1) End-to-end models are usually more reliable. The risks from each single stage accumulate in multistage systems, which can eventually become unmanageable. (2) End-to-end models are usually easy to train. Their end-to-end nature enables training of the whole system in accordance with a single objective function. In contrast, the optimal protocols for the individual modules in a multistage system may vary significantly. Tuning a multistage system is thus significantly more difficult than tuning an end-to-end system.
- As a family of representation learners, deep learning models can be easily integrated with other approaches. The raw data can be easily replaced with their representations learned from a deep neural network. This algorithm design approach has been extensively utilized in computer vision and natural language processing to considerably accelerate the pace of innovation.

1.3 Applications of Deep Learning

Significant developments have emerged in the field of deep learning over the past decade. The rapid progress in deep learning is empowering abundant developments in many, possibly almost all, areas of both engineering and science and is further driving a technological revolution in a variety of industry sectors.

- *Computer vision* (CV) is the field of technology that seeks to help machines "see" and understand a digital form of the world, including images and videos captured by cameras and point clouds generated by LiDAR sensors. The primary subareas of CV include recognition, detection, segmentation, and tracking. Potential applications involve a variety of industry sectors, including autonomous vehicles, medical diagnosis, and entertainment. The outstanding approximation capabilities and trainability of deep learning models enable deep learning to vastly enhance performance in CV tasks, providing improved speed and accuracy.
- *Natural language processing* (NLP) is the field of training machines to read, understand, and derive meaning from natural languages. Machines are good at dealing with machine languages or programming languages, which follow clear and concise logic. Unfortunately, this characteristic does not apply to natural languages. Natural languages exhibit considerable ambiguity, vagueness, and violations of the *grammar* rules as summarized in linguistics;[1] this, however, is exactly the wisdom and charm of natural languages. Deep learning models, particularly long short-term memory (LSTM) networks and recurrent neural networks (RNNs), enable data-driven approaches to the discovery of knowledge from natural languages. Prevalent subareas of NLP include information retrieval, text mining, dialogue

[1] Theoretically, it would be possible to summarize the grammar of a natural language in its entirety, as per finite length, the capacity of all potential sentences is finite. However, such a grammar book would be unacceptably large.

and interactive systems, machine translation, question answering, syntax analysis (tagging, chunking, and parsing), sentiment analysis, stylistic analysis, and argument mining.

- *Speech processing* is the field of understanding the contents of speech and translating speech into text or other forms. One conventional method is a hidden Markov model, which models every spoken sentence as a *sequence of observations*. This method assumes the presence of a *state sequence*, which is assumed to be a Markov chain, underlying the observation sequence and carrying the desired information. However, the Markov chain assumption is rarely stringent and is rather flexible. Fortunately, deep learning models, such as LSTM and RNN models, do not rely on such assumptions and have therefore emerged as essential techniques in dealing with speech processing tasks. Prevalent applications include automatic translation, real-time speech writing, real-time captioning, and virtual assistants (such as Apple's Siri).
- *Recommender systems* seek to rank products or service options in accordance with consumer preference. Such systems been extensively deployed by advertising providers and e-commerce corporations, including Google, Amazon, and JD.com. Formally, recommender systems aim to solve ranking problems.
- *Autonomous vehicles* are automobiles that can travel either under limited human control or entirely without human control. It is estimated that by 2040, autonomous vehicles will occupy over 40% of the entire automobile market. Based on the level of autonomy, a hierarchy has been introduced: Level 0: humans take full responsibility for driving; Level 1, "*hands on*": machines and humans share vehicle control; Level 2, "*hands off*": machines normally have complete control, but humans need to oversee the driving process and constantly remain ready to intervene; Level 3, "*eyes off*": machines can override human control in response to emergencies, while humans need to oversee the driving process and "advise" the machines in taking action; Level 4, "*mind off*": machines take complete responsibility for vehicle control without any human oversight in limited scenarios; Level 5, "*steering wheel optional*": machines have complete vehicle control at all times and human intervention is not needed. Autonomous vehicles seamlessly integrate technologies from several areas, including manufacturing, recognition, reasoning, and cybernetics. Deep learning contributes significantly to the last three of these areas. Specifically, techniques for various CV tasks are needed, such as segmentation, object detection, tracking, and depth estimation. In addition, deep learning has emerged as an efficient representation learning tool that enables the deployment of RL to control vehicles.
- *Autonomous medical diagnosis* aims to teach machines to diagnose disease and thereby improve the affordability and accuracy of medical diagnosis. The methods involved heavily rely on many key CV and NLP technologies, including classification, detection, and segmentation. The extensive applications of deep learning in CV have significantly accelerated the development of autonomous medical diagnosis. However, the U.S. Food and Drug Administration (FDA) has yet to authorize any autonomous diagnosis system for direct medical use. This lack of approval can be attributed to the fact that in the current state of research on deep learning,

theoretical foundations and rigorous empirical verifications, such as hypothesis tests, are largely missing.[2] This situation highlights that investigations of deep learning theory and rigorous empirical verifications are essential.
- *Drug discovery* refers to the discovery of novel medications. Neural networks, particularly *graph neural networks* (GNNs), show promise in the discovery of knowledge from data expressed in a graph-style format, which is closely analogous in form to the molecular formulas of medications.

1.4 The Status Quo of Deep Learning Theory

Many applications of deep learning lie in extremely security-critical areas, such as autonomous vehicles and medical diagnosis. In these areas, even a small algorithmic error can lead to fatal disasters. Consequently, deep learning is required to be transparent and accountable. Intuitively, the theoretical foundations of deep learning algorithms should serve as a major source of transparency and accountability.

To date, however, the development of deep learning has relied heavily on experiments that have not thoroughly explored or sought an understanding of its theoretical foundations. Although unstable at times, heuristic approaches in deep learning have achieved significant performance in many areas. The success of deep learning has been both astonishing and baffling, as researchers have yet to comprehend the mechanism that governs deep learning's behavior. The black-box nature of deep learning introduces unmanageable risk into its applications since we do not know why deep learning works, when it may fail, or how to prevent failures. This knowledge gap considerably compromises our ability to identify, manage, and prevent algorithm-led disasters, severely undermines confidence in the application of recent advances in many industrial sectors, and restricts the further development of innovative machine learning algorithms. In particular, deep learning faces three major problems:

- *Approximation* refers to the ability of a machine learning model to fit the training data. Conventional machine learning models exhibit a relatively limited ability to approximate data. For example, a vanilla SVM can fit only linearly separable data. Nevertheless, several works in the 1980s and 1990s have proven universal approximation theorems for neural networks: *if the width is sufficiently large, even a neural network with only one hidden layer can approximate any continuous function at any precision.* What is the bound on the approximation ability of a feed-forward neural networks if the network is of regular width? What are the approximation bounds for RNNs and GANs?

[2] The lack of hypothesis tests is not merely due to negligence. It is honestly doubtful that current deep learning algorithms could pass hypothesis tests if they were performed. Most deep learning algorithms exhibit significant variability in performance, and partial reporting is common in current experimental practice, which severely undermines any confidence in the reproducibility of deep learning algorithms.

- *Optimization* refers to the ability to search for the optimal solution to a problem. Machine learning (including deep learning) algorithms are usually formulated to solve optimization problems. Thus, their optimization performance is of vital importance. Unfortunately, the optimization problems in deep learning tend to be highly nonconvex. Nevertheless, they can be well solved in practice by means of stochastic-gradient-based methods, such as stochastic gradient descent (SGD), which is apparently a stochastic convex optimization method. Why does SGD, a convex optimization method, work for a nonconvex optimization problem?
- *Generalization* refers to the ability to perform well on unseen data. Good generalizability ensures that a trained model can robustly handle unseen circumstances. Generalizability is particularly important when long-tail events regularly appear and can cause catastrophic failure. Statistical learning theory has established a rigorous upper bound on the generalization error depending on the hypothesis complexity; however, major difficulties arise in analysing deep learning.

Theoretical foundations have been established for conventional machine learning algorithms. However, significant difficulties arise when we attempt to apply these theoretical foundations to deep learning. These difficulties are listed as follows:

- *Overparameterization.* Statistical learning theory is grounded in *Occam's razor principle*:

Plurality should not be posited without necessity.

This principle emphasizes the importance of simplicity in model design to prevent unnecessary complexity. According to Occam's razor, a model should be as simple as possible while still effectively capturing the patterns present in the training data. This principle ensures that generalization bounds in statistical learning are tied to the complexity of the hypothesis space used in machine learning scenarios. However, deep learning models often appear to diverge from Occam's razor due to their inherent complexity. Deep neural networks typically comprise numerous trainable parameters, sometimes exceeding the size of the training dataset. This phenomenon challenges conventional generalization guarantees because excessively complex models can memorize specific examples from the training data rather than learning generalizable patterns. As a result, deep learning models are suspected to be susceptible to overfitting, where they perform well on training data but fail to generalize to unseen examples.

Remark 1.4 Before the era of foundation models (large-scale deep neural networks trained on broad data), the complexity of deep learning models necessitates careful regularization techniques and model selection strategies to mitigate overfitting and promote generalization. Techniques such as dropout, weight regularization, and early stopping are employed to simplify and regularize deep neural networks, encouraging them to learn essential features while suppressing noise and irrelevant details. Balancing model complexity with generalization capability remains a fundamental challenge in deep learning research, highlighting the need for continued exploration of regularization methods and theoretical insights to improve model robustness and performance.

Remark 1.5 In the era of foundation models, state-of-the-art models often do not impose strict constraints on model complexity. On the contrary, larger models have shown greater effectiveness on testing benchmarks, which may exceed the size of the training set by a significant margin. The underlying reason for this phenomenon can be attributed to the capacity of these larger models to capture and learn intricate patterns and nuances present in vast and diverse datasets. Additionally, larger models possess more parameters and computational power, enabling them to represent complex relationships and dependencies within the data more comprehensively. This enhanced capacity often leads to improved performance and generalization capabilities, especially in tasks involving extensive and diverse data inputs. Surely, it is still important to balance model size with considerations of efficiency, interpretability, and practical deployment in real-world applications.

- *Highly complex empirical risk landscape.* Highly complex empirical risk landscape. The landscape of empirical risk in deep learning is characterized by significant complexity, posing unique challenges compared to traditional machine learning paradigms. In conventional settings, learnability and optimizability are typically ensured through regularization techniques that leverage properties like convexity, Lipschitz continuity, and differentiability. However, these conventional guarantees do not directly translate to deep learning due to the intrinsic nature of neural networks. Deep neural networks are composed of multiple layers with nonlinear activation functions, leading to loss surfaces that exhibit extreme nonsmoothness and nonconvexity. This inherent complexity renders traditional optimization guarantees ineffective and raises significant obstacles in understanding the underlying principles of deep learning. The complex and rugged nature of deep learning loss surfaces has long been a barrier to developing comprehensive theories and analytical frameworks for neural network behavior. Recent progress in deep learning research has shifted the focus towards investigating the geometric properties of loss surfaces as a means to decode the behavior of neural networks. It is increasingly recognized that the intricate geometry of loss surfaces directly reflects the dynamics of learning and generalization in deep learning models. Exploring and characterizing the geometric features of loss landscapes holds promise as a pathway to gaining deeper insights into the working mechanisms of neural networks. By delving into the geometric intricacies of loss surfaces, researchers aim to uncover fundamental principles governing optimization, generalization, and model performance in deep learning. This evolving perspective highlights the importance of advancing our understanding of loss surface geometry to enhance the robustness, efficiency, and interpretability of deep learning systems.

1.5 Notations

Let $S = \{(x_1, y_1), \ldots, (x_m, y_m) | x_i \in \mathcal{X} \subset \mathbb{R}^{d_X}, y_i \in \mathcal{Y} \subset \mathbb{R}^{d_Y}, i = 1, \ldots, m\}$ be a training sample set, where d_X and d_Y are the dimensionalities of the feature X and the label Y, respectively. All $\{z_i\}_{i=1}^m$ are i.i.d. observations of the random variable

$Z = (X, Y) \in \mathcal{Z}$. Deep learning algorithms aim to learn a hypothesis (denoted by h) from training data. All possible such hypotheses form the hypothesis space, denoted by $\mathcal{H}$. The mathematical structure of a neural network is usually defined in the following form:

$$x \mapsto W_D\sigma_{D-1}(W_{D-1}\sigma_{D-2}(\ldots\sigma_1(W_1x))), \ x \in \mathcal{X}$$

where D is the depth of the neural network, W_j is the j-th weight matrix, and σ_j is the j-th nonlinear activation function for any $j = 1, \ldots, D$. Usually, we assume that the activation function σ_j is a continuous function between Euclidean spaces and that $\sigma_j(0) = 0$. Popular activation functions include the softmax, sigmoid, and tanh functions.

The generalization bound measures the generalizability of an algorithm. For any hypothesis h learned by an algorithm $\mathcal{A}$, the expected risk $R(h)$ and the empirical risk $\hat{R}_S(h)$ with respect to the training sample set S are respectively defined as follows:

$$R(h) = \mathbb{E}_z l(h, z), \ \hat{R}_S(h) = \frac{1}{m}\sum_{i=1}^{m} l(h, z_i),$$

where $l(\cdot)$ is the loss function. Furthermore, when the output hypothesis h learned by algorithm $\mathcal{A}$ is stochastic (i.e., exhibits randomness), we typically calculate the expectations of both the expected risk, $\hat{R}(h)$, and the empirical risk $\hat{R}_S(h)$. This involves averaging over the randomness introduced by algorithm $\mathcal{A}$.

$$R(\mathcal{A}) = \mathbb{E}_{\mathcal{A}(S)} R(\mathcal{A}(S)), \ \hat{R}_S(\mathcal{A}) = \mathbb{E}_{\mathcal{A}(S)} \hat{R}_S(\mathcal{A}(S)),$$

where $\mathcal{A}(S)$ is the hypothesis learned by algorithm $\mathcal{A}$ on the sample S.

Part I
Background

Chapter 2
Statistical Learning Theory

This chapter introduces mathematical tools for establishing the foundations of deep learning and, more broadly, machine learning. We first introduce Glivenko-Cantelli theorem and concentration inequalities, which characterizes the convergence of an empirical process in the context of convergence in probability. The Glivenko-Cantelli theorem and concentration inequalities also inspire the concept of the sample complexity required to ensure a desirable level of generalizability on unseen data. We then discuss the Probably Approximately Correct (PAC) learning framework, which characterizes learning algorithms that can learn a target concept in an appropriate amount of time given a sufficiently high sample complexity. The PAC framework has become the foundation of statistical learning. Usually, PAC is categorized in the 'frequentist' sense, which suffers from significant deficiency in studying stochastic algorithms. To address this issue, the PAC-Bayes framework is also discussed, which integrates the PAC framework with Bayesian methods to characterize randomness.

2.1 Glivenko-Cantelli Theorem and Concentration Inequalities

This section discusses the convergence of an empirical process. We first introduce Glivenko-Cantelli theorem based on measure theory, which characterizes whether an empirical process asymptotically converges. Then, concentration inequalities are introduced as a tool for evaluating the convergence speed. Please note that some concentration inequalities also imply asymptotic convergence of a sequence.

F. He and D. Tao, *Foundations of Deep Learning*, Machine Learning: Foundations, Methodologies, and Applications, https://doi.org/10.1007/978-981-16-8233-9_2

2.1.1 Glivenko-Cantelli Theorem

This section introduces Glivenko-Cantelli theorem (or central limit theorem) based on the measure theory. Measure theory offers a portfolio of rigorous tools for mathematically characterizing the convergence of a sequence of hypotheses. However, this rigour is usually accompanied by technical difficulty in practice. We thus also introduce concentration inequalities in the next section to simplify the reader's understanding, which may also characterize the convergence.

We first recall some necessary definitions in the measure theory. Readers can also consult Stein and Shakarchi (2009) for reference. Measure is defined as follows.

Definition 2.1 Let $\mathcal{X}$ be the sample space. A collection $\mathcal{A}$ of subsets of $\mathcal{X}$ is called a $\sigma-$field if it satisfies the following conditions:

- $\varnothing \in \mathcal{A}$;
- if $A_1, A_2, \cdots \in \mathcal{A}$ then $\bigcup_{i=1}^{\infty} A_i \in \mathcal{A}$;
- if $A \in \mathcal{A}$ then $A^c \in \mathcal{A}$.

The tuple $(\mathcal{X}, \mathcal{A})$ is called a measurable space. A subset A of $\mathcal{X}$ is called measurable if $A \in \mathcal{A}$. Moreover, a function $f : \mathcal{X} \to \mathbb{R}$ is called measurable, if for all $a \in \mathbb{R}$, the set $f^{-1}([-\infty, a)) = \{x \in \mathcal{X} : f(x) < a\}$ is measurable.

We then may define probability measure.

Definition 2.2 A probability measure $\mathbb{P}$ on a measurable space $(\mathcal{X}, \mathcal{A})$ is a function $\mathbb{P} : \mathcal{A} \to [0, 1]$ satisfying

- $\mathbb{P}(\varnothing) = 0, \mathbb{P}(\mathcal{X}) = 1$;
- if $A_1, A_2, \ldots$ is a collection of disjoint members of $\mathcal{A}$, then

$$\mathbb{P}\left(\bigcup_{i=1}^{\infty} A_i\right) = \sum_{i=1}^{\infty} \mathbb{P}(A_i).$$

The triple $(\mathcal{X}, \mathcal{A}, \mathbb{P})$ is called a probability space.

The empirical measure $\mathbb{P}_m$ of a sample of random elements $X_1, \ldots, X_m$ on a measurable space $(\mathcal{X}, \mathcal{A})$ is defined as $\mathbb{P}_m(C) = m^{-1}\#(1 \le i \le m : X_i \in C)$. Alternatively, if the points are measurable, this measure is characterized as a random measure that imposes a mass $1/m$ at each point.

The following defined Dirac measure is critical in probability theory.

Definition 2.3 The Dirac measure δ_x is defined by

$$\delta_x(A) = \begin{cases} 1 \text{ if } x \in A \\ 0 \text{ if } x \notin A \end{cases}$$

Given the definition of the Dirac measure, the empirical measure is usually represented as a linear combination of the Dirac measures at the observations, $\mathbb{P}_m = m^{-1} \sum_{i=1}^{m} \delta_{x_i}$.

Given a collection $\mathcal{F}$ of measurable functions $f : \mathcal{X} \to \mathbb{R}$, the empirical measure induces a map from $\mathcal{F}$ to $\mathbb{R}$ given by

$$f \mapsto \mathbb{P}_m f = \frac{1}{m}\sum_{i=1}^{m} f(x_i).$$

Let P be the shared distribution of all X_i. We define the notation $\|P\|_{\mathcal{F}}$ as $\|P\|_{\mathcal{F}} = \sup\{|Pf| : f \in \mathcal{F}\}$. We now recall the law of large numbers.

Theorem 2.1 (Law of large numbers) *Let $X_1, X_2, \ldots X_m$ be independently identically distributed random variables with $\mathbb{E}|X_1| < \infty$. Then*

$$\frac{1}{m}\sum_{i=1}^{m} x_i \to \mathbb{E}[X_1] \text{ almost surely, as } m \to \infty.$$

We can write the uniform version of the law of large numbers as follows:

$$\|\mathbb{P}_m - P\|_{\mathcal{F}} \to 0.$$

The above equation is guaranteed to converge almost surely in the outer norm, *i.e.*, the event $\{\|\mathbb{P}_m - P\|_{\mathcal{F}} \to 0\}$ holds with probability 1. A Glivenko-Cantelli class is defined as a class $\mathcal{F}$ that satisfies the law of large numbers. It is also referred to as a P-Glivenko-Cantelli class since the class is dependent on the underlying measure P.

Let l and u be real functions with finite norms on a measurable space $\mathcal{X}$. We define a bracket $[l, u]$ as the set of all functions $f \in \mathcal{F}$ satisfying $l(x) \le f(x) \le u(x)$ for every $x \in \mathcal{X}$. It should be noted that the functions l and u are not necessarily in the Glivenko-Cantelli class $\mathcal{F}$. An ϵ-bracket is a bracket $[l, u]$ that satisfies $\|u - l\| \le \epsilon$, where ϵ is a positive real number and $\|\cdot\|$ is a given norm. The bracketing number $\mathcal{N}_{[]}(\epsilon, \mathcal{F}, \|\cdot\|)$ is the minimum number of ϵ-brackets that can contain the class $\mathcal{F}$. The bracketing entropy is the natural logarithm of the bracketing number. The occasional stars as superscripts refer to outer measures in the first case, and minimal measureable envelopes in the second case.

Theorem 2.2 *Let $\mathcal{F}$ be a class of measurable functions. If $\mathcal{N}_{[]}(\epsilon, \mathcal{F}, \|\cdot\|) < \infty$ for all $\epsilon > 0$, then $\mathcal{F}$ is a P-Glivenko-Cantelli class. We have that*

$$\|\mathbb{P}_m - P\|^*_{\mathcal{F}} \underset{a.s.}{\to} 0,$$

where P is a shared distribution and $\mathbb{P}_m$ is the empirical observation.

We present a brief proof of this theorem as follows. For any $\epsilon > 0$, we choose finitely many ϵ-brackets $\{[l_i, u_i]\}_{i=1}^{m}$ and argue that, by selecting a bound on $|(\mathbb{P}_m - P)f|$ (for each f) in terms of the $[l_i, u_i]$ that contains it, we obtain

$$\sup_{f\in\mathcal{F}}|(\mathbb{P}_m - P)f| \le \left\{\max_{1\le i\le m}(\mathbb{P}_m - P)u_i \vee \max_{1\le i\le m}(P - \mathbb{P}_m)l_i\right\} + \epsilon.$$

To conclude, according to the strong law for random variables, the right-hand side of the above inequality is guaranteed to be less than 2ϵ.

The following is the definition of the Glivenko-Cantelli theorem for a continuous distribution function on a line. Let F be a continuous cumulative distribution function (CDF), and let P be the corresponding measure. By making the continuity of F on the line uniform, we can find that for every $\epsilon > 0$, $-\infty = t_0 < t_1 < t_2 < \ldots < t_k < t_{k+1} = \infty$, where k is a positive integer such that the union of the brackets $\left[\mathbf{I}_{x\le t_i}, \mathbf{I}_{x\le t_{i+1}}\right]$ for $i = 0, 1, \ldots, k$ contains $\{\mathbf{I}_{x\le t} : t \in \mathbb{R}\}$ and satisfies the equation $F(t_{i+1}) - F(t_i) \le \epsilon$. Then, the above theorem applies. Notably, the continuity of the distribution function F is essential, although the Glivenko-Cantelli theorem on a line holds for arbitrary distribution functions. This more general result will be considered a subsequent corollary of the Glivenko-Cantelli theorem.

The following lemma characterizes a setting that guarantees a finite bracketing number for appropriate classes of functions and characterizes ready application in inference for parametric statistical models.

Lemma 2.1 *Suppose that $\mathcal{F} = \{f(\cdot, t) : t \in T\}$, where T is a compact subset of a metric space (D, d) and the functions $f : \mathcal{X} \times T \to \mathbb{R}$ are continuous in t for P-almost $x \in \mathcal{X}$. If the envelope function F defined by $F(x) = \sup_{t\in T}|f(x, t)|$ satisfies the condition $PF < \infty$, then $\mathcal{N}_{[]}(\epsilon, \mathcal{F}, L_1(P)) < \infty$ for each $\epsilon > 0$.*

We will derive the consistency in parametric statistical models.

Consistency in parametric models: Let $\{p(x, \theta) : \theta \in \Theta\}$, $\Theta \subset \mathbb{R}^d$ be a class of parametric densities. Suppose that $X_1, X_2, \ldots$ are generated from some $p(x, \theta_0)$. Additionally, assume that Θ is compact and $p(x, \theta)$ is continuous w.r.t. θ for P_{θ_0}-almost x. We define $M(\theta) = \mathbb{E}_{\theta_0} l(X_1, \theta)$, where $l(x, \theta) = \log p(x, \theta)$. Eventually, we assume that $\sup_{\theta\in\Theta}|l(x, \theta)| \le B(x)$ for some B with $\mathbb{E}_{\theta_0} B(X_1) < \infty$.

Note that $M(\theta)$ is continuous on Θ if it is finite for all θ. If P_{θ_0} denotes the measure corresponding to θ_0, then $M(\theta) = P_{\theta_0} l(\cdot, \theta)$, and the maximum likelihood estimate (MLE) of θ is given by $\hat{\theta}_m = \text{argmax}_\theta \mathbb{M}_m(\theta)$, where $\mathbb{M}_m(\theta) = \mathbb{P}_m l(\cdot, \theta)$. Under the assumption that the model is identifiable (i.e., the probability distributions corresponding to different θs are different), it can be observed that $M(\theta)$ is uniquely (and globally) maximized at θ_0.

Eventually, θ_0 is a well-separated maximizer in the sense that for any $\eta > 0$, $\sup_{\theta\in\Theta\cap B_\eta(\theta_0)^c} M(\theta) < M(\theta_0)$, while $B_\eta(\theta_0)$ is the open ball of radius η centred at θ_0. Let $\psi(\eta) = M(\theta_0) - \sup_{\theta\in\Theta\cap B_\eta(\theta_0)^c} M(\theta)$. Then, we have $\psi(\eta) > 0$.

Our goal is to show that $\hat{\theta}_n \to_{P^*_{\theta_0}} \theta_0$. Given $\epsilon > 0$, we have that

$$\hat{\theta}_m \in B_\epsilon(\theta_0)^c \Rightarrow M\left(\hat{\theta}_m\right) \le \sup_{\theta\in\Theta\cap B_\eta(\theta_0)^c} M(\theta) \qquad \text{(Property of the expectation)}$$

$$\Rightarrow M\left(\hat{\theta}_m\right) - M(\theta_0) \le -\psi(\epsilon)$$

$$\Rightarrow M\left(\hat{\theta}_m\right) - M(\theta_0) + \mathbb{M}_m(\theta_0) - \mathbb{M}_m\left(\hat{\theta}_m\right) \leq -\psi(\epsilon)$$
$$\Rightarrow 2\sup_{\theta\in\Theta}|\mathbb{M}_m(\theta) - M(\theta)| \geq \psi(\epsilon). \quad \text{(Property of the limit superior)}$$

Thus,

$$P^*\left(\hat{\theta}_m \in B_\epsilon(\theta_0)^c\right) \leq P^*\left(\sup_{\theta\in\Theta}|\mathbb{M}_m(\theta) - M(\theta)| \geq \psi(\epsilon)/2\right)$$
$$= P^*\left(\sup_{\theta\in\Theta}\left|\left(\mathbb{P}_m - P_{\theta_0}\right)l(\cdot,\theta)\right| \geq \psi(\epsilon)/2\right).$$

This equation converges to 0 because $\left(\sup_{\theta\in\Theta}\left|\left(\mathbb{P}_m - P_{\theta_0}\right)l(\cdot,\theta)\right|\right)^* \to 0$ a.s.; based on our assumptions on the parametric model, we can conclude from Lemma 2.1 that $\mathcal{N}_{[]}\left(\eta, \{l(\cdot,\theta) : \theta\in\Theta\}, L_1\left(P_{\theta_0}\right)\right) < \infty$ for every $\eta > 0$ and then invoke Theorem 2.2.

Next, we provide the necessary and sufficient conditions for a class of functions $\mathcal{F}$ to be a Glivenko-Cantelli class in terms of covering numbers Wellner et al. (2013).

Theorem 2.3 *Let $\mathcal{F}$ be a P-measurable class of measurable functions bounded in $L_1(P)$. Then, $\mathcal{F}$ is a P-Glivenko-Cantelli class if and only if*
(a) $P^*F < \infty$ and
(b)

$$\lim_{m\to\infty}\frac{\mathbb{E}^*\log\mathcal{N}(\epsilon,\mathcal{F}_M, L_2(\mathbb{P}_m))}{m} = 0$$

for all $M < \infty$ and $\epsilon > 0$. Here, $\mathcal{F}_M = \{f\mathbf{I}_{F\leq M} : f\in\mathcal{F}\}$.

We will only consider the 'if' part of the proof here due to space limitations. We first note that L_2 can be replaced by any L_r with $r \geq 1$. Then, for the 'if' part, the second condition can be replaced by the weaker condition that $\log\mathcal{N}(\epsilon,\mathcal{F}_M, L_2(\mathbb{P}_m))/m \to_{P^*} 0$. Since $\mathcal{N}(\epsilon,\mathcal{F}_M, L_2(\mathbb{P}_m)) \leq \mathcal{N}(\epsilon,\mathcal{F}, L_2(\mathbb{P}_m))$ for all $M > 0$, condition (b) in the theorem can be replaced by the alternative condition that $\mathbb{E}^*(\log\mathcal{N}(\epsilon,\mathcal{F}, L_2(\mathbb{P}_m)/m) \to 0$ (or a condition involving convergence in probability for the 'if' part). Eventually, if $\mathcal{F}$ has a measurable and integrable envelope F, then $\mathbb{P}_m F$ is most certainly finite (as per the simple strong law). We then may argue the equivalence relation between

$$\forall\epsilon > 0, (\log\mathcal{N}(\epsilon,\mathcal{F}, L_1(\mathbb{P}_m)))^* = o_p(m)$$

and

$$\forall\epsilon > 0, \left(\log\mathcal{N}\left(\epsilon\|F\|_{\mathbb{P}_m,1},\mathcal{F}, L_1(\mathbb{P}_m)\right)\right)^* = o_p(m).$$

We can use the characterization of in-probability convergence on the likely convergence along the subsequences to verify this. Thus, there is a large class of functions, called a Vapnik-Chervonenkis (VC) class of functions, for which the quantity

$\log \mathcal{N}\left(\epsilon \|F\|_{\mathbb{P}_m,1}, \mathcal{F}, L_1\left(\mathbb{P}_m\right)\right)$ is bounded; in fact, for such a class $\mathcal{F}$ of functions, we have that

$$\sup_Q \mathcal{N}\left(\epsilon \|F\|_{Q,r}, \mathcal{F}, L_r(Q)\right) \leq K_1 \left(\frac{1}{\epsilon^r}\right)^M$$

for an integer $M \geq 1$ that depends solely on $\mathcal{F}$ and a constant K_1 that also depends solely only on $\mathcal{F}$, where the supremum is taken over all probability measures for which $\|F\|_{Q,r} > 0$. Thus, a VC class of functions with an integrable envelope F is a Glivenko-Cantelli class for any probability measure on the corresponding sample space. Fortunately, functions formed by combining VC classes of functions via various mathematical operations often satisfy entropy bounds similar to those illustrated above. Therefore, classes of such (greater) complexity remain Glivenko-Cantelli classes under the integrability hypothesis.

As a unique case, consider $\mathcal{F} = \left\{f_t(x) = \mathbf{I}_{-\infty<x\leq t} : t \in \mathbb{R}^d\right\}$. Thus, $f_t(x)$ is the indicator of the infinite rectangle to the 'southwest' of the point t. For any probability measure Q on the d-dimensional Euclidean space, we have that

$$\mathcal{N}(c, F, L_1(Q)) \leq M_d \left(\frac{K}{\epsilon}\right)^d,$$

which immediately implies the classical Glivenko-Cantelli theorem in $\mathbb{R}^d$.

Proof of Theorem 2.2 We now prove the 'if' part. Applying the P-measurability of the class $\mathcal{F}$ and Corollary 1.1 of the symmetrization notes, we find that

$$\begin{aligned}
\mathbb{E}^* \|\mathbb{P}_m - P\|_{\mathcal{F}} &\leq 2\mathbb{E} \left\| \frac{1}{m} \sum_{i=1}^m \epsilon_i f(X_i) \right\|_{\mathcal{F}} \\
&= 2\mathbb{E}_X \mathbb{E}_\epsilon \left\| \frac{1}{m} \sum_{i=1}^m \epsilon_i f(X_i) \right\|_{\mathcal{F}} \\
&\leq 2\mathbb{E}_X \mathbb{E}_c \left\| \frac{1}{m} \sum_{i=1}^m \epsilon_i f(X_i) \right\|_{\mathcal{F}_M} + 2P^*(F\mathbf{I}_{F>M}).
\end{aligned}$$

Given any $\epsilon > 0$, an appropriate choice of M ensures that the second term is not larger than ϵ. It is sufficient that, for this choice of M, the first term will eventually be smaller than ϵ. To this end, we fix $X_1, X_2, \ldots, X_m$. An ϵ-net $\mathcal{G}$ (assumed to be of minimal size) over $\mathcal{F}_M$ in $L_2\left(\mathbb{P}_m\right)$ is also an ϵ-net in $L_1\left(\mathbb{P}_m\right)$. Then, we have

$$\mathbb{E}_\epsilon \left\| \frac{1}{m} \sum_{i=1}^m \epsilon_i f(X_i) \right\|_{\mathcal{F}_M} \leq \mathbb{E}_\epsilon \left\| \frac{1}{m} \sum_{i=1}^m \epsilon_i f(X_i) \right\|_{\mathcal{G}} + \epsilon.$$

Each $g \in \mathcal{G}$ is assumed to be uniformly bounded in absolute value by M since each f in $\mathcal{F}_M$ is bounded in absolute value by M. Therefore, given an arbitrary ϵ-net $\mathcal{G}$, we perturb each g to a $\tilde{g}$ that coincides with g whenever $|g| \leq M$, and the complement of this set is equal to $(g) \times M$. These perturbed functions constitute an ϵ-net over $\mathcal{F}_M$.

Let us analyze the first term on the right-hand side of the equation. Since the L_1 norm is bounded up to a constant by the ψ_1 Orlicz norm, which is bounded up to a constant by the ψ_2 Orlicz norm, we can use Lemma 2.1 in the chaining notes to bound the first term up to the following constant:

$$B_m = \sqrt{1 + \log \mathcal{N}\left(\epsilon, \mathcal{F}_M, L_2\left(\mathbb{P}_m\right)\right)} \max_{f \in \mathcal{G}} \left\| \frac{1}{m} \sum_{i=1}^{m} \epsilon_i f\left(X_i\right) \right\|_{\psi_2 | X}.$$

Applying Hoeffding's inequality, we obtain

$$\left\| \frac{1}{m} \sum_{i=1}^{m} \epsilon_i f\left(X_i\right) \right\|_{\psi_2 | X} \leq \sqrt{6} \frac{1}{\sqrt{m}} \left(\mathbb{P}_m f^2\right)^{1/2} \leq \sqrt{6} \frac{1}{\sqrt{m}} M,$$

and thus,

$$B_m \leq \sqrt{6} M \sqrt{\frac{1 + \log \mathcal{N}\left(\epsilon, \mathcal{F}_M, L_2\left(\mathbb{P}_m\right)\right)}{m}} \to 0$$

by condition (b) of the theorem. We conclude that

$$\mathbb{E}_\epsilon \left\| \frac{1}{m} \sum_{i=1}^{m} \epsilon_i f\left(X_i\right) \right\|_{\mathcal{F}_M} \to_P 0.$$

Since the above random variable is bounded, we have

$$\mathbb{E}_X \mathbb{E}_c \left\| \frac{1}{m} \sum_{i=1}^{m} \epsilon_i f\left(X_i\right) \right\|_{\mathcal{F}_M} \to 0.$$

It follows that $\mathbb{E}^* \left\| \mathbb{P}_m - P \right\|_{\mathcal{F}} \to 0$. However, our goal is to show almost sure convergence. This is deduced through a submartingale argument, a simplified version of which is presented at the end of these notes. The idea here is to show that $\left\| \mathbb{P}_m - P \right\|_{\mathcal{F}}^*$ is a reverse submartingale with respect to a (decreasing) filtration that converges to the symmetric sigma field and therefore has an almost sure limit. This almost sure limit, measurable with respect to the symmetric sigma field, must be a nonnegative constant almost surely. The fact that the expectation converges to 0 forces this constant to become 0.

2.1.2 Concentration Inequality

Concentration inequality, which describes the deviation between the sum (or sample mean) of a group of random variables and their expected value, is a very useful concept in the analysis of PAC learning. In addition to characterizing the asymptotic convergence of an iterative machine learning algorithm, concentration inequalities also offer practical tools for evaluating the convergence rate. We will give a brief summary of various concentration inequalities in this subsection.

Markov's inequality. Let X be a nonnegative random variable, and $a > 0$. Then,

$$\mathbb{P}(X \geq a) \leq \frac{\mathbb{E}X}{a}.$$

Chebyshev's inequality. Let X be a random variable with mean μ and variance σ^2. Then, for any constant $k > 0$,

$$\mathbb{P}(|X - \mu| \geq k\sigma) \leq \frac{1}{k^2}.$$

Chernoff bounds. Let $X = \sum_{i=1}^{m} X_i$, and t be a positive constant. Then,

$$\mathbb{P}\left(X_i > a\right) = \mathbb{P}\left(e^{tX_i} > e^{ta}\right) \leq e^{-ta}\mathbb{E}\left[e^{tX_i}\right],$$

$$\mathbb{P}(X > a) \leq e^{-ta}\mathbb{E}\left[\Pi_i^m e^{tX_i}\right],$$

$$\mathbb{P}(X > a) \leq \min_{t>0}\left[e^{-ta}\mathbb{E}\left[\Pi_i^m e^{tX_i}\right]\right].$$

Hoeffding's inequality. Let $X_1, X_2, \cdots, X_m$ be a family of independent random variables, and $0 \leq X_i \leq 1$. Then,

$$\mathbb{P}(\frac{1}{m}\sum_{i=1}^{m} X_i - \frac{1}{m}\sum_{i=1}^{m} \mathbb{E}X_i \geq \epsilon) \leq \exp(-2m\epsilon^2),$$

$$\mathbb{P}(|\frac{1}{m}\sum_{i=1}^{m} X_i - \frac{1}{m}\sum_{i=1}^{m} \mathbb{E}X_i| \geq \epsilon) \leq 2\exp(-2m\epsilon^2).$$

McDiarmid's inequality. Let $X_1, X_2, \cdots, X_m \in \mathcal{X}$ be a family of independent random variables, and for any $1 \leq i \leq m$, the function f satisfy

$$\sup_{X_1, X_2, \cdots, X_m, X_i' \in \mathcal{X}} |f(X_1, \cdots, X_m) - f(X_1, \cdots, X_{i-1}, X_i', X_{i+1}, \cdots, X_m)| \leq c_i.$$

Then, for any $\epsilon > 0$,

$$\mathbb{P}(f(X_1, \cdots, X_m) - \mathbb{E}(f(X_1, \cdots, X_m)) \geq \epsilon) \leq \exp\left(\frac{-2\epsilon^2}{\sum_{i=1}^m c^2}\right),$$

$$\mathbb{P}(|f(X_1, \cdots, X_m) - \mathbb{E}(f(X_1, \cdots, X_m))| \geq \epsilon) \leq 2\exp\left(\frac{-2\epsilon^2}{\sum_{i=1}^m c_i^2}\right).$$

Bennett's inequality. Let $X_1, X_2, \cdots, X_m$ be a family of independent random variables, and the following relations hold: $a_i < X_i < b_i$, $b_i - a_i = c_i \leq C$, $\bar{X} = \frac{1}{m}(X_1 + \cdots + X_m)$, and $\sigma^2 = \sum_{i=1}^m \mathbb{E}[X_i^2] - \mathbb{E}[X]^2$. Then, for any $t \in \mathbb{R}$,

$$\mathbb{P}(\bar{X} - E[\bar{X}] \geq t) \leq \exp\left(-\frac{\sigma^2}{C^2} h\left(\frac{Ct}{m\sigma^2}\right)\right),$$

where $h(u) = (1 + u)\log(1 + u) - u$.

Bernstein's inequality. Let $X_1, X_2, \cdots, X_m$ be a family of independent random variables, and the following relations hold: $a_i < X_i < b_i$, $b_i - a_i = c_i \leq C$, $\bar{X} = \frac{1}{m}(X_1 + \cdots + X_m)$, and $\sigma^2 = \sum_{i=1}^m \mathbb{E}[X_i^2] - \mathbb{E}[X]^2$. Then, we have

$$\mathbb{P}(\bar{X} - \mathbb{E}[\bar{X}] \geq t) \leq \exp\left(-\frac{m^2t^2/2}{\sigma^2 + Ctm/3}\right), \forall t \in \mathbb{R}.$$

The inequalities discussed above require the random variables to be independent. In contrast, for Azuma's inequality below, the sequence of random variables can be correlated.

Azuma's inequality. Let $X_1, X_2, \cdots, X_m$ be a martingale sequence, i.e., $\mathbb{E}[|X_m|] \leq \infty$ and $\mathbb{E}[X_{m+1}|X_1, \cdots, X_m] = X_m$. Then, under the assumption that $|X_i - X_{i-1}| < c_i$, for any $t \in \mathbb{R}$, we have

$$\mathbb{P}(X_m - X_0 \geq t) \leq \exp\left(-\frac{t^2}{2\sum_{i=1}^m c_i^2}\right).$$

Doob's martingale inequality. Let $\{X_t, \mathcal{F}_t, 0 \leq t < \infty\}$ be a submartingale, i.e., for every $0 \leq s < t < \infty$, $\mathbb{E}(X_t|\mathcal{F}_s) \geq X_s$. Then, for any positive constant C,

$$\mathbb{P}\left(\sup_{0\leq t\leq T} X_t \geq C\right) \leq \frac{\mathbb{E}[\max(X_T, 0)]}{C}.$$

2.2 Probably Approximately Correct (PAC) Learning

Before designing algorithms to learn from training data, we should consider the following issues. What is the most essential consideration in a learning task? How many examples are required to successfully train a model? Additionally, is there a generalizable model that can be learned for the current problem? To formalize and solve these issues, this section introduces the PAC learning framework. The PAC learning framework can help determine an algorithm's sample complexity (how many examples are required to train a desirable approximator for the given problem) and whether a problem is learnable. In this section, we first present the PAC framework, then provide a theoretical assurance for learning algorithms, and eventually hope to guide algorithm design in accordance with the results of our analysis.

2.2.1 *The PAC Learning Model*

We first present some relevant definitions and notations before introducing the PAC model. We consider a binary classification task, i.e., the label $y \in \{0, 1\}$. We express a concept c as a mapping from a feature x to the label y. We use C to denote a concept class, which represents the set of concepts we hope to learn. For instance, it may be the set of all triangles on a plane.

Suppose that all examples are ii.i.d. according to a fixed but unknown distribution $\mathcal{D}$. Then, the framework of the learning problem is constructed as follows. The learner considers a fixed set of possible concepts $\mathcal{H}$, which is termed the hypothesis set. The hypothesis set does not need to coincide with C. The learner receives a sample set $S : S = (x_1, \ldots, x_m)$ with known labels $(c(x_1), \ldots, c(x_m))$ that depend on the concept $c \in C$ to be learned (here, c represents an optimal mapping). The examples in S are independently sampled from the distribution $\mathcal{D}$. The learning task requires the selection of a hypothesis $h_S \in \mathcal{H}$ based on the labelled sample S, which has a small generalization error with respect to c. The generalization error (or risk, or true error, or simply error) of a hypothesis $h \in \mathcal{H}$ is denoted by $R(h)$ and is defined as follows.

Definition 2.4 (Generalization risk) The generalization risk of a hypothesis $h \in \mathcal{H}$ is defined as

$$R(h) = \mathop{\mathbb{P}}_{x \sim \mathcal{D}} [h(x) \neq c(x)], \tag{2.1}$$

where $c \in C$ is the target concept.

The generalization error of h cannot be obtained directly by the learner because both the distribution $\mathcal{D}$ and the target concept c are unknown. Nevertheless, the learner can learn the empirical error of a mapping on the labelled sample S to approximate the generalization error.

Definition 2.5 (Empirical error) The empirical risk of a hypothesis $h \in \mathcal{H}$ is defined as

$$\hat{R}_S(h) = \frac{1}{m}\sum_{i=1}^{m} \mathbf{I}_{h(x_i)\neq c(x_i)}, \tag{2.2}$$

where $\mathbf{I}_w$ is the indicator function of event w. The generalization error of $h \in \mathcal{H}$ is the expected error based on the distribution $\mathcal{D}$. The empirical error is the average error over the sample S. Under certain general assumptions, there exist a series of guarantees with high probability for these two quantities. It is also important to note that the expected empirical error based on the i.i.d. sample set S is equivalent to the generalization error for a fixed $h \in \mathcal{H}$:

$$\underset{S\sim\mathcal{D}^m}{\mathbb{E}}\left[\hat{R}_S(h)\right] = R(h). \tag{2.3}$$

By linearizing the expectation and considering that the sample S is drawn in an i.i.d. manner, we obtain

$$\underset{S\sim\mathcal{D}^m}{\mathbb{E}}\left[\hat{R}_S(h)\right] = \frac{1}{m}\sum_{i=1}^{m}\underset{S\sim\mathcal{D}^m}{\mathbb{E}}\left[\mathbf{I}_{h(x_i)\neq c(x_i)}\right] = \underset{x\sim\mathcal{D}}{\mathbb{E}}\left[\mathbf{I}_{h(x)\neq c(x)}\right] = R(h).$$

Next, we present the PAC learning framework. We use n to denote a number, size(c) to denote the maximal cost of the computational representation of $c \in C$, and h_S to denote the hypothesis that is obtained by algorithm $\mathcal{A}$ when applied to the labelled sample S. The computational cost of representing the element $x \in X$ is at most $O(n)$. To simplify the notation, we do not explicitly express the dependence of h_S on $\mathcal{A}$.

Definition 2.6 (PAC learning) For any $\epsilon > 0$, any $\delta > 0$, any distributions $\mathcal{D}$ on X and any target concept $c \in C$, the concept class C is PAC learnable if there exist an algorithm $\mathcal{A}$ and a polynomial function poly$(\cdot,\cdot,\cdot,\cdot)$ such that the following inequality holds for any sample size $m \geq \text{poly}(1/\epsilon, 1/\delta, n, \text{size}(c))$:

$$\mathbb{P}\left[R\left(h_S\right) \leq \epsilon\right] \geq 1-\delta. \tag{2.4}$$

Furthermore, C is efficiently PAC learnable if $\mathcal{A}$ runs in poly$(1/\epsilon, 1/\delta, n, \text{size}(c))$ time. The algorithm $\mathcal{A}$ is said to be a PAC learning algorithm for C.

Hence, if the hypothesis acquired by the algorithm after receiving a series of points that are polynomial in $1/\epsilon$ and $1/\delta$ is approximately correct with high probability, then the concept class C is PAC learnable. The parameter $\epsilon > 0$ defines the accuracy $1-\epsilon$, while $\delta > 0$ defines the confidence $1-\delta$.

We need to emphasize a certain point regarding the definition of PAC learning. No particular assumption is made about the distribution $\mathcal{D}$ from which examples are drawn in the PAC framework. Therefore, it is a distribution-free model. Additionally, the training and test examples are drawn from the same distribution $\mathcal{D}$ when defining the error. This is a natural assumption to make generalization possible. Furthermore, the PAC framework is suitable for solving the learnability question for a concept class C rather than for a particular concept. It is important to note that the target concept $c \in C$ is unknown to the algorithm, whereas the concept class C is known. Hence, we can focus only on the sample complexity in the definition of PAC learning and omit the polynomial dependency on n and size(c) without explicitly discussing the concepts' computational representation.

2.2.2 Generalities

Next, we discuss some general aspects of learning scenarios.

2.2.2.1 Deterministic Versus Stochastic Scenarios

In supervised learning, the distribution $\mathcal{D}$ is usually defined over $\mathcal{X} \times \mathcal{Y}$, and the training data consist of a labelled sample S drawn in an i.i.d. manner according to $\mathcal{D}$:

$$S = ((x_1, y_1), \ldots, (x_m, y_m)) .$$

The target of learning is to find a hypothesis $h \in \mathcal{H}$ with a small generalization error

$$R(h) = \underset{(x,y)\sim\mathcal{D}}{\mathbb{P}}[h(x) \neq y] = \underset{(x,y)\sim\mathcal{D}}{\mathbb{E}}\left[\mathbf{I}_{h(x)\neq y}\right] .$$

This general scenario is stochastic. Under such circumstances, the output label is a probabilistic function of the input. This scenario can describe a number of real-world problems. For instance, when the goal is to predict gender using input pairs consisting of the weight and height of a person, the labels of the input sample generally may not be unique. Both 'man' and 'woman' will be possible genders for most pairs, although there may exist distinct probability distributions for the label being 'man' or 'woman' for each fixed pair. An extension of the PAC learning framework, known as agnostic PAC learning, is able to deal with this setting.

Definition 2.7 (Agnostic PAC learning) Let $\mathcal{H}$ be a hypothesis set. The algorithm $\mathcal{A}$ is said to be an agnostic PAC learning algorithm if there exists a polynomial function poly$(\cdot, \cdot, , \cdot\cdot)$ such that for any $\epsilon > 0$ and $\delta > 0$, for distribution $\mathcal{D}$ over $X \times Y$, and for any sample size $m \geq \text{poly}(1/\epsilon, 1/\delta, n, \text{size}(c))$, the following expression holds:

$$\mathbb{P}\Big(R\left(h_S\right) - \min_{h\in\mathcal{H}} R(h) \leq \epsilon\Big) \geq 1 - \delta, \tag{2.5}$$

Furthermore, if $\mathcal{A}$ runs in poly$(1/\epsilon, 1/\delta, n)$ time, then it is called an efficient agnostic PAC learning algorithm.

The scenario is considered deterministic if the label of an input sample can be determined with probability one by a function $f : x \to y$. In such a scenario, it is sufficient to consider the distribution $\mathcal{D}$ in the input space. The training sample can be acquired by drawing $(x_1, \ldots, x_m)$ according to $\mathcal{D}$. The corresponding labels can be obtained from $f : y_i = f(x_i)$ for all $i \in [m]$.

2.2.2.2 Bayes Error and Noise

In the deterministic case, there exists a target function f with no generalization error ($R(h) = 0$), while in the stochastic case, there is a minimal nonzero error for any hypothesis.

Definition 2.8 (Bayes error) Given a distribution $\mathcal{D}$ over $\mathcal{X} \times Y$, the Bayes error R^* is defined as the infimum of the errors obtained by a measurable function $h : \mathcal{X} \to$ y:

$$R^* = \inf_h \quad R(h). \tag{2.6}$$

A hypothesis h with $R(h) = R^*$ is called a Bayes hypothesis or Bayes classifier.

$R^* = 0$ in the deterministic case, while $R^* \neq 0$ in the stochastic case. The Bayes classifier h_{Bayes} can be defined in terms of conditional probabilities as follows:

$$\forall x \in \mathcal{X}, \quad h_{\text{Bayes}}(x) = \underset{y \in \{0,1\}}{\text{argmax}} \, \mathbb{P}(y \mid x). \tag{2.7}$$

Therefore, the average error generated by h_{Bayes} on $x \in \mathcal{X}$ is $\min\{\mathbb{P}(0 \mid x), \mathbb{P}(1 \mid x)\}$, the minimum possible error. On this basis, the definition of noise is given as follows.

Definition 2.9 (Noise) Given a distribution $\mathcal{D}$ over $\mathcal{X} \times Y$, the noise at point $x \in \mathcal{X}$ is defined as

$$\text{noise}(x) = \min\{\mathbb{P}(1 \mid x), \mathbb{P}(0 \mid x)\}. \tag{2.8}$$

The average noise, or the noise associated with $\mathcal{D}$, is $\mathbb{E}[\text{noise}(x)]$.

It is easy to see that the average noise is precisely the Bayes error: noise $= \mathbb{E}[\text{noise}(x)] = R^*$. A point $x \in \mathcal{X}$ with noise(x) close to $1/2$ is sometimes called noisy and is a challenge to predict accurately.

2.2.2.3 Guarantees for Finite Hypothesis Sets–Consistent Case

We consider that the hypotheses are consistent and that the target concept c is within the cardinality $|\mathcal{H}|$ of the finite hypothesis set. In this section, we will present a general sample complexity bound for such consistent hypotheses.

Theorem 2.4 (Learning bound–finite $\mathcal{H}$, consistent case) *Let $\mathcal{A}$ be an algorithm that, for any target concept $c \in \mathcal{H}$ and i.i.d. sample S, returns a consistent hypothesis $h_S : \hat{R}_S(h_S) = 0$, where $\mathcal{H}$ is a finite set of functions mapping from $\mathcal{X}$ to $\mathcal{Y}$. For any $\epsilon, \delta > 0$, if*

$$m \geq \frac{1}{\epsilon}\left(\log|\mathcal{H}| + \log\frac{1}{\delta}\right), \tag{2.9}$$

then the inequality $\mathbb{P}_{S\sim\mathcal{D}^m}\left[R(h_S) \leq \epsilon\right] \geq 1-\delta$ holds. The following generalization bound can be regarded as the equivalent statement according to this sample complexity. That is, for any $\epsilon, \delta > 0$, with a probability of at least $1-\delta$,

$$R(h_S) \leq \frac{1}{m}\left(\log|\mathcal{H}| + \log\frac{1}{\delta}\right). \tag{2.10}$$

Proof For any $\epsilon > 0$, we define $\mathcal{H}_\epsilon$ as $\mathcal{H}_\epsilon = \{h \in \mathcal{H} : R(h) > \epsilon\}$. The probability that a hypothesis h is consistent on an i.i.d. sample set S, i.e., that it has no error on any point in S, can be bounded as follows:

$$\mathbb{P}\left[\hat{R}_S(h) = 0\right] \leq (1-\epsilon)^m.$$

Applying the union bound yields the following expression:

$$\begin{aligned}
\mathbb{P}\left[\exists h \in \mathcal{H}_\epsilon : \hat{R}_S(h) = 0\right] &= \mathbb{P}\left[\hat{R}_S(h_1) = 0 \vee \cdots \vee \hat{R}_S\left(h_{|\mathcal{H}_\epsilon|}\right) = 0\right] \\
&\leq \sum_{h\in\mathcal{H}_\epsilon} \mathbb{P}\left[\hat{R}_S(h) = 0\right] \qquad \text{(Triangle inequality)} \\
&\leq \sum_{h\in\mathcal{H}_\epsilon} (1-\epsilon)^m \leq |\mathcal{H}|e^{-m\epsilon}.
\end{aligned}$$

Let the right-hand side be equal to δ; then, solving for ϵ completes the proof. □

This theorem illustrates that a consistent algorithm $\mathcal{A}$ is PAC learnable when the hypothesis set $\mathcal{H}$ is finite since the sample complexity in (2.9) is dominated by a polynomial in $1/\epsilon$ and $1/\delta$. From expression (2.10), we know that the generalization error of consistent hypotheses has an upper bound, which decreases with an increasing sample size m.

The cost of proposing a consistent algorithm is the necessity of using a larger hypothesis set $\mathcal{H}$ that contains the target concepts. The upper bound of (2.10) increases with increasing $|\mathcal{H}|$, although the dependency is only logarithmic. The term $\log|\mathcal{H}|$, which is not a constant factor, can be interpreted as the number of bits needed to represent $\mathcal{H}$. As a result, the ratio between the number of bits ($\log_2|\mathcal{H}|$) and the sample size m determines this theorem's generalization guarantee.

2.2.2.4 Guarantees for Finite Hypothesis Sets–Inconsistent Case

Difficult learning problems may arise in which there exist concept classes that are more complex than the hypotheses used in the learning algorithm. In other words, there may not exist a hypothesis in $\mathcal{H}$ that is consistent with the labelled training sample. Nevertheless, such inconsistent hypotheses with minor errors in the training sample can also be useful. Under certain assumptions, they may even benefit from favourable guarantees. In this section, we introduce learning guarantees for finite hypotheses in the inconsistent case by using either Hoeffding's inequality or the following corollary, which establishes the relationship between the empirical error and generalization error of a single hypothesis.

Corollary 2.1 *Let $\epsilon > 0$ be fixed. Then, for any hypothesis $h : x \to \{0, 1\}$, the following expressions hold:*

$$\mathbb{P}\Big(\hat{R}_S(h) - R(h) \geq \epsilon\Big) \leq \exp\left(-2m\epsilon^2\right), \tag{2.11}$$

$$\mathbb{P}\Big(\hat{R}_S(h) - R(h) \leq -\epsilon\Big) \leq \exp\left(-2m\epsilon^2\right). \tag{2.12}$$

Applying the union bound yields the following two-sided inequality:

$$\mathbb{P}\Big(\Big|\hat{R}_S(h) - R(h)\Big| \geq \epsilon\Big) \leq 2\exp\left(-2m\epsilon^2\right). \tag{2.13}$$

Let the right-hand side of (2.13) be equal to δ; then, solving for ϵ immediately produces the following bound.

Corollary 2.2 (Generalization bound–single hypothesis) *Consider a fixed hypothesis $h : x \to \{0, 1\}$. For any $\delta > 0$, the following expression holds with a probability of at least $1 - \delta$:*

$$R(h) \leq \hat{R}_S(h) + \sqrt{\frac{\log \frac{2}{\delta}}{2m}}. \tag{2.14}$$

Because h_S is a random variable that depends on the training sample S, we cannot utilize Corollary 2.2 to bound its generalization error. Additionally, the generalization error $R\left(h_S\right)$ is a random variable and differs from the expectation $\mathbb{E}\left[\hat{R}_S\left(h_S\right)\right]$, which is a constant. This is distinct from the case of a given hypothesis, for which the expectation of the empirical error is the generalization error (Eq. 2.3). Thus, for the consistent case, it is necessary to derive a uniform convergence bound that is satisfactory for all hypotheses $h \in \mathcal{H}$ with high probability.

Theorem 2.5 (Learning bound–finite $\mathcal{H}$, inconsistent case) *Let $\mathcal{H}$ be a finite hypothesis set. Then, for any $\delta > 0$, the following expression holds with a probability of at least $1 - \delta$:*

$$\forall h \in \mathcal{H}, \quad R(h) \leq \hat{R}_S(h) + \sqrt{\frac{\log|\mathcal{H}| + \log\frac{2}{\delta}}{2m}}. \tag{2.15}$$

Proof We denote the elements of $\mathcal{H}$ by $h_1, \ldots, h_{|\mathcal{H}|}$. Applying the union bound and Corollary 2.2 to each hypothesis yields

$$\begin{aligned}
&\mathbb{P}\Big(\exists h \in \mathcal{H} \Big| R(h) - \hat{R}_S(h)\Big| > \epsilon\Big) \\
&= \mathbb{P}\Big(\Big(\Big|\hat{R}_S(h_1) - R(h_1)\Big| > \epsilon\Big) \vee \ldots \vee \Big(\Big|\hat{R}_S\left(h_{|\mathcal{H}|}\right) - R\left(h_{|\mathcal{H}|}\right)\Big| > \epsilon\Big)\Big) \\
&\leq \sum_{h \in \mathcal{H}} \mathbb{P}\Big(\Big|\hat{R}_S(h) - R(h)\Big| > \epsilon\Big) \qquad \text{(Triangle inequality)} \\
&\leq 2|\mathcal{H}| \exp\left(-2m\epsilon^2\right).
\end{aligned}$$

Setting the right-hand side equal to δ completes the proof. □

Therefore, for a finite hypothesis set $\mathcal{H}$, we have

$$R(h) - \hat{R}_S(h) \leq O\left(\sqrt{\frac{\log_2|\mathcal{H}|}{m}}\right).$$

A larger sample size m provides a better generalization guarantee, and the bound increases logarithmically with $|\mathcal{H}|$. However, the bound is a less favourable function of $\log_2|\mathcal{H}|/m$, varying with the square root of $\log_2|\mathcal{H}|/m$. A quadratically larger labelled sample is required for a given $|\mathcal{H}|$ to obtain the same guarantee as in the consistent case. The values of the bounds show that a balance is sought between the empirical error and the size of the hypothesis set. Although a larger hypothesis set is penalized by the second term, increasing the size of the hypothesis set helps reduce the first term, i.e., the empirical error. For a similar empirical error, however, the use of a smaller hypothesis set is recommended; this can be regarded as an example of the Occam's razor principle, named after the theologian William of Occam, which states that the simplest explanation is the best. In this context, this principle can be understood to mean that when all other things are equal, a hypothesis set of a smaller size is better.

2.2.3 Insights from Glivenko-Cantelli Theorem

A fundamental result of statistical learning theory states that a concept class is PAC learnable if and only if it is a uniform Glivenko-Cantelli class. However, the theorem is valid only under special assumptions regarding the class's measurability, in which case even the PAC learnability becomes consistent. Otherwise, a classical example can be constructed, under the continuum hypothesis developed by Dudley and Durst and further adapted by Blumer, Ehrenfeucht, Haussler, and Warmuth, of a concept class of VC dimension one that is neither uniformly Glivenko-Cantelli nor consistently PAC learnable. We show that, under an additional set-theoretic hypothesis that is much milder than the continuum hypothesis (Martin's axiom), PAC learnability is equivalent to a finite VC dimension for every concept class.

A class $\mathcal{F}$ of measurable functions on a measurable space $(\mathcal{X}, \mathcal{A})$ is uniformly Glivenko-Cantelli in $P \in \mathcal{P}$ for a class $\mathcal{P}$ of probability measures on $(\mathcal{X}, \mathcal{A})$ if

$$\sup_{P \in \mathcal{P}} \mathbb{P}^* \left(\sup_{m \geq a} \|\mathbb{P}_m - P\|_{\mathcal{F}} > \varepsilon \right) \to 0, \forall \varepsilon > 0, \quad \text{as } a \to \infty.$$

Theorem 2.6 *For a concept class $\mathcal{F}$, the following two conditions are equivalent: 1. $\mathcal{F}$ is distribution-free PAC learnable; 2. $\mathcal{F}$ is a uniform Glivenko-Cantelli class.*

2.3 PAC-Bayesian Learning

PAC-Bayesian theory was initially developed by McAllester as an attempt to explain Bayesian learning from a learning theory perspective (McAllester 1999a).

Let $\mathcal{X} \in \mathbb{R}^p$ be the input space and $\mathcal{Y} \in \{-1, 1\}$ be the response space, respectively. We use $\mathcal{D}$ to denote the unknown joint distribution on $\mathcal{Z} := \mathcal{X} \times \mathcal{Y}$. Suppose that the estimators in the hypothesis space $\mathcal{H}$ can be indexed by a parameter set Θ, i.e., $\mathcal{H} = \{h_\theta : \theta \in \Theta\}$. Let $\mathcal{Q}$ be the space of probability distributions on Θ. The output of an algorithm here refers to a distribution $Q \in \mathcal{Q}$ (named the posterior distribution) instead of a single estimator $h_\theta \in \mathcal{H}$. Here, we consider only a 0, 1-valued loss function $l(h_\theta, z)$, i.e., $l(h_\theta, z) = 1$ if $h_\theta(x) = y$ and $l(h_\theta, z) = 0$ otherwise. We present the definitions of the expected risk and the empirical risk as follows:

Definition 2.10 Suppose that there exists a probability distribution $Q \in \mathcal{Q}$ on Θ. Given a sample $S = \{(x_i, y_i)\}_{i=1}^m$, the expected and empirical risk are

$$R(Q) = \mathbb{E}_{z \sim \mathcal{D}} \mathbb{E}_{\theta \sim Q}[l(h_\theta, z)]$$

and

$$\hat{R}_S(Q) = \frac{1}{m} \sum_{i=1}^{m} \mathbb{E}_{\theta \sim Q}[l(h_\theta, z_i)].$$

To establish the PAC-Bayesian bound, we introduce the following definition for the Kullback-Leibler divergence.

Definition 2.11 Given two probability measures $D, G \in \mathcal{Q}$, the Kullback-Leibler (KL) divergence between G and D is

$$\mathrm{KL}(D\|G) = \int D(\theta)\log(\frac{D(\theta)}{G(\theta)})d\theta.$$

In particular, if the space Θ is finite, then we have

$$\mathrm{KL}(D\|G) = \sum_{\theta\in\Theta} D(\theta)\log(\frac{D(\theta)}{G(\theta)}).$$

Note that for any probability measures D and G, $\mathrm{KL}(D\|G) \geq 0$, where the equality holds if and only if $D = G$.

In this section, we will focus on a bound as proposed in (Catoni 2007). Let us fix a probability measure $\pi \in \mathcal{Q}$, which is usually called the prior.

Theorem 2.7 *We assume that $l(h_\theta, z) \leq C$, $\forall h_\theta \in \mathcal{H}$ and $\forall z \in \mathcal{Z}$. For any $\lambda > 0$ and $\delta \in (0, 1)$, it holds that*

$$\mathbb{P}\Big(\mathbb{E}_{\theta\sim Q}R(\theta) \leq \mathbb{E}_{\theta\sim Q}\hat{R}_S(\theta) + \frac{\lambda C^2}{8m} + \frac{\mathrm{KL}(Q\|\pi) + \log 1/\delta}{\lambda}\Big) \geq 1-\delta,\ \forall Q \in \mathcal{Q}.$$

Proof We first recall Donsker and Varadhan's variational formula (Donsker and Varadhan 1976). For any measurable and bounded function $g : \Theta \to \mathbb{R}$, we have

$$\log \mathbb{E}_{\theta\sim\pi}[e^{g(\theta)}] = \sup_{Q\in\mathcal{Q}}[\mathbb{E}_{\vartheta\sim Q}[g(\vartheta)] - \mathrm{KL}(Q\|\pi)]. \tag{2.16}$$

Moreover, given a risk $R(\cdot)$, the supremum with respect to Q on the right-hand side is reached for the Gibbs measure π_R, where its density with respect to π is

$$\frac{d\pi_R}{d\pi}(\theta) = \frac{e^{g(\theta)}}{\mathbb{E}_{\vartheta\sim\pi}[e^{g(\vartheta)}]}.$$

With the help of Hoeffding's lemma, for a fixed $h_\theta \in \mathcal{H}$ and any $t > 0$, we have

$$\mathbb{E}_S[e^{tm[R(\theta)-\hat{R}_S(\theta)]}] \leq e^{\frac{mt^2C^2}{8}}.$$

By taking $t = \lambda/m$, we obtain

$$\mathbb{E}_S[e^{tm|R(\theta)-\hat{R}_S(\theta)|}] \leq e^{\frac{\lambda^2C^2}{8m}}.$$

Clearly, this bound with respect to π is

$$\mathbb{E}_{\theta\sim\pi}\mathbb{E}_S[e^{tm|R(\theta)-\hat{R}_S(\theta)|}] \leq e^{\frac{\lambda^2C^2}{8m}}.$$

We can exchange the integrations with respect to π and the sample S:

$$\mathbb{E}_S \mathbb{E}_{\theta\sim\pi}[e^{tm|R(\theta)-\hat{R}_S(\theta)|}] \le e^{\frac{\lambda^2 C^2}{8m}}.$$

Donsker and Varadhan's variational formula gives

$$\mathbb{E}_S[e^{\sup_{Q\in\mathcal{Q}} \lambda \mathbb{E}_{\theta\in Q}[R(\theta)-\hat{R}_S(\theta)]-\mathrm{KL}(Q\|\pi)}] \le e^{\frac{\lambda^2 C^2}{8m}}.$$

Direct computation shows that

$$\mathbb{E}_S[e^{\sup_{Q\in\mathcal{Q}} \lambda \mathbb{E}_{\theta\in Q}[R(\theta)-\hat{R}_S(\theta)]-\mathrm{KL}(Q\|\pi)-\frac{\lambda^2 C^2}{8m}}] \le 1.$$

The following result is obtained using the Chernoff bound:

$$\begin{aligned}&\mathbb{P}\Big(\sup_{Q\in\mathcal{Q}} \lambda \mathbb{E}_{\theta\in Q}[R(\theta)-\hat{R}_S(\theta)] - \mathrm{KL}(Q\|\pi) - \frac{\lambda^2 C^2}{8m} > s\Big)\\ \le & \mathbb{E}_S\Big[e^{\sup_{Q\in\mathcal{Q}} \lambda \mathbb{E}_{\theta\in Q}[R(\theta)-\hat{R}_S(\theta)]-\frac{\lambda^2 C^2}{8m}}\Big]e^{-s} \le e^{-s}.\end{aligned}$$

By taking $aligns = \log(1/\delta)$, we deduce that

$$\mathbb{P}\Big(\sup_{Q\in\mathcal{Q}} \lambda \mathbb{E}_{\theta\in Q}[R(\theta)-\hat{R}_S(\theta)] - \mathrm{KL}(Q\|\pi) - \frac{\lambda^2 C^2}{8m} > \log 1/\delta\Big) \le \delta.$$

We complete the proof by rearranging the above expression. □

The above error bound motivates the study of a data-dependent learning problem defined as

$$\hat{Q} = \arg\min_{Q\in\mathcal{Q}}\{\mathbb{E}_{\theta\sim Q}\hat{R}_S(\theta) + \frac{\mathrm{KL}(Q\|\pi)}{\lambda}\}.$$

Equivalently, this optimization problem can be rewritten as

$$\hat{Q} = \arg\max_{Q\in\mathcal{Q}}\{-\lambda\mathbb{E}_{\theta\sim Q}\hat{R}_S(\theta) - \mathrm{KL}(Q\|\pi)\}.$$

According to Donsker and Varadhan's variational formula (2.16), the minimum is reached for the Gibbs measure $\hat{Q} = \pi_{-\lambda\hat{R}_S}$.

Corollary 2.3 *Let $\hat{Q}$ be the Gibbs measure $\hat{Q} = \pi_{-\lambda\hat{R}_S}$. For any $\lambda > 0$ and $\delta \in (0, 1)$,*

$$\mathbb{P}\Big(\mathbb{E}_{\theta\sim\hat{Q}}R(\theta) \le \inf_{Q\in\mathcal{Q}}\Big[\mathbb{E}_{\theta\sim Q}\hat{R}_S(\theta) + \frac{\lambda C^2}{8m} + \frac{\mathrm{KL}(Q\|\pi)+\log 1/\delta}{\lambda}\Big]\Big) \ge 1-\delta.$$

We will present two examples related to the general PAC-Bayes bound above.

Example 2.1 (Finite case) Let us consider a special case in which the set $\mathcal{Q}$ is finite. In such a case, the Gibbs posterior $\hat{Q}$ is a probability on the finite set Θ given by

$$\hat{Q}(\theta) = \frac{e^{-\lambda \hat{R}_S(\theta)}\pi(\theta)}{\sum_{\vartheta\in\Theta} e^{-\lambda \hat{R}_S(\vartheta)}\pi(\vartheta)},$$

and, with a probability of at least $1 - \delta$, we have

$$\mathbb{E}_{\theta\sim\hat{Q}} R(\theta) \le \inf_{Q\in\mathcal{Q}} \left[\mathbb{E}_{\vartheta\sim Q}\hat{R}_S(\vartheta) + \frac{\lambda C^2}{8m} + \frac{\mathrm{KL}(Q\|\pi) + \log 1/\delta}{\lambda}\right].$$

The above upper bound holds for all $Q \in \mathcal{Q}$. Thus, it holds for the Q in the set of the Dirac masses $\{\delta_\theta, \theta \in \Theta\}$. Then, we have $\mathbb{E}_{\vartheta\sim\delta_\theta} \hat{R}_S(\vartheta) = \hat{R}_S(\theta)$ and

$$\mathrm{KL}(\delta_\theta\|\pi) = \sum_{\vartheta\in\Theta} \delta_\theta(\vartheta)\log\frac{\delta_\theta(\vartheta)}{\pi(\vartheta)} = \log\frac{1}{\pi(\theta)}.$$

Furthermore, if the prior π follows a uniform probability distribution, then $\log(1/\pi(h)) = \log(|\Theta|)$, and the bound becomes

$$\mathbb{P}\left(\mathbb{E}_{\theta\sim\hat{Q}} R(\theta) \le \inf_{Q\in\mathcal{Q}} \hat{R}_S(\theta) + \frac{\lambda C^2}{8m} + \frac{\log|\Theta| + \log 1/\delta}{\lambda}\right) \ge 1 - \epsilon.$$

By taking $\lambda = \sqrt{8m/(C^2\log(|\Theta|/\epsilon))}$, we further obtain

$$\mathbb{P}\left(\mathbb{E}_{\theta\sim\hat{Q}} R(\theta) \le \inf_{Q\in\mathcal{Q}} \hat{R}_S(\theta) + C\sqrt{\frac{\log\frac{|\Theta|}{\epsilon}}{2m}}\right) \ge 1 - \epsilon.$$

Example 2.2 (Lipschitz loss and Gaussian priors) We assume that $l(h_\theta, z) \le C$, $\forall h_\theta \in \mathcal{H}$ and $\forall z \in \mathcal{Z}$. Let the loss function $l(h, z)$ be L-Lipschitz for any $z \in \mathcal{Z}$, and let the prior π follow a standard Gaussian distribution $\pi = N(0, \sigma^2 I_{\dim(\theta)})$. Then, for any $\delta \in (0, 1)$, with a probability of at least $1 - \delta$, it holds that

$$\mathbb{E}_{\theta\sim\hat{Q}} R(\theta) \le \inf_{\gamma\in\Theta,\ \gamma:\|\gamma\|\le B} \hat{R}_S(\gamma) + L\sigma\sqrt{\frac{\dim(\gamma)}{m}} + \sqrt{\frac{\frac{B^2}{2\sigma^2} + \frac{\dim(\gamma)}{2}\log m + \log\frac{1}{\epsilon}}{2m}}.$$

Proof Let the posterior $\hat{Q}$ be a Gibbs measure. For any $\delta \in (0, 1)$, it holds that

$$\mathbb{E}_{\theta\sim\hat{Q}} R(\theta) \le \inf_{Q\in\mathcal{Q}} \left[\mathbb{E}_{\vartheta\sim Q}\hat{R}_S(\vartheta) + \frac{\lambda C^2}{8m} + \frac{\mathrm{KL}(Q\|\pi) + \log 1/\delta}{\lambda}\right]$$

with a probability of at least $1 - \delta$. If the probability distribution $Q \in \mathcal{Q}$ is restricted to Gaussian distributions of the form $N(\gamma, s^2 I_{\dim(\theta)})$, where $\gamma \in \mathbb{R}^{\dim(\theta)}$, we have

$$\mathbb{E}_{\theta\sim\hat{Q}}R(\theta) \leq \inf_{Q=N(\gamma,s^2 I_{\dim(\theta)})}\Big[\mathbb{E}_{\vartheta\sim Q}\hat{R}_S(\vartheta) + \frac{\lambda C^2}{8m} + \frac{\mathrm{KL}(Q\|\pi)+\log 1/\delta}{\lambda}\Big].$$

When $Q = N(\gamma, s^2 I_{\dim(\theta)})$ and $\pi = N(0, \sigma^2 I_{\dim(\theta)})$, we obtain

$$\mathrm{KL}(Q\|\pi) = \frac{\|\gamma\|_2^2}{2\sigma^2} + \frac{\dim(\theta)}{2}[\frac{s^2}{\sigma^2} + \log\frac{\sigma^2}{s^2} - 1].$$

Moreover, according to the Jensen's inequality, we obtain

$$\begin{aligned}\mathbb{E}_{\vartheta\sim Q}\hat{R}_S(\vartheta) \leq \hat{R}_S(\gamma) + L\mathbb{E}_{\theta\sim Q}[\|\theta-\gamma\|] \leq& \hat{R}_S(\gamma) + L\sqrt{\mathbb{E}_{\theta\sim Q}[\|\theta-\gamma\|^2]}\\ =&\hat{R}_S(\gamma) + Ls\sqrt{\dim(\theta)}.\end{aligned}$$

Combining the above results, we have

$$\begin{aligned}\mathbb{E}_{\theta\sim\hat{Q}}R(\theta) \leq \inf_{\gamma}\Big[&\hat{R}_S(\gamma) + Ls\sqrt{\dim(\gamma)}\\ &+\frac{\lambda C^2}{8m} + \frac{\frac{\|\gamma\|_2^2}{2\sigma^2} + \frac{\dim(\gamma)}{2}[\frac{s^2}{\sigma^2} + \log\frac{\sigma^2}{s^2} - 1] + \log 1/\delta}{\lambda}\Big].\end{aligned}$$

Taking $s = \sigma/\sqrt{m}$ yields

$$\begin{aligned}\mathbb{E}_{\theta\sim\hat{Q}}R(\theta) \leq \inf_{\gamma}\Big[&\hat{R}_S(\gamma) + L\sigma\sqrt{\frac{\dim(\gamma)}{m}} + \frac{\lambda C^2}{8m}\\ &+\frac{\frac{\|\gamma\|_2^2}{2\sigma^2} + \frac{\dim(\gamma)}{2}[\frac{1}{m} - 1 + \log m] + \log 1/\delta}{\lambda}\Big].\end{aligned}$$

By further assuming that $\|\gamma\| \leq B$ for some $B > 0$, we obtain

$$\begin{aligned}\mathbb{E}_{\theta\sim\hat{Q}}R(\theta) \leq \inf_{\gamma:\|\gamma\|\leq B}\Big[&\hat{R}_S(\gamma) + L\sigma\sqrt{\frac{\dim(\gamma)}{m}} + \frac{\lambda C^2}{8m}\\ &+\frac{\frac{\|\gamma\|_2^2}{2\sigma^2} + \frac{\dim(\gamma)}{2}[\frac{1}{m} - 1 + \log m] + \log 1/\delta}{\lambda}\Big].\end{aligned}$$

In this case, we can see that the optimal λ is $\frac{1}{C}\sqrt{8m(\frac{B^2}{2\sigma^2} + \frac{\dim(\gamma)}{2}\log m + \log\frac{1}{\delta})}$, which indicates that, with a probability of at least $1-\delta$, it holds that

$$\mathbb{E}_{\theta\sim\hat{Q}}R(\theta) \leq \inf_{\gamma:\|\gamma\|\leq B}\hat{R}_S(\gamma) + L\sigma\sqrt{\frac{\dim(\gamma)}{m}} + \sqrt{\frac{\frac{B^2}{2\sigma^2} + \frac{\dim(\gamma)}{2}\log m + \log\frac{1}{\epsilon}}{2m}}.$$

This completes the proof. □

References

David, A McAllester. 1999. PAC-Bayesian model averaging. *In Annual Conference of Learning Theory* 99: 164–170.

Donsker, M.D., S.R. Srinivasa, and Varadhan. 1976. On the principal eigenvalue of second-order elliptic differential operators. *Communications on Pure and Applied Mathematics* 29 (6): 595–621.

Elias, M Stein and Rami Shakarchi. 2009. *Real Analysis: Measure Theory, Integration, and Hilbert Spaces*. Princeton University Press.

Jon Wellner et al. 2013. *Weak Convergence and Empirical Processes: With Applications to Statistics*. Springer Science & Business Media.

Olivier, Catoni. 2007. Pac-bayesian supervised classification: the thermodynamics of statistical learning. *arXiv preprint* arXiv:0712.0248.

Chapter 3
Hypothesis Complexity

This chapter presents the hypothesis complexity concept frequently used in statistical learning theory, which is important for deriving generalization bounds for deep learning models. The hypothesis complexity characterizes the complexity of a machine learning algorithm and can be measured in terms of the Vapnik-Chervonenkis (VC) dimension, Rademacher complexity, and covering number. Intuitively, a more complex algorithm has worse generalizability. In this way, we may study the generalizability via the hypothesis complexity.

3.1 Worst-Case Bounds Based on the Rademacher Complexity

A major measure for evaluating the hypothesis complexity in conventional statistical learning theory is the Rademacher complexity (Bartlett and Mendelson 2002), defined below.

Definition 3.1 (Empirical Rademacher complexity and Rademacher complexity) For a real-valued function class denoted by $\mathcal{H}$ and a dataset S, the empirical Rademacher complexity is defined as follows:

$$\hat{\mathfrak{R}}(\mathcal{H}) = \mathbb{E}_\epsilon \left[\sup_{h \in \mathcal{H}} \frac{1}{m} \sum_{i=1}^{m} \epsilon_i h(x_i) \right], \tag{3.1}$$

where $\epsilon = \{\epsilon_1, \ldots, \epsilon_m\}$ and the ϵ_i, $i = 1, \ldots, m$, are independently and uniformly drawn from $\{-1, +1\}$. In turn, the Rademacher complexity can be defined as follows:

$$\mathfrak{R}(\mathcal{H}) = \mathbb{E}_S \hat{\mathfrak{R}}(\mathcal{H}). \tag{3.2}$$

F. He and D. Tao, *Foundations of Deep Learning*, Machine Learning: Foundations, Methodologies, and Applications, https://doi.org/10.1007/978-981-16-8233-9_3

One can obtain uniform generalization bounds via the Rademacher complexity.

Theorem 3.1 (see Mohri et al. (2018)) *Let $\mathcal{H}$ be a family of functions taking values in {-1,+1}, and let $\mathcal{D}$ be the distribution over the input space $\mathcal{X}$. Then, for any $\delta > 0$, with probability at least $1-\delta$ over samples S of size m drawn according to $\mathcal{D}^m$, the following holds for any $h \in \mathcal{H}$:*

$$R(h) \leq \hat{R}_S(h) + \mathfrak{R}(\mathcal{H}) + \sqrt{\frac{\log \frac{1}{\delta}}{2m}}. \tag{3.3}$$

Proof For any sample $S = (x_1, \ldots, x_m)$, we define a function D as follows:

$$D(S) = \sup_{h \in \mathcal{H}} \left(R(h) - \hat{R}_S(h) \right). \tag{3.4}$$

If S and S' have only one different point, we have

$$D(S) - D(S') \leq \sup_{h \in \mathcal{H}} \left(\hat{R}_S(h) - \hat{R}_{S'}(h) \right) \leq \frac{1}{m}. \tag{3.5}$$

Then, by McDiarmid's inequality, for any $\delta > 0$, with probability at least $1-\delta$, the following holds:

$$D(S) \leq \mathbb{E}_S[D(S)] + \sqrt{\frac{\log \frac{1}{\delta}}{2m}}. \tag{3.6}$$

In addition, the expectation of the right-hand side can be bounded as follows:

$$\begin{aligned}
\mathbb{E}_S[D(S)] =& \mathbb{E}_S\left[\sup_{h \in \mathcal{H}} \left(R(h) - \hat{R}_S(h) \right) \right] \\
\leq& \mathbb{E}_{S,S'}\left[\sup_{h \in \mathcal{H}} \left[\hat{R}_{S'}(h) - \hat{R}_S(h) \right] \right] \\
=& \mathbb{E}_{\sigma,S,S'}\left[\sup_{h \in \mathcal{H}} \frac{1}{m} \sum_{i=1}^m \sigma_i \left(\frac{1 - y_i' h(x_i')}{2} - \frac{1 - y_i h(x_i)}{2} \right) \right] \\
\leq& \mathbb{E}_{\sigma,S'}\left[\sup_{h \in \mathcal{H}} \frac{1}{m} \sum_{i=1}^m \sigma_i \frac{h(x_i')}{2} \right] + \mathbb{E}_{\sigma,S}\left[\sup_{h \in \mathcal{H}} \frac{1}{m} \sum_{i=1}^m \sigma_i \frac{h(x_i)}{2} \right] \\
=& \mathfrak{R}(\mathcal{H}).
\end{aligned} \tag{3.7}$$

As a result,

$$R(h) \leq \hat{R}_S(h) + \mathfrak{R}(\mathcal{H}) + \sqrt{\frac{\log \frac{1}{\delta}}{2m}}. \tag{3.8}$$

The proof is complete. □

Moreover, an important bound of $\mathfrak{R}(\mathcal{H})$ is given based on Massart's lemma.

Lemma 3.1 *(Massart's lemma) Let $\mathcal{G} \subset \mathbb{R}^m$ be a finite set, with $r = \max_{\mathbf{x}\in\mathcal{G}} \|x\|_2$; then, the following holds:*

$$\mathbb{E}_\sigma\left[\frac{1}{m}\sup_{\mathbf{x}\in\mathcal{G}}\sum_{i=1}^{m}\sigma_i x_i\right] \leq \frac{r\sqrt{2\log|\mathcal{G}|}}{m}, \tag{3.9}$$

where the σ_i are independent uniform random variables taking values in $\{-1, +1\}$ and $x_1, \ldots, x_m$ are the components of the vector $\mathbf{x}$.

Proof Let $t > 0$ be a number to be chosen later.

$$\begin{aligned}
\exp\mathbb{E}_\sigma\left[\sup_{\mathbf{x}\in\mathcal{G}}\sum_{i=1}^{m}t\sigma_i x_i\right] &\leq \mathbb{E}_\sigma\left[\exp\sup_{\mathbf{x}\in\mathcal{G}}\sum_{i=1}^{m}t\sigma_i x_i\right] = \mathbb{E}_\sigma\left[\sup_{\mathbf{x}\in\mathcal{G}}\exp\sum_{i=1}^{m}t\sigma_i x_i\right]\\
&\leq \sum_{\mathbf{x}\in\mathcal{G}}\mathbb{E}_\sigma\left[\exp\sum_{i=1}^{m}t\sigma_i x_i\right] = \sum_{\mathbf{x}\in\mathcal{G}}\prod_{i=1}^{m}\mathbb{E}_\sigma[\exp(t\sigma_i x_i)]\\
&\leq \sum_{\mathbf{x}\in\mathcal{G}}\prod_{i=1}^{m}\exp\left(\frac{(tx_i)^2}{2}\right) = \sum_{\mathbf{x}\in\mathcal{G}}\exp\left(\sum_{i=1}^{m}\frac{(tx_i)^2}{2}\right)\\
&\leq \sum_{\mathbf{x}\in\mathcal{G}}\exp(\frac{t^2r^2}{2}) = |\mathcal{G}|\exp(\frac{t^2r^2}{2}).
\end{aligned}$$

The first inequality comes from Jensen's inequality, and the third inequality is from Hoeffding's inequality. By the result above, we have

$$\mathbb{E}_\sigma\left[\sup_{\mathbf{x}\in\mathcal{G}}\sum_{i=1}^{m}\sigma_i x_i\right] \leq \frac{1}{t}\log|\mathcal{G}| + \frac{tr^2}{2}. \tag{3.10}$$

Let t be $\sqrt{2\log|\mathcal{G}|/r^2}$. Then, we can obtain the result that we have claimed.

The proof is complete. □

One can prove a margin bound on the basis of the *covering number*, defined as follow, or the *Rademacher complexity* of the hypothesis space.

Definition 3.2 (Covering number) The covering number $\mathcal{N}(\mathcal{H}, \epsilon, \|\cdot\|)$ of a space $\mathcal{H}$ is defined as the minimal cardinality of any subset $V \subset \mathcal{H}$ that covers $\mathcal{H}$ at scale ϵ under metric $\|\cdot\|$, i.e.,

$$\sup_{A\in\mathcal{H}}\min_{B\in V}\|A - B\| \leq \epsilon. \tag{3.11}$$

Intuitively, the covering number reflects the richness of the hypothesis space. A larger covering number indicates a space with more balls needed to represent it, suggesting a potentially more complex set of hypotheses.

The covering number indicates how many balls are needed to cover the hypothesis space, whereas the Rademacher complexity measures the maximal correlation between a hypothesis and noise and thus characterizes the "goodness-of-fit" of the hypothesis space to noise. A higher Rademacher complexity suggests that hypotheses in the class might be overly susceptible to noise, leading to poorer generalization performance on unseen data.

The relationship between hypothesis complexity and these measures becomes evident when we consider techniques like the Dudley entropy integral.

Theorem 3.2 (Dudley entropy integral) *Let $\mathcal{H}$ be a real-valued function class taking values in* $[0, 1]$, *and assume that* $0 \in \mathcal{H}$. *Then,*

$$\mathfrak{R}(\mathcal{H}) \leq \inf_{\alpha>0} \left(\frac{4\alpha}{\sqrt{m}} + \frac{12}{m} \int_{\alpha}^{\sqrt{m}} \sqrt{\log \mathcal{N}(\mathcal{H}, \epsilon, \|\cdot\|_2)} d\epsilon \right). \tag{3.12}$$

Proof Let N be an arbitrary positive integer in $\mathbb{N}$, and let $\epsilon_i = \sqrt{m} 2^{-(i-1)}$ for each $i \in [N]$. Let B_i denote the cover achieving $\mathcal{N}(\mathcal{H}, \epsilon, \|\cdot\|_2)$. By the definition of the covering number, for any $f \in \mathcal{H}$, there exists a $b^i[f] \in B_i$ such that

$$\|f - b^i[f]\|_2 \leq \epsilon_i.$$

Moreover, we have the following:

$$\begin{aligned}
&\mathbb{E}_\sigma \left[\sup_{f\in\mathcal{H}} \sum_{t=1}^{m} \sigma_t f(x_t) \right] \\
=&\mathbb{E}_\sigma \sup_{f\in\mathcal{H}} \left[\sum_{t=1}^{m} \sigma_t (f(x_t) - b_t^N[f]) + \sum_{i=1}^{N-1} \sum_{t=1}^{m} \sigma_t (b_t^{i+1}[f] - b_t^i[f]) + \sum_{t=1}^{m} \sigma_t b_t^1[f] \right] \\
\leq&\mathbb{E}_\sigma \sup_{f\in\mathcal{H}} \left[\sum_{t=1}^{m} \sigma_t (f(x_t) - b_t^N[f]) \right] + \sum_{i=1}^{N-1} \mathbb{E}_\sigma \sup_{f\in\mathcal{H}} \left[\sum_{t=1}^{m} \sigma_t (b_t^{i+1}[f] - b_t^i[f]) \right] \\
&+ \mathbb{E}_\sigma \sup_{f\in\mathcal{H}} \left[\sum_{t=1}^{m} \sigma_t b_t^1[f] \right]. \qquad \text{(Triangle inequality)}
\end{aligned}$$

Let $b^1 = \mathbf{0}$; then, the third term becomes 0. For the first term, we can use the Cauchy-Schwarz inequality to obtain

$$\begin{aligned}
\mathbb{E}_\sigma \sup_{f\in\mathcal{H}} \left[\sum_{t=1}^{m} \sigma_t (f(x_t) - b_t^N[f]) \right] \leq& \sqrt{\mathbb{E}_\sigma \left[\sum_{t=1}^{m} \sigma_t^2 \right] \mathbb{E}_\sigma \left[\sup_{f\in\mathcal{H}} \sum_{t=1}^{m} (f(x_t) - b_t^N[f])^2 \right]} \\
\leq& \sqrt{m} \epsilon_N.
\end{aligned}$$

Let $W_i = \{b_t^{i+1}[f] - b_t^i[f] : f \in \mathcal{H}\}$. Then, we have

$$|W_i| \le |B_i||B_{i+1}| \le |B_{i+1}|^2$$

and

$$\sup_{w_i \in W_i} \|w_i\|_2 \le \sup_{f \in \mathcal{H}} \|f - b^i[f]\|_2 + \sup_{f \in \mathcal{H}} \|f - b^{i+1}[f]\|_2 \le \epsilon_i + \epsilon_{i+1} \le 3\epsilon_{i+1}.$$

Thus, by Massart's lemma (Lemma 3.1), we have

$$\mathbb{E}_\sigma \sup_{f \in \mathcal{H}} \left[\sum_{t=1}^m \sigma_t (b_t^{i+1}[f] - b_t^i[f]) \right] \le \mathbb{E}_\sigma \sup_{w_i \in W_i} \left[\sum_{t=1}^m \sigma_t w_t \right] \le 6\sqrt{\log |B_{i+1}|}\epsilon_{i+1}.$$

Collecting all terms, we have

$$\begin{aligned}
\mathbb{E}_\sigma \left[\sup_{f \in \mathcal{H}} \sum_{t=1}^m \sigma_t f(x_t) \right] &\le \sqrt{m}\epsilon_N + \sum_{i=1}^{N-1} 6\sqrt{\log |B_{i+1}|}\epsilon_{i+1} \\
&\le \sqrt{m}\epsilon_N + 12 \sum_{i=1}^{N} (\epsilon_i - \epsilon_{i+1}) \sqrt{\log \mathcal{N}(\mathcal{H}, \epsilon_i, \|\cdot\|_2)} \\
&\le \sqrt{m}\epsilon_N + 12 \int_{\epsilon_{N+1}}^{\sqrt{m}} \sqrt{\log \mathcal{N}(\mathcal{H}, \epsilon, \|\cdot\|_2)} d\epsilon.
\end{aligned}$$

Finally, for any $\alpha > 0$, let $\epsilon_{N+1} > \alpha$ and $\epsilon_N \le 4\alpha$ for some N; then,

$$\mathfrak{R}(\mathcal{H}) \le \frac{4\alpha}{\sqrt{m}} + \frac{12}{m} \int_{\alpha}^{\sqrt{m}} \sqrt{\log \mathcal{N}(\mathcal{H}, \epsilon, \|\cdot\|_2)} d\epsilon, \tag{3.13}$$

which completes the proof. □

3.2 Worst-Case Bounds Based on the Vapnik-Chervonenkis (VC) Dimension

Another major measure of hypothesis complexity is the VC dimension (Vapnik and Chervonenkis 2015), which is defined as follows.

Definition 3.3 (Growth function, shattering, and Vapnik-Chervonenkis (VC) dimension) For any nonnegative integer m, the growth function of a hypothesis space $\mathcal{H}$ is defined as follows:

$$\Pi_{\mathcal{H}}(m) = \max_{x_1,\ldots,x_m \in \mathcal{X}} |\{(h(x_1), \ldots, h(x_m)) : h \in \mathcal{H}\}|. \tag{3.14}$$

If $\Pi_{\mathcal{H}}(m) = 2^m$, we say that $\mathcal{H}$ shatters the dataset $\{x_1, \ldots, x_m\}$. We define the VC dimension VCdim($\mathcal{H}$) as the size of the largest shattered set.

We can obtain the following theorem regarding the relationship between the growth function and the VC dimension.

Theorem 3.3 (Sauer's lemma; Shelah (1972); Sauer (1972)) *Let $\mathcal{H}$ be a hypothesis set with VCdim($\mathcal{H}$) $= d$. Then, for all $m \in \mathbb{N}$, the following inequality holds:*

$$\Pi_{\mathcal{H}}(m) \leq \sum_{i=0}^{d} \binom{m}{i}. \tag{3.15}$$

Proof We will prove this theorem by induction on $m + d$.

First, the statement obviously holds for $m = 1$ and $d = 0$ or $d = 1$.

Second, we assume that the statement holds for $(m-1, d-1)$ and $(m-1, d)$. Then, we will show that the statement is also true for (m, d). Let $\mathcal{S} = \{x_1, \ldots, x_m\}$ be a set with $\Pi_{\mathcal{H}}(m)$ dichotomies, and let $\mathcal{G} = \mathcal{H}_{|\mathcal{S}}$ be the set of concepts $\mathcal{H}$ induced by restriction to $\mathcal{S}$, that is, $\mathcal{G} = \{\{x_i : h(x_i) = 1\} : h \in \mathcal{H}\}$. Thus, $\Pi_{\mathcal{H}}(m) = |\mathcal{G}|$.

To use our assumptions, similarly, we consider $\mathcal{S}' = \{x_1, \ldots, x_{m-1}\}$ and $\mathcal{G}_1 = \mathcal{H}_{|\mathcal{S}_1}$ as the set of concepts induced by restriction to $\mathcal{S}_1$.

Now, we consider the gap between $\mathcal{G}$ and $\mathcal{G}_1$. That is, in $\mathcal{S}'$, we omit only the value $h(x_m)$. Based on this observation, we define $\mathcal{G}_2$ as

$$\mathcal{G}_2 = \{g' \subset S' : (g' \in \mathcal{G}) \wedge (g' \cup \{x_m\} \in \mathcal{G})\}. \tag{3.16}$$

This implies that $|\mathcal{G}| = |\mathcal{G}_1| + |\mathcal{G}_2|$.

Next, we will calculate the scales of $\mathcal{G}_1$ and $\mathcal{G}_2$. We can view each point g in $\mathcal{G}_1$ as a concept satisfying $h(x_i) = 1$ if and only if $x_i \in g$. In this sense, $\mathcal{G}_1$ and $\mathcal{G}_2$ can be viewed as consisting of concepts induced by restriction to $\mathcal{S}_1$.

Note that VCdim($\mathcal{G}_1$) $\leq$ VCdim($\mathcal{G}$) $\leq d$; then, from the assumption for $(m-1, d)$, we know that

$$|\mathcal{G}_1| \leq \Pi_{\mathcal{G}_1}(m-1) \leq \sum_{i=0}^{d} \binom{m-1}{i}. \tag{3.17}$$

Moreover, if $\mathcal{Z} \subset \mathcal{S}'$ is shattered by $\mathcal{G}_2$, then the set $\mathcal{Z} \cup \{x_m\}$ is shattered by $\mathcal{G}$. Hence, we have

$$\text{VCdim}(\mathcal{G}_2) \leq \text{VCdim}(\mathcal{G}) - 1 = d - 1, \tag{3.18}$$

and from the assumption for $(m-1, d-1)$,

$$|\mathcal{G}_2| \leq \Pi_{\mathcal{G}_2}(m-1) \leq \sum_{i=0}^{d-1} \binom{m-1}{i}. \tag{3.19}$$

Finally, we have

$$|\mathcal{G}| = |\mathcal{G}_1| + |\mathcal{G}_2| \le \sum_{i=0}^{d} \binom{m-1}{i} + \sum_{i=0}^{d-1} \binom{m-1}{i} = \sum_{i=0}^{d} \binom{m}{i}, \tag{3.20}$$

which completes the inductive proof. □

One can obtain uniform generalization bounds in terms of the VC dimension as follows (Mohri et al. 2018).

Theorem 3.4 (see (Mohri et al. 2018)) *Suppose that a hypothesis space $\mathcal{H}$ has VC dimension $VCdim(\mathcal{H})$. Then, for any $\delta > 0$, with probability $1 - \delta$, the following inequality holds for any $h \in \mathcal{H}$:*

$$R(h) \le \hat{R}_S(h) + \sqrt{\frac{2VCdim(\mathcal{H}) \log \frac{em}{VCdim(\mathcal{H})}}{m}} + \sqrt{\frac{\log \frac{1}{\delta}}{2m}}. \tag{3.21}$$

This theorem is based on two lemmas, as follows.

Lemma 3.2 *Let $\mathcal{H}$ be a family of functions taking values in $\{-1, +1\}$. Then, the following inequality holds:*

$$\mathfrak{R}(\mathcal{H}) \le \sqrt{\frac{2 \log \Pi_{\mathcal{H}}(m)}{m}}. \tag{3.22}$$

Proof Lemma 3.2 follows from Massart's lemma (Lemma 3.1). Let the set of vectors of function values $(h(x_1), \ldots, h(x_m))^T$ be denoted by $\mathcal{H}_{|S}$; then, we have

$$\mathfrak{R}(\mathcal{H}) \le \mathbb{E}_S\left[\frac{\sqrt{m}\sqrt{2\log|\mathcal{H}_{|S}|}}{m}\right] \le \sqrt{\frac{2\log \Pi_{\mathcal{H}}(m)}{m}}. \tag{3.23}$$

The first inequality follows from Massart's lemma, and the second inequality follows from the definition of $\Pi_{\mathcal{H}}(m)$. □

The other lemma is a foundational lemma in combinatorial mathematics and extremal set theory that was independently proven by (Shelah 1972) and (Sauer 1972).

We now may prove Theorem 3.4.

Proof of Theorem 3.4 Theorem 3.3 yields

$$\Pi_{\mathcal{H}}(m) \le \sum_{i=0}^{d} \binom{m}{i} \le \sum_{i=0}^{m} \binom{m}{i} \left(\frac{m}{d}\right)^{d-i} = \left(\frac{m}{d}\right)^{d} \left(1 + \frac{d}{m}\right)^{m} \le \left(\frac{m}{d}\right)^{d} e^{d}. \tag{3.24}$$

By combining this result with Lemma 3.2 above, we complete the proof of Theorem 3.4.

References

Mehryar, Mohri, Afshin Rostamizadeh, and Ameet Talwalkar. 2018. *Foundations of Machine Learning*. MIT press.

Nesterov, Yurii E. 1983. A method for solving the convex programming problem with convergence rate o (1/k∧ 2). *In Dokl. Akad. Nauk Sssr* 269: 543–547.

Peter, L Bartlett, and Shahar Mendelson. 2002. Rademacher and Gaussian complexities: Risk bounds and structural results. *Journal of Machine Learning Research* 3: 463–482.

Sauer, Norbert. 1972. On the density of families of sets. *Journal of Combinatorial Theory, Series A* 13 (1): 145–147.

Shelah, Saharon. 1972. A combinatorial problem; stability and order for models and theories in infinitary languages. *Pacific Journal of Mathematics* 41 (1): 247–261.

Vladimir, N Vapnik and A Ya Chervonenkis. 2015. On the uniform convergence of relative frequencies of events to their probabilities. In *Measures of complexity*, 11–30. Springer.

Chapter 4
Algorithmic Stability

Recall the classical uniform generalization bounds, which essentially depend on a measure of the capacity of the hypothesis space $\mathcal{H}$ (e.g., the Rademacher complexity or covering number). This type of upper bound is independent of any specific algorithm because it provides a guarantee for all hypotheses $h \in \mathcal{H}$ simultaneously. In contrast to the above approaches, generalization bounds associated with a specific learning algorithm can be established based on the concept of algorithmic stability, which refers to the property that the hypothesis output by an algorithm does not change significantly when one training point is perturbed. In this chapter, we introduce algorithmic stability to characterize the impact of randomness in the training process on the generalizability of an algorithm.

4.1 Definition of Algorithmic Stability

Algorithm stability is defined in terms of the difference in the output of an algorithm on a training set S and a perturbed version S^i, where $S^i = (z_1, ..., z_i', ..., z_m)$ is obtained by replacing the i-th point of S with a new point $z_i' \in \mathcal{Z}$. We regard a learning algorithm as a function $\mathcal{A}(S) : \mathcal{Z}^m \to \mathcal{H}$ that takes the training sample S as input and produces the hypothesis $h \in \mathcal{H}$ as output. In this section, we consider only a deterministic learning algorithm $\mathcal{A}$, which does not depend on the ordering of the points in the training set.

We introduce several widely used definitions of algorithmic stability below (Bousquet and Elisseeff 2002).

F. He and D. Tao, *Foundations of Deep Learning*, Machine Learning: Foundations, Methodologies, and Applications, https://doi.org/10.1007/978-981-16-8233-9_4

Definition 4.1 *(Hypothesis stability)* An algorithm $\mathcal{A}$ has hypothesis stability β with respect to the loss function l if the following holds:

$$\forall i \in \{1, \ldots, m\}, \mathbb{E}_S[|l(\mathcal{A}(S), z_i) - l(\mathcal{A}(S^{\setminus i}), z_i)|] \leq \beta, \tag{4.1}$$

where $S^{\setminus i} = \{z_1, .., z_{i-1}, z_{i+1}, ..., z_m\}$.

Definition 4.2 *(Pointwise hypothesis stability)* An algorithm $\mathcal{A}$ has pointwise hypothesis stability β with respect to the loss function l if the following holds:

$$\forall i \in \{1, \ldots, m\}, \mathbb{E}_S[|l(\mathcal{A}(S), z_i) - l(\mathcal{A}(S^{\setminus i}), z_i)|] \leq \beta. \tag{4.2}$$

Definition 4.3 *(Error stability)* An algorithm $\mathcal{A}$ has error stability β with respect to the loss function l if the following holds:

$$\forall S \in \mathcal{Z}^m, \forall i \in \{1, \ldots, m\}, |\mathbb{E}_z[l(\mathcal{A}(S), z)] - \mathbb{E}_z[l(\mathcal{A}(S^{\setminus i}), z)]| \leq \beta. \tag{4.3}$$

Definition 4.4 *(Uniform stability)* Given a training sample $S = \{(x_i, y_i)\}_{i=1}^m$ of size m, an algorithm $\mathcal{A}$ is β-uniformly stable if $\forall m \in \mathbb{N}, i \in [m], S \in \mathcal{Z}^m, z_i' \in \mathcal{Z}$, we have

$$\sup_{z \in \mathcal{Z}}[|l(\mathcal{A}(S), z) - l(\mathcal{A}(S^i), z)|] \leq \beta.$$

Definition 4.5 *(Classification stability)* A real-valued classification algorithm $\mathcal{A}$ has classification stability β if the following holds:

$$\forall S \in \mathcal{Z}^m, \forall i \in \{1, \ldots, m\}, \|\mathcal{A}(S)(z) - \mathcal{A}(S^{\setminus i})(z)\| \leq \beta. \tag{4.4}$$

4.2 Algorithmic Stability and Generalization Error Bounds

Given that an algorithm $\mathcal{A}$ has uniform stability β, an upper bound on its generalization error can be established with the help of McDiarmid's inequality.

Definition 4.6 (McDiarmid's inequality) Suppose that a function $f : \mathcal{Z}^m \to \mathbb{R}$ satisfies $\sup_{S,S^i} |f(S) - f(S^i)| \leq c_i$, $i = 1, ..., m$; then, the following statement holds:

$$\mathbb{P}(|f(S) - \mathbb{E}_S[f(S)]| > \epsilon) \leq 2\exp(-\frac{2\epsilon^2}{\sum_{i=1}^m c_i^2}).$$

We can then prove the following polynomial generalization bound in terms of the hypothesis stability (Bousquet and Elisseeff 2002).

Theorem 4.1 (Polynomial generalization bound in terms of hypothesis stability) *For a learning algorithm $\mathcal{A}$ with hypothesis stability β_1 and pointwise hypothesis* *stability β_2 with respect to a loss function l such that $0 \le l(y, y') \le M$, we have, with probability $1 - \delta$,*

$$R(\mathcal{A}(S)) \le \hat{R}_S(\mathcal{A}(S)) + \sqrt{\frac{M^2 + 12Mm\beta_2}{2m\delta}} \tag{4.5}$$

and

$$R(\mathcal{A}(S)) \le \hat{R}_S(\mathcal{A}(S^{\backslash i})) + \sqrt{\frac{M^2 + 6Mm\beta_1}{2m\delta}}. \tag{4.6}$$

Proof Note that

$$\begin{aligned}\mathbb{E}_{\mathcal{S},z_i'}[|l(\mathcal{A}(S), z_i) - l(\mathcal{A}(S^i), z_i)|] \le & \mathbb{E}_{\mathcal{S}}[|l(\mathcal{A}(S), z_i) - l(\mathcal{A}(S^{\backslash i}), z_i)|] \\ & + \mathbb{E}[|l(\mathcal{A}(S^{\backslash i}), z_i) - l(\mathcal{A}(S^i), z_i)|] \\ \le & 2\beta_2.\end{aligned}$$

Thus, we have

$$\begin{aligned}\mathbb{E}[(R(\mathcal{A}(S)) - \hat{R}_S(\mathcal{A}(S))^2] &\le \frac{M^2}{2m} + 3M\mathbb{E}_{\mathcal{S},z_i'}[|l(\mathcal{A}(S), z_i) - l(\mathcal{A}(S^i), z_i)|] \\ &= \frac{M^2}{2m} + 6M\beta_2\end{aligned}$$

and

$$\begin{aligned}\mathbb{E}[(R(\mathcal{A}(S)) - \hat{R}_S(\mathcal{A}(S^{\backslash i})))^2] &\le \frac{M^2}{2m} + 3M\mathbb{E}_{S,z}[|l(\mathcal{A}(S), z) - l(\mathcal{A}(S^{\backslash i}), z)|] \\ &= \frac{M^2}{2m} + 3M\beta_1.\end{aligned}$$

Finally, by Chebyshev's inequality $\mathbb{P}[X \ge \delta] \le \mathbb{E}[X^2]/\delta^2$, we have that, with probability at least $1 - \delta$,

$$X \le \sqrt{\frac{\mathbb{E}[X^2]}{\delta}}. \tag{4.7}$$

By combining this result with

$$X = R(\mathcal{A}(S)) - \hat{R}_S(\mathcal{A}(S)) \tag{4.8}$$

and

$$X = R(\mathcal{A}(S)) - \hat{R}_S(\mathcal{A}(S^{\backslash i})), \tag{4.9}$$

we obtain Theorem (4.1).

The proof is complete. □

We introduce a modified cost function

$$l_\gamma(f, z) = \begin{cases} 1 & \text{if } f(x)y < 0 \\ 1 - f(x)y/\gamma & \text{if } 0 \le f(x)y \le \gamma \\ 0 & \text{if } f(x)y \ge \gamma \end{cases} \tag{4.10}$$

and then define

$$\hat{R}_S^\gamma(\mathcal{A}(S)) = \frac{1}{m}\sum_{i=1}^{m} l_\gamma(\mathcal{A}(S), z_i). \tag{4.11}$$

Then, we may prove the following theorem based on this new notion.

Theorem 4.2 *Let $\mathcal{A}$ be a real-valued classification algorithm with stability β. Then, for all $\gamma > 0$, any $m \ge 1$, and any $\delta \in (0, 1)$, with probability at least $1 - \delta$ over the random draw of sample S,*

$$R(\mathcal{A}(S)) \le \hat{R}_S^\gamma(\mathcal{A}(S)) + 2\frac{\beta}{\gamma} + \left(4m\frac{\beta}{\gamma} + 1\right)\sqrt{\frac{\ln(1/\delta)}{2m}} \tag{4.12}$$

and

$$R(\mathcal{A}(S)) \le \hat{R}_S^\gamma(\mathcal{A}(S^{\setminus i})) + \frac{\beta}{\gamma} + \left(4m\frac{\beta}{\gamma} + 1\right)\sqrt{\frac{\ln(1/\delta)}{2m}}. \tag{4.13}$$

Proof The loss function l_γ is bounded by $M = 1$, and the algorithm is β/γ-stable. By using the fact that $R(\mathcal{A}(S)) \le R^\gamma = \mathbb{E}_z[l_\gamma(\mathcal{A}(S), z)]$ and applying Theorem 4.1, we complete the proof. □

Based on the uniform stability, we give an exponential generalization bound.

Theorem 4.3 (Exponential generalization bound in terms of uniform stability) *Let algorithm $\mathcal{A}$ be β-uniformly stable, and let the loss function satisfy $l(h, z) \le M$, $\forall h \in \mathcal{H}$ and $\forall z \in \mathcal{Z}$. Given a training sample $S = \{(x_i, y_i)\}_{i=1}^m$, for any $\delta \in (0, 1)$, with probability at least $1 - \delta$, it holds that*

$$R(\mathcal{A}(S)) \le \hat{R}_S(\mathcal{A}(S)) + \beta + (m\beta + M)\sqrt{\frac{2\ln(2/\delta)}{m}}.$$

Proof We adopt the notation $D(\mathcal{A}(S)) = R(\mathcal{A}(S)) - \hat{R}_S(\mathcal{A}(S))$ for notational simplicity. In the rest of the proof, we will prove that $D(\mathcal{A}(S))$ is close to its expectation $\mathbb{E}_S[D(\mathcal{A}(S))]$ and that both have uniform β-stability:

$$\begin{aligned} \mathbb{E}_S[D(\mathcal{A}(S))] &= \mathbb{E}_{S,z}[l(\mathcal{A}(S), z) - \frac{1}{m}\sum_{i=1}^{m} l(\mathcal{A}(S), z_i)] \\ &= \mathbb{E}_{S,z'}[l(\mathcal{A}(S), z') - \frac{1}{m}\sum_{i=1}^{m} l(\mathcal{A}(S^i), z')] \\ &\le \beta, \end{aligned}$$

where the third equality is based on the "symmetry" of the expectation. Let us verify the conditions required in McDiarmid's inequality (Definition 4.6). Given that an algorithm $\mathcal{A}$ with uniform stability β, for any $m \in \mathbb{N}$ and $S, S^i \in \mathcal{Z}^m$, it holds that

$$\begin{aligned}|D(\mathcal{A}(S)) - D(\mathcal{A}(S^i))| &\leq |R(\mathcal{A}(S)) - R(\mathcal{A}(S^i))| + |\hat{R}_S(\mathcal{A}(S)) - \hat{R}_{S^i}(\mathcal{A}(S^i))| \\ &\leq \beta + \frac{1}{m}|l(\mathcal{A}(S), z_i) - l(\mathcal{A}(S^i), z_i')| \\ &\quad + \frac{1}{m}\sum_{j \neq i}|l(\mathcal{A}(S), z_j) - l(\mathcal{A}(S^i), z_j)| \\ &\leq 2\beta + \frac{M}{m}.\end{aligned}$$

Applying McDiarmid's inequality to $D(\mathcal{A}(S))$, we find that for any $\epsilon \in (0, 1)$,

$$\mathbb{P}(|D(\mathcal{A}(S)) - \mathbb{E}D(\mathcal{A}(S))| > \epsilon) \leq 2\exp(-\frac{m\epsilon^2}{2(m\beta + M)^2}).$$

Recall that $\mathbb{E}D(\mathcal{A}(S)) \leq \beta$; then, we have

$$\mathbb{P}(D(\mathcal{A}(S)) > \beta + \epsilon) \leq \mathbb{P}(|D(\mathcal{A}(S)) - \mathbb{E}D(\mathcal{A}(S))| > \epsilon).$$

Hence,

$$\mathbb{P}(D(\mathcal{A}(S)) > \beta + \epsilon) \leq 2\exp(-\frac{m\epsilon^2}{2(m\beta + M)^2}).$$

Eventually, we take

$$\delta = 2\exp(-\frac{m\epsilon^2}{2(m\beta + M)^2}),$$

i.e.,

$$\epsilon = (m\beta + M)\sqrt{\frac{2\ln(2/\delta)}{m}}.$$

Thus, we obtain the claimed result. □

Deep learning algorithms usually have an extremely large "total" hypothesis complexity, whereas optimizers may explore only a small part of the hypothesis space. The following notion helps characterize the "effective" hypothesis complexity, which depends on both the learning algorithm and the training data.

Definition 4.7 *(Algorithmic Rademacher complexity)* The Rademacher complexity of a hypothesis class $\mathcal{H}$ on the feature space $\mathcal{X}$ is defined as

$$\mathfrak{R}(\mathcal{H}) = \mathbb{E}\left[\sup_{h \in \mathcal{H}} \frac{1}{m}\sum_{i=1}^{m} \sigma_i \langle h, x_i \rangle\right], \tag{4.14}$$

where $\sigma_1, \ldots, \sigma_m$ are i.i.d. Rademacher variables that are uniformly distributed in $\{-1, +1\}$.

With the help of algorithmic stability, one may prove a generalization bound in terms of the algorithmic Rademacher complexity (Liu et al. 2017).

Theorem 4.4 (Generalization bound in terms of algorithmic Rademacher complexity) *Let $\mathfrak{B}$ be a separable Hilbert space. Suppose that $\|x_i\|_* \leq B$ with probability one for some $B > 0$ and that the loss function is bounded and Lipschitz, that is, $l(h, z) \leq M$ with probability one for some $M > 0$ and $|l(h, z) - l(h', z)| \leq L|\langle h, x\rangle - \langle h', x\rangle|$ for all $z \in \mathcal{Z}$ and $h, h' \in \mathcal{H}$ ($\mathcal{H}$ is a subset of the separable Hilbert space $\mathfrak{B}$). If a learning algorithm is $\alpha(m)$-uniformly argument stable, then its generalization error is bounded as follows. With probability at least $1 - 2\delta$,*

$$R(h_S) - \hat{R}_S(h_S) \leq 2LB\sqrt{2\log(2/\delta)}\alpha(m) + M\sqrt{\frac{\log(1/\delta)}{2m}}.$$

Proof For a sample size m and confidence $\delta > 0$, let

$$r := r(m, \delta) = D\alpha(m)\sqrt{2m\log(2/\delta)}, \quad D > 0$$

and

$$B_r = \{h \in \mathcal{H} | \|h - \mathbb{E}h_S\| \leq r(m, \delta)\}.$$

Our proof consists of two parts. First, we show that $\mathfrak{R}(B_r) \leq B\sqrt{2\log(2/\delta)}\alpha(m)$. This is because

$$\begin{aligned}
\mathfrak{R}(B_r) &= \mathbb{E}\left[\sup_{h \in B_r} \frac{1}{n} \sum_{i=1}^{m} \sigma_i \langle h, x_i \rangle\right] \\
&= \mathbb{E}\left[\sup_{h \in B_r} \frac{1}{m} \sum_{i=1}^{m} \sigma_i (\langle h, x_i \rangle - \mathbb{E}[\langle h_S, x_i \rangle])\right] \\
&\leq \mathbb{E}\left[\sup_{h \in B_r} \frac{1}{m} \|h - \mathbb{E}[h_S]\| \left\|\sum_{i=1}^{m} \sigma_i x_i\right\|_*\right] \\
&\leq \frac{r}{m} \left\|\sum_{i=1}^{m} \sigma_i x_i\right\|_* \\
&\leq \frac{1}{m} \alpha(m) D\sqrt{2m\log(2/\delta)} C_p \left(\sum_{i=1}^{m} \|x_i\|_*^p\right)^{1/p} \\
&\leq DC_p B\sqrt{2\log(2/\delta)}\alpha(m) m^{-1/2+1/p}.
\end{aligned}$$

Note that $\mathfrak{B}$ is a Hilbert space; thus, the first step is complete. Next, let us consider the Doob-Meyer difference

$$D_t = \mathbb{E}[h_S|Z_1, \ldots, Z_t] - \mathbb{E}[h_S|Z_1, \ldots, Z_{t-1}],$$

and further,

$$\begin{aligned}
\sum_{t=1}^{m} \|D_t\|_\infty^2 &= \sum_{t=1}^{m} \|\mathbb{E}[h_S|Z_1, \ldots, Z_t] - \mathbb{E}[h_S|Z_1, \ldots, Z_{t-1}]\|_\infty^2 \\
&= \sum_{t=1}^{m} \|\mathbb{E}[h_S - h_{S'}|Z_1, \ldots, Z_t]\|_\infty^2 \\
&\leq \sum_{t=1}^{m} \Big[\mathbb{E}[\|h_S - h_{S'}\|_\infty|Z_1, \ldots, Z_t]\Big]^2 \\
&\leq m\alpha(m)^2.
\end{aligned}$$

Then, we know that with probability at least $1 - \delta$,

$$R(h_S) - \hat{R}_S(h_S) \leq \sup_{h \in B_r}(R(h) - \hat{R}_S(h)).$$

Since the loss function is bounded by M, with probability at least $1 - \delta$, it holds

$$\begin{aligned}
\sup_{h \in B_r}(R(h) - \hat{R}_S(h)) &\leq \mathbb{E} \sup_{h \in B_r}(R(h) - \hat{R}_S(h)) + M\sqrt{\frac{\log(1/\delta)}{2m}} \\
&\leq 2L\mathfrak{R}(B_r) + M\sqrt{\frac{\log(1/\delta)}{2m}}.
\end{aligned}$$

By combining the two results above, we complete the proof. □

Furthermore, we may prove a generalization bound in terms of the uniform argument stability as follows.

Theorem 4.5 (Generalization bound in terms of uniform argument stability; see *(Liu et al. 2017)*) *Let $\mathfrak{B}$ be a separable Hilbert space. Suppose that the marginal distribution of the X_i is such that $\|X_i\|_* \leq B$ with probability one for some $B > 0$ and that the loss function is bounded and Lipschitz, that is, $l(h, z) \leq M$ with probability one for some $M > 0$ and $|l(h, z) - l(h', z)| \leq L|\langle h, x\rangle - \langle h', x\rangle|$ for all $z \in \mathcal{Z}$ and $h, h' \in \mathcal{H}$. Let $a > 1$. If a learning algorithm is $\alpha(m)$-uniformly argument stable, then with probability at least $1 - 2\delta$,*

$$R(h_S) - \frac{a}{a-1}\hat{R}_S(h_S) \leq 8LB\sqrt{2\log(2/\delta)}\alpha(m) + \frac{(6a+8)M\log(1/\delta)}{3m}.$$

This theorem relies on the following two lemmas.

Lemma 4.1 (Theorem 2.1 in *(Bartlett et al. 2005)*) *Let $\mathcal{F}$ be a class of functions that map $\mathcal{X}$ to $[0, M]$. Assume that there exists some $\rho > 0$ such that for every $f \in \mathcal{F}$,* $\mathrm{var}(f(X)) \leq \rho$. *Then, with probability at least $1 - \delta$, we have*

$$\sup_{f \in \mathcal{F}} \left(\mathbb{E}[f(x)] - \frac{1}{m} \sum_{i=1}^{m} f(x_i) \right) \leq 4\mathfrak{R}(\mathcal{F}) + \sqrt{\frac{2\rho \log(1/\delta)}{m}} + \frac{4M \log(1/\delta)}{3m}.$$

Proof Based on the concept of concentration inequality, let $V = \sup_{f \in \mathcal{F}} \Big(\mathbb{E}[f(x)] - \frac{1}{m} \sum_{i=1}^{m} f(x_i) \Big)$. Since $\sup_{f \in \mathcal{F}} \mathrm{var}[f(x_i)] \leq \rho$, it holds, with probability at least $1 - e^{-x}$,

$$V \leq \mathbb{E}[V] + \sqrt{\frac{2x\rho}{m} + \frac{4xM\mathbb{E}[V]}{m}} + \frac{Mx}{3m}.$$

By the inequalities $\sqrt{x+y} \leq \sqrt{x} + \sqrt{y}$ and $2\sqrt{xy} \leq \alpha x + \frac{y}{\alpha}$, with probability at least $1 - e^{-x}$,

$$V \leq \inf_{\alpha > 0} \left((1+\alpha)\mathbb{E}[V] + \sqrt{\frac{2\rho x}{m}} + \left(\frac{1}{3} + \frac{1}{\alpha}\right) \frac{Mx}{m} \right).$$

Finally, let $\alpha = 1$ and consider $\mathbb{E}[V] \leq 2\mathfrak{R}(\mathcal{F})$; this completes the proof. □

Lemma 4.2 (*Bartlett et al. 2005*) *Let*

$$\mathcal{G}_r(z) = \left\{ \frac{r}{\max\{r, \mathbb{E}l(h, z)\}} l(h, z) | h \in B_r \right\}$$

and

$$V_r = \sup_{g \in \mathcal{G}_r} \left(\mathbb{E}[g(z)] - \frac{1}{m} \sum_{i=1}^{m} g(z_i) \right)$$

be defined. For any $r > 0$ and $a > 1$, if $V_r \leq r/a$, then every $h \in B_r$ satisfies

$$\mathbb{E}[l(h, z)] \leq \frac{a}{a-1} \frac{1}{m} \sum_{i=1}^{m} l(h, z_i) + V_r.$$

Proof Let $l(h, z)$ be denoted by $f(z)$, and let $g = rf/\omega(f)$. Then, by the definition of V_r, we have $\mathbb{E}[g(z)] \leq \frac{1}{m} \sum_{i=1}^{m} g(z_i) + V_r$. We note that when $r \geq \mathbb{E}[f(z)]$, $g = f$. Otherwise, when $r < \mathbb{E}[f(z)]$, we have

$$\mathbb{E}[f(z)] \leq \frac{1}{m}\sum_{i=1}^{m} f(z_i) + \frac{\mathbb{E}[f(z)]}{r}V_r \leq \frac{1}{m}\sum_{i=1}^{m} f(z_i) + \frac{\mathbb{E}[f(z)]}{a}.$$

Finally, the statement is true under both conditions, as shown. □

Now, we will prove Theorem 4.5.

Proof First, with probability at least $1 - \delta$, we have

$$R(h_S) - \frac{a}{a-1}\hat{R}_S(h_S) \leq \sup_{h \in B_r}\left(R(h) - \frac{a}{a-1}\hat{R}_S(h)\right).$$

Furthermore, we note that

$$\text{var}(g(z)) \leq \mathbb{E}[g(z)^2] \leq M\mathbb{E}[g(z)] \leq Mr.$$

Then, by Lemma 4.1, we have

$$V_r \leq 4\mathfrak{R}(\mathcal{G}_r) + \sqrt{\frac{2Mr\log(1/\delta)}{m}} + \frac{4M\log(1/\delta)}{3m}.$$

Let the right-hand side be r/a; then, by applying Lemma 4.2, we obtain

$$\mathbb{E}[l(h,z)] \leq \frac{a}{a-1}\frac{1}{m}\sum_{i=1}^{m} l(h,z_i) + \frac{r}{a} \tag{4.15}$$

$$\leq \frac{a}{a-1}\frac{1}{m}\sum_{i=1}^{m} l(h,z_i) + \frac{2Ma\log(1/\delta)}{m} + 8\mathfrak{R}(\mathcal{G}_r) + \frac{8M\log(1/\delta)}{3m}. \tag{4.16}$$

Then, we have

$$\sup_{h \in B_r}\left(\mathbb{E}[l(h,z) - \frac{a}{a-1}\frac{1}{m}\sum_{i=1}^{m} l(h,z_i)\right) \leq \frac{2Ma\log(1/\delta)}{m} + 8\mathfrak{R}(l \circ B_r) + \frac{8M\log(1/\delta)}{3m}.$$

By combining the two results above, the proof is completed. □

In general, the stability β is a function of the sample size m. For instance, if $\beta = \frac{k}{m}$ for some positive constant k, then we have that, for any $\delta \in (0,1)$,

$$R(\mathcal{A}(S)) \leq \hat{R}_S(\mathcal{A}(S)) + \frac{k}{m} + (2k+M)\sqrt{\frac{2\ln(2/\delta)}{m}}$$

with probability at least $1 - \delta$.

We have used McDiarmid's inequality to prove a generalization bound for a uniformly β-stable algorithm. This generalization bound tells us that with a high probability, the expected error will be close to the empirical error. We thus conclude that a low empirical error serves as evidence of a low expected error.

In the next section, we will show that the Tikhonov regularization algorithm possesses this property.

4.3 Uniform Stability of Regularized Learning

In this section, we provide an example to illustrate the uniform stability of the kernel-based Tikhonov regularization algorithm. Recall the definition of the Tikhonov regularization scheme:

$$\min_{h\in\mathcal{H}}\{\frac{1}{m}\sum_{i=1}^{m} l(h, z_i) + \lambda\Omega(h)\},$$

where $\Omega(h)$ is a convex and differentiable function mapping from $\mathcal{H}$ to $\mathbb{R}$ and λ is a positive regularization parameter. We discuss the stability properties and generalization error bound of a regularized algorithm that has a reproducing kernel Hilbert space (RKHS) and a kernel-based norm regularizer.

Let $K : \mathcal{X} \times \mathcal{X} \to \mathbb{R}$ be a symmetric and positive definite kernel function. For each $u, x \in \mathcal{X}$, we adopt the notation $K_u(v) = K(u, v)$. The RKHS is defined as

$$\mathcal{H}_K = \{f : \mathcal{X} \to \mathbb{R} | f(x) = \sum_{i=1}^{m} \alpha_i K_{x_i}(x), \alpha_i \in \mathbb{R}\},$$

with inner product

$$\left\langle \sum_{i=1}^{m} \alpha_i K_{x_i}, \sum_{j=1}^{m} \beta_j K_{x_j} \right\rangle = \sum_{i=1}^{m}\sum_{j=1}^{m} \alpha_i \beta_j K(x_i, x_j).$$

According to the reproducing property, we have that, for any $h \in \mathcal{H}_K$ and $x \in \mathcal{X}$,

$$\langle f, K_x\rangle = \left\langle \sum_{i=1}^{m} \alpha_i K_{x_i}, K_x \right\rangle = \sum_{i=1}^{m} \alpha_i K_{x_i}(x) := f(x).$$

Let $\mathcal{A}$ be a kernel-based regularized learning algorithm that outputs a function $h_S \in \mathcal{H}_K$ such that

$$h_S = \arg\min_{f \in \mathcal{H}_K} \frac{1}{m} \sum_{i=1}^{m} l(h, z_i) + \lambda \|f\|_K^2,$$

where the loss function l is convex and satisfies the L-Lipschitz continuity condition.

Theorem 4.6 *Suppose that the kernel function is bounded, i.e.,* $K(x, x) \leq \kappa^2 < \infty$, $\forall x \in \mathcal{X}$. *The kernel-based Tikhonov regularization algorithm* $\mathcal{A}$ *is uniformly* β*-stable, with probability at least* $1 - \delta$*, we have*

$$R(\mathcal{A}(S)) \leq \hat{R}_S(\mathcal{A}(S)) + \frac{2L\kappa^2}{\lambda m} + (2k + M)\sqrt{\frac{2\ln(2/\delta)}{m}}.$$

Proof Let $D_f = \mathcal{A}(S^i) - \mathcal{A}(S)$. According to the property of a convex function, we have that $\forall t \in [0, 1]$, it is easy to obtain

$$\hat{R}_S(\mathcal{A}(S) + tD_f) + \hat{R}_S(\mathcal{A}(S^i) - tD_f) \leq \hat{R}_S(\mathcal{A}(S)) + \hat{R}_S(\mathcal{A}(S^i)).$$

The definitions of $\mathcal{A}(S)$ and $\mathcal{A}(S^i)$ indicate that

$$\hat{R}_S(\mathcal{A}(S)) + \lambda\|\mathcal{A}(S)\|_K^2 \leq \hat{R}_S(\mathcal{A}(S) + tD_f) + \lambda\|\mathcal{A}(S) + tD_f\|_K^2$$

and

$$\hat{R}_{S^i}(\mathcal{A}(S^i)) + \lambda\|\mathcal{A}(S^i)\|_K^2 \leq \hat{R}_{S^i}(\mathcal{A}(S^i) - tD_f) + \lambda\|\mathcal{A}(S^i) - tD_f\|_K^2.$$

Furthermore, we can deduce that

$$\begin{aligned}
&\frac{\lambda}{2}\|D_f\|_K^2 \\
&= \lambda[\|\mathcal{A}(S)\|_K^2 + \|\mathcal{A}(S^i)\|_K^2 - \|\mathcal{A}(S) + tD_f\|_K^2 - \|\mathcal{A}(S^i) - tD_f\|_K^2] \\
&\leq [\hat{R}_{S^i}(\mathcal{A}(S^i) - tD_f) - \hat{R}_S(\mathcal{A}(S^i) - tD_f)] + [\hat{R}_S(\mathcal{A}(S^i)) - \hat{R}_{S^i}(\mathcal{A}(S^i))] \\
&= \frac{1}{m}[l(\mathcal{A}(S^i) - tD_f, z_i') - l(\mathcal{A}(S^i) - tD_f, z_i)] + \frac{1}{m}[l(\mathcal{A}(S^i), z_i) - l(\mathcal{A}(S^i), z_i')] \\
&= \frac{1}{m}[l(\mathcal{A}(S^i) - tD_f, z_i') - l(\mathcal{A}(S^i), z_i')] + \frac{1}{m}[l(\mathcal{A}(S^i), z_i) - l(\mathcal{A}(S^i) - tD_f, z_i)] \\
&\leq \frac{tL}{m}(|D_f(x_i')| + |D_f(x_i)|) \leq \frac{2tL\kappa}{m}\|D_f\|_K \\
&\leq \frac{L\kappa}{m}\|D_f\|_K \quad (\text{by taking } t = \frac{1}{2}),
\end{aligned}$$

where the last line follows from the reproducing property, the Cauchy-Schwartz inequality, and the boundedness assumption on the kernel K, i.e., $|f| \leq \kappa\|f\|_K$ for any f in the RKHS. Moreover, by taking $t = 1/2$, we then have

$$|D_f| \leq \kappa \|D_f\|_K^2 \leq \frac{2L\kappa^2}{\lambda m}.$$

Finally, we show that the algorithm has a uniform stability of $2L\kappa^2/\lambda m$. We complete the proof by combining this result with the result of Theorem 4.3. □

References

Olivier, Bousquet, and André Elisseeff. 2002. Stability and generalization. *Journal of Machine Learning Research* 2: 499–526.

Peter, L Bartlett, Olivier Bousquet, and Shahar Mendelson. 2005. Local rademacher complexities. *The Annals of Statistics* 33 (4): 1497–1537.

Tongliang, Liu, Gábor Lugosi, Gergely Neu, and Dacheng Tao. 2017. Algorithmic stability and hypothesis complexity. In *International Conference on Machine Learning*, 2159–2167.

Yurii, E Nesterov. 1983. A method for solving the convex programming problem with convergence rate o $(1/k^2)$. *In Dokl. Akad. Nauk Sssr* 269: 543–547.

Part II
Deep Learning Theory

Chapter 5
Capacity and Complexity

This chapter presents the results for deep learning theory from the perspective of hypothesis complexity, including the Vapnik-Chervonenkis (VC) dimension, the Rademacher complexity, and the covering number. By determining upper and lower bounds for the VC dimension of neural networks, we can better understand their generalizability. Additionally, we discuss margin bounds, which offer more robust generalization assurances compared to worst-case bounds based on the VC dimension. These bounds ensure that trained models can achieve a small empirical margin loss with high confidence. Furthermore, we examine the effect of residual connections on hypothesis complexity by analyzing the covering number of the hypothesis space. We propose an upper bound for the covering number, providing insights into how residual connections influence model complexity.

5.1 Worst-Case Bounds Based on the VC Dimension

One can upper bound or lower bound the VC dimension of a neural network in order to characterize its generalizability. Goldberg and Jerrum (1995) gave an $O(W^2)$ upper bound for the VC dimension of a neural network with W parameters. Bartlett et al. (1999) improved this bound to $O((WL \log W + WL^2)$, where L is the number of layers. The tightest bound thus far has been proven by Harvey et al. (2017) as follows.

Theorem 5.1 (see Harvey et al. (2017)) *Consider a neural network with W parameters and U units. These units have activation functions that are piecewise polynomials with at most p pieces and a degree of at most d. Let $\mathcal{F}$ be the set of (real-valued) functions computed by this network. Then, the VC dimension of the sign function (sgn) applied to elements of $\mathcal{F}$ is upper bounded by $O(WU \log((d+1)p))$.*

F. He and D. Tao, *Foundations of Deep Learning*, Machine Learning: Foundations, Methodologies, and Applications, https://doi.org/10.1007/978-981-16-8233-9_5

The following theorem is based on a theorem presented by Goldberg and Jerrum (1993), which states that any class of functions that can be expressed using a small number of distinct polynomial inequalities has a smaller VC dimension.

Theorem 5.2 *Let k and n be positive integers, and let $f : \mathbb{R}^n \times \mathbb{R}^k \to \{0, 1\}$ be a function that can be expressed as a Boolean formula containing s distinct atomic predicates, where each atomic predicate is a polynomial inequality or equality in $k + n$ variables with a degree of at most d. Let $\mathcal{F} = \{f(\cdot, \omega) : \omega \in \mathbb{R}^k\}$. Then, $VCdim(\mathcal{F}) \leq 2k \log_2(8eds)$.*

Proof The output signal of a neural network with piecewise activation can be expressed as a Boolean formula. In detail, in each layer, the input to each computation unit must lie in one of the p pieces of the activation function ψ, which can be written as a Boolean formula. Accordingly, we can express the output signal of the network as a Boolean function consisting of fewer than $2(1 + p)^U$ atomic predicates, each of which is a polynomial inequality with a degree of at most $\max\{U + 1, 2d^U\}$. □

We can now prove Theorem 5.1.

Proof By Theorem 5.2, we prove an upper bound of $2W \log(16e \cdot \max\{U+1, 2d^U\} \cdot (1 + p)^U) = O(WU \log(1 + d)p)$ for the VC dimension. This completes the proof. □

Harvey et al. (2017) also derived a lower bound for the VC dimension, presented as the following theorem.

Theorem 5.3 (see (Harvey et al. 2017)) *Let C be an universal constant. Given any W and L satisfying $W > CL > C^2$, there exists a rectified linear unit (ReLU) network with fewer than L layers and fewer than W parameters that has a VC dimension lower bounded by $WL \log(W/L)/C$.*

This theorem has a more general version. Theorem 5.3 is a special case in which $r = \log_2(W/L)/2$, $m = rL/8$, and $n = W - 5m2^r$.

Theorem 5.4 *Let r, m, and n be positive integers, and let $k = \lceil m/r \rceil$. There exists a ReLU network with $3 + 5d$ layers, $2 + n + 4m + k((11 + r)2^r + 2r + 2)$ parameters, $m + n$ input nodes and $m + 2 + k(5 \times 2^r + 1)$ computational nodes with VC dimension $\geq mn$.*

5.2 Rademacher Complexity and Covering Number

Margin bounds (Vapnik 2006; Schapire et al. 1998; Bartlett and Shawe-Taylor 1999; Koltchinskii and Panchenko 2002; Taskar et al. 2004) represent a distinct category of generalization bounds that offer stronger assurances compared to worst-case bounds based on the VC dimension. Unlike traditional bounds, margin bounds focus on the concept of margin, which refers to the difference between the predicted score of the correct label and the highest score of incorrect labels for a given example. These bounds provide robust guarantees that trained models can achieve a small empirical margin loss within a large confidence margin. Essentially, margin bounds quantify the degree of separation between different classes in the model's decision space, reflecting the model's ability to generalize beyond the training data. Furthermore, similar generalization guarantees can be derived based on a measure of "luckiness" (Schapire et al. 1998; Koltchinskii and Panchenko 2002), which captures the notion of how likely a model is to perform well on unseen data due to favorable properties of the training set. By incorporating margin-based metrics and measures of luckiness, researchers gain valuable insights into the behavior and performance of machine learning models, enhancing our understanding of their generalization capabilities and reliability in real-world applications.

Margin bound

For any distribution $\mathcal{D}$ and margin $\gamma > 0$, we define the expected margin loss for a hypothesis h as follows:

$$L_\gamma(h) = \mathbb{P}_{(x,y)\sim\mathcal{D}}\left(h(x)[y] \leq \gamma + \max_{j=y} h(x)[j]\right),$$

where $h(x)[j]$ is the j-th component of the vector $h(x)$. The generalization bounds obtained under the margin loss are called margin bounds.

Based on the covering number, the Dudley entropy integral, and the Rademacher complexity, Bartlett et al. (2017) obtained the following spectrally normalized upper bound for the generalization error:

$$\tilde{O}\left(\frac{\|X\|_2 \log Q}{\gamma m}\left(\prod_{i=1}^{D}\rho_i \|W_i\|_\sigma\right)\left(\sum_{i=1}^{D}\frac{\|W_i^\top - M_i^\top\|_{2,1}^{2/3}}{\|W_i\|_\sigma^{2/3}}\right)^{3/2} + \sqrt{\frac{1/\delta}{m}}\right),$$

where $W_i, i = 1, \ldots, D$, is the weight matrix, Q is an upper bound on the number of output units in each layer, $\|\cdot\|_\sigma$ is the spectral norm, $\rho_i > 0$ is the Lipschitz constant and $M_i, i = 1, \ldots, D$, is a reference matrix. This bound is strictly tighter than that of Neyshabur et al. (2015). Moreover, this bound suggests that one can boost the generalizability of a network by controlling its spectral norm. This is consistent with spectral normalization (Miyato et al. 2018), an effective regularization approach mainly used in generative adversarial networks.

Concurrently, Neyshabur et al. (2017) proved a similar generalization bound via Probably Approximately Correct (PAC)-Bayesian theory, as shown below:

$$O\left(\sqrt{\frac{B^2D^2Q\log(DQ)\prod_{i=1}^{D}\frac{\|W_i\|_F^2}{\|W_i\|_2^2}+\log\frac{Dm}{\delta}}{\gamma^2 m}}\right). \tag{5.1}$$

The authors of (Neyshabur et al. 2017) pursued a two-pronged approach in their research endeavor. First, they developed a perturbation bound aimed at controlling the changes in the network's output resulting from perturbations introduced into the model's weights. This perturbation bound allowed them to effectively quantify the sensitivity or sharpness of the neural network's response to variations in its parameters. Building upon the perturbation analysis, Neyshabur et al. (2017) then employed PAC-Bayesian theory to derive a generalization bound. This bound was intended to provide insights into the model's ability to generalize from the training data to unseen examples. However, it was noted that this derived generalization bound proved to be weaker compared to a similar bound proposed by Bartlett et al. (2017). This observation underscores the importance of exploring different theoretical frameworks and bounds in assessing the robustness and performance of machine learning models in various contexts. Further investigations may be warranted to elucidate the factors contributing to the variance in generalization guarantees across different theoretical approaches.

By introducing some norm constraints on the weights, Golowich et al. (2018) improved the exponential depth dependence of Eq. (5.1) in (Neyshabur et al. 2015) to a polynomial dependence, as follows:

$$O\left(\frac{B\sqrt{D}\prod_{j=1}^{D}M_F(j)}{\sqrt{m}}\right) \text{ or } O\left(\frac{B\sqrt{D+\log m}\prod_{j=1}^{D}M_F(j)}{\sqrt{m}}\right), \tag{5.2}$$

where $M_F(1), \ldots, M_F(D)$ mean the maximum of Frobenius norm of networks parameter matrices. Besides, Golowich et al. (2018) further improved the bound to a depth-independent bound, as follows:

$$\tilde{O}\left(B\left(\prod_{j=1}^{D}M_F(j)\right)\min\left\{\sqrt{\frac{\log\left(\frac{1}{\Gamma}\prod_{j=1}^{D}M_F(j)\right)}{\sqrt{m}}}, \sqrt{\frac{D}{m}}\right\}\right), \tag{5.3}$$

where Γ is an upper bound on the spectral norm of all weight matrices.

Moreover, Golowich et al. (2018) also proved a lower bound, as follows:

$$\Omega\left(\frac{B\prod_{j=1}^{D}M_F(j)Q^{\max\{0,\frac{1}{2}-\frac{1}{p}\}}}{\sqrt{m}}\right), \tag{5.4}$$

where Q is the network width. This bound suggests a possible tighter bound of $O\left(m^{-1/2}\right)$.

Tu et al. (2020) developed a gradient measure based on the Fisher-Rao norm, which gives rise to a generalization bound as follows.

Theorem 5.5 (see Tu et al. (2020)) *If the margin parameter α is fixed, then for any $\delta > 0$, with probability at least $1 - \delta$, the following holds for every recurrent neural network (RNN) whose weight matrices $\theta = (U, W, V)$ satisfy $||V^T||_1 \leq \beta_V$, $||W^T||_1 \leq \beta_W$, $||U^T||_1 \leq \beta_U$ and $||\theta||_{fs} \leq r$:*

$$\mathbb{E}[\mathbf{I}_{\mathcal{M}_{y_L}(x,y)\leq 0}] \leq \frac{4k}{\alpha}\left(\frac{r||X||_F}{2m}\sqrt{\frac{1}{\lambda_{min}(\mathbb{E}(XX^T))}} + \frac{1}{n}\beta_V\beta_U||X^T||_1\Lambda\right) + \frac{1}{m}\sum \mathbf{I}_{\mathcal{M}_{y_L}(x_i,y_i)\leq a} + 3\sqrt{\frac{\log\frac{2}{\delta}}{2m}}. \tag{5.5}$$

The gradient measure plays a crucial role in refining generalization bounds and illuminating the intricate relationship between a model's generalizability and its trainability. By analyzing the impact of gradients during training, researchers gain insights into how noise can be strategically introduced into input data to enhance a model's ability to generalize to unseen examples. This observation underscores the value of robust training strategies that incorporate regularization techniques like gradient clipping (Merity et al. 2018; Gehring et al. 2017; Peters et al. 2018). Furthermore, the gradient measure provides theoretical justification for gradient clipping, a technique used to stabilize training by constraining the magnitude of gradients during optimization. By controlling the scale of gradients, gradient clipping helps prevent exploding gradients and facilitates more stable and effective learning. Overall, understanding the implications of gradient measures enriches our understanding of model behavior and informs practical strategies to improve model performance and generalization capabilities.

(Novak et al. 2018) conducted comprehensive experiments to analyze the generalization capacity of deep neural networks. Their empirical investigations revealed pivotal factors influencing generalization, particularly focusing on the input-output Jacobian norm and the linear region counting within the networks. The input-output Jacobian norm measures the sensitivity of network outputs to changes in input, providing insights into how robustly the network behaves across different input variations. Meanwhile, the linear region counting quantifies the number of linear decision boundaries that can be formed within the network's architecture, offering clues about the complexity and flexibility of its decision-making process. Furthermore, the study highlighted the critical relationship between the output hypothesis of a model and the underlying data manifold. The generalization bound, which defines the model's ability to perform well on unseen data, is heavily influenced by how closely the model's output hypothesis aligns with the distribution of the training data in the input space. This suggests the importance of designing models that not only fit the training data well but also generalize effectively to new, unseen instances. Understanding these fundamental aspects contributes to the development of more robust and reliable deep learning models capable of adapting to diverse real-world scenarios.

5.3 Generalization Bound of ResNet

In recent years, deep neural networks have repeatedly demonstrated their significant value in practical applications and have led to breakthroughs in image recognition, language translation, and predictive analytics, driving innovation across various industries (LeCun et al. 2015; Greff et al. 2016; Shi et al. 2018; Silver et al. 2016; Litjens et al. 2017; Chang et al. 2018), particularly with the introduction of ResNet (He et al. 2016a) and the wide adoption of residual connections to induce top performed neural network architectures (He et al. 2016a; Huang et al. 2017; Xie et al. 2017). These advancements have catalyzed transformations in computer vision (Krizhevsky et al. 2012; Lin et al. 2017; He et al. 2017; Liu et al. 2019) and data analysis (Witten et al. 2016). Residual connections link non-neighboring layers, diverging from the traditional chain-like stacking of layers.

Empirical studies have consistently proven that incorporating residual connections can facilitate the training of deep neural networks by largely alleviating issues related to vanishing gradients and exploding gradients. Despite their effectiveness in training, there remains a notable gap in theoretical analysis regarding the impact of residual connections on the generalization ability of deep neural networks.

While residual connections have shown instrumental in improving training dynamics and model convergence, their influence on generalization-i.e., a model's performance on unseen data in the inference stage-remains less explored from a theoretical perspective. Comprehensive research is critical to elucidate the underlying mechanisms by which residual connections impact on the generalization capabilities of deep neural networks, thereby providing deeper insights into the theoretical foundations of this powerful addon.

Residual connections introduce a novel approach to neural network architecture by connecting non-neighboring layers, diverging from the traditional chain-like stacking of layers. This departure from conventional network structures introduces loops within the neural network, fundamentally altering its topology. Intuitively, the incorporation of residual connections significantly expands the hypothesis space of deep neural networks. This increased complexity, however, raises concerns regarding the generalization ability of the network, as per the principle of Occam's razor, which suggests a negative correlation between algorithmic generalization and hypothesis complexity.

The lack of comprehensive analysis on this aspect poses challenges for deploying residual-connected neural networks in safety-critical domains, such as autonomous vehicles and medical diagnosis, where algorithmic errors could have severe consequences. Without addressing this theoretical gap, leveraging the advancements made possible by residual connections in critical applications, from autonomous vehicles (Janai et al. 2017) to medical diagnose (Esteva et al. 2017), remains constrained. Efforts to bridge this gap are crucial for ensuring the reliability and safety of neural network-based systems in real-world scenarios where accuracy and generalization are paramount.

In this section, we investigate the impact of residual connections on hypothesis complexity by examining the covering number of the hypothesis space. We propose an upper bound for the covering number that reveals insights into the influence of residual connections on model complexity. Our analysis demonstrates that when the total number of weight matrices in a neural network is fixed, the upper bound on the covering number remains consistent regardless of whether the weight matrices are part of the residual connections or the main "stem"[1] of the network. This finding suggests that residual connections do not necessarily increase the hypothesis space complexity compared to a traditional chain-like neural network, provided that the total number of weight matrices and non-linearities is held constant. Leveraging this understanding, we establish an $O(1/\sqrt{m})$ generalization bound for ResNet as a representative case of neural networks with residual connections, where m represents the training sample size.

Furthermore, our framework enables the derivation of generalization bounds for similar architectures created by incorporating residual connections into chain-like neural networks. This approach facilitates the assessment and comparison of hypothesis complexity across different network configurations, offering valuable insights into the theoretical underpinnings of deep learning models with residual connections. Our generalization bound is intricately linked to the product of norms of all weight matrices within the neural network. Specifically, as the product of these norms increases, there tends to be a detrimental impact on the network's ability to generalize effectively. This observation validates the importance of employing regularization techniques aimed at controlling the magnitudes of these weight matrix norms to enhance generalization performance.

To achieve robust generalization, it is common practice to incorporate regularization methods such as weight decay and spectral normalization during the training of deep neural networks. Weight decay, also known as L_2 regularization, imposes a penalty based on the L_2 norm of the weights, encouraging smaller and more stable weight values. Similarly, spectral normalization constrains the spectral norm of weight matrices, leading to more controlled and stable learning dynamics.

These regularization techniques (Krogh and Hertz 1992) play a critical role in managing model complexity, mitigating overfitting, and improving generalization capabilities across a variety of deep learning tasks. The endorsement of these methods by Bartlett et al. (2017) suggests their effectiveness in promoting robust and reliable performance of neural networks, particularly in scenarios where generalization is paramount. By incorporating these strategies, practitioners can optimize neural network architectures for improved performance and reliability in real-world applications.

[1] The "stem" is defined to denote the chain-like part of the neural network besides all the residuals. For more details, please refer to Sect. 5.3.1.

5.3.1 Stem-Vine Framework

This section provides a notation system for deep neural networks with residual connections. Motivated by the topological structure, we call it the *stem-vine framework*.

Deep neural networks are typically assembled by linking numerous weight matrices with nonlinear operators, such as ReLU, sigmoid, and max-pooling functions. In this discussion, we focus on a neural network design that integrates multiple residual connections into a traditional "chain-like" architecture, which sequentially stacks layers of weight matrices and nonlinearities. Inspired by the topological arrangement of this network, we refer to the sequential layers as the "stem" of the neural network and designate the added residual connections as the "vines".

Both the stem and vines consist of stacked layers of weight matrices and nonlinear functions. The stem represents the core structure of the neural network, comprising the primary sequence of layers, while the vines introduce additional connections that bypass certain layers, enhancing the network's depth and flexibility. This configuration allows for more complex transformations and feature representations within the neural network, leveraging both the traditional sequential architecture and the introduced residual connections to enhance learning capabilities and model expressiveness.

We denote the weight matrices and the nonlinearities in the stem S respectively as

$$A_i \in \mathbb{R}^{n_{i-1} \times n_i} \,, \tag{5.6}$$

$$\sigma_j : \mathbb{R}^{n_j} \to \mathbb{R}^{n_j} \,, \tag{5.7}$$

where $i = 1, \ldots, L$, L is the number of weight matrices in the stem, $j = 1, \ldots, L_N$, L_N is the number of nonlinearities in the stem, n_i is the dimension of the output of the i-th weight matrix, n_0 is the dimension of the input data to the network, and n_L is the dimension of the output of the network. Thus we can write the stem S as a vector to express the chain-like structure. Here for the simplicity and without any loss of the generality, we give an example that the numbers of weight matrices and nonlinearities are equal[2], i.e., $L_N = L$, as the following equation,

$$S = (A_1, \sigma_1, A_2, \sigma_2, \ldots, A_L, \sigma_L) \,. \tag{5.8}$$

---

[2] If two weight matrices, A_i and A_{i+1}, are connected directly without a nonlinearity between them, we define a new weight matrix $A = A_i \cdot A_{i+1}$. The situations that nonlinearities are directly connected are similar, as the composition of any two nonlinearities is still a nonlinearity.

Meanwhile, the number of weight matrices does not necessarily equal the number of nonlinearities. Sometimes, if a vine connects the stem at a vertex between two weight matrices (or two nonlinearities), the number of the weight matrices (nonlinearities) would be larger than the number of nonlinearities (weight matrices). Taken the 34-layer ResNet as an example, a vine connects the stem between two nonlinearities σ_{33} and σ_{34}. In this situation, we cannot merge the two nonlinearities, so the number of nonlinearities is larger than the number of weight matrices.

For the brevity, we give an index j to each vertex between a weight matrix and a nonlinearity and denote the j-th vertex as $N(j)$. Specifically, we give the index 1 to the vertex that receives the input data and $L + L_N + 1$ to the vertex after the last weight matrix/nonlinearity. Taken Eq. (5.8) as an example, the vertex between the nonlinearity σ_{i-1} and the weight matrix A_i is denoted as $N(2i - 1)$ and the vertex between the weight matrix A_i and the nonlinearity σ_i is denoted as $N(2i)$.

Vines are constructed to connect the stem at two different vertexes. And there could be over one vine connecting a same pair of the vertexes. Therefore, we use a triple vector (s, t, i) to index the i-th vine connecting the vertexes $N(s)$ and $N(t)$ and denote the vine as $V(s, t, i)$. All triple vectors (s, t, i) constitute an index set I_V, i.e., $(s, t, i) \in I_V$. Similar to the stem, each vine $V(s, t, i)$ is also constructed by a series of weight matrices $A_1^{s,t,i}, \ldots, A_{L^{s,t,i}}^{s,t,i}$ and nonlinearities $\sigma_1^{s,t,i}, \ldots, \sigma_{L_N^{s,t,i}}^{s,t,i}$, where $L^{s,t,i}$ is the number of weight matrices in the vine, while $L_N^{u,v,i}$ is the number of the nonlinearities.

Multiplying by a weight matrix corresponds to an affine transformation on the data matrix. Also, nonlinearities induce nonlinear transformations. Through a series of affine transformations and nonlinear transformations, hierarchical features are extracted from the input data by neural networks. Usually, we use the *spectrum norms* of weight matrices and the *Lipschitz constants* of nonlinearities to express the intensities respectively of the affine transformations and the nonlinear transformations. We call a function $f(x)$ is ρ-Lipschitz continuous if for any x_1 and x_2 in the support domain of $f(x)$, it holds that

$$\|f(x_1) - f(x_2)\|_f \leq \rho \|x_1 - x_2\|_x , \tag{5.9}$$

where $\|\cdot\|_f$ and $\|\cdot\|_x$ are respectively the norms defined on the spaces of $f(x)$ and x. Fortunately, almost all nonlinearities normally used in neural networks are Lipschitz continuous, such as ReLU, max-pooling, and sigmoid (see (Bartlett et al. 2017)).

Many important tasks for deep neural networks can be categorized into multi-class classification. Suppose input examples $z_1 \ldots, z_m$ are given, where $z_i = (x_i, y_i), x_i \in \mathbb{R}^{n_0}$ is an instance, $y \in \{1, \ldots, n_L\}$ is the corresponding label, and n_L is the number of the classes. Collect all instances $x_1, \ldots, x_m$ as a matrix $X = (x_1, \ldots, x_m)^T \in \mathbb{R}^{m \times n_0}$ that each row of X represents a data point. By employing optimization methods (usually stochastic gradient decent, SGD), neural networks are trained to fit the training data and then predict on test data. In mathematics, a trained deep neural network with all parameters fixed computes a hypothesis function $F : \mathbb{R}^{n_0} \to \mathbb{R}^{n_L}$. And a natural way to convert F to a multi-class classifier is to select the coordinate of $F(x)$ with the largest magnitude. In other words, for an instance x, the classifier is $x \mapsto \arg\max_i F(x)_i$. Correspondingly, the *margin* for an instance x labelled as y_i is defined as $F(x)_y - \max_{i \neq y} F(x)_i$. It quantitatively expresses the confidence of assigning a label to an instance.

To express F, we first define the functions respectively computed by the stem and vines. Specifically, we denote the function computed by a vine $V(s, t, i)$ as:

$$F_V^{s,t,i}(X) = \sigma^{u,v,i}_{L_{u,v,i}}(A^{u,v,i}_{L_{u,v,i}}\sigma^{u,v,i}_{L_{u,v,i}-1}(\ldots\sigma_1(A_1^{u,v,i}X)\ldots)) . \tag{5.10}$$

Similarly, the stem computes a function as the following equation:

$$F_S(X) = \sigma_L(A_L\sigma_{L-1}(\ldots\sigma_1(A_1X)\ldots)) . \tag{5.11}$$

Furthermore, we denote the output of the stem at the vertex $N(j)$ as the following equation:

$$F_S^j(X) = \sigma_j(A_j\sigma_{j-1}(\ldots\sigma_1(A_1X)\ldots)) . \tag{5.12}$$

$F_S^j(X)$ is also the input of the rest part of the stem. Eventually, with all residual connections, the output hypothesis function $F^j(X)$ at the vertex $N(j)$ is expressed by the following equation:

$$F^j(X) = F_S^j(X) + \sum_{(u,j,i)\in I_V} F_V^{u,j,i}(X) . \tag{5.13}$$

Apparently,

$$F_S(X) = F_S^L(X),\ F(X) = F^L(X) . \tag{5.14}$$

Naturally, we call this notation system as the *stem-vine framework*, and Fig. 5.1 gives an example.

5.3.2 Generalization Bound

In this section, we investigate the generalization capabilities of deep neural networks with residual connections and establish a specific generalization bound for ResNet as a case study. This bound is derived from the margin-based multi-class bound detailed in Lemmas 3.1 and 3.2. Guided by these lemmas, which highlight a natural approach to deriving the generalization bound through exploration of the covering number of the corresponding hypothesis space, we first propose an upper bound for the covering number (referred to as a covering bound) applicable to any deep neural network within the stem-vine framework. We then apply this approach to ResNet, providing a specific calculation of its covering bound. By leveraging Lemmas 3.1 and 3.2, we subsequently present a detailed generalization bound tailored specifically for ResNet. The proofs supporting these covering bounds will be provided in Sect. 5.3.3, enhancing our understanding of the theoretical underpinnings behind the generalization capabilities of neural networks with residual connections.

In practice, when introducing new structures to improve training performance, such as boosting accuracy or reducing training time, it's essential to proceed with caution to avoid overfitting, which can lead to significant generalization errors. ResNet, with its introduction of "loops" via residual connections in traditional chain-like neu-

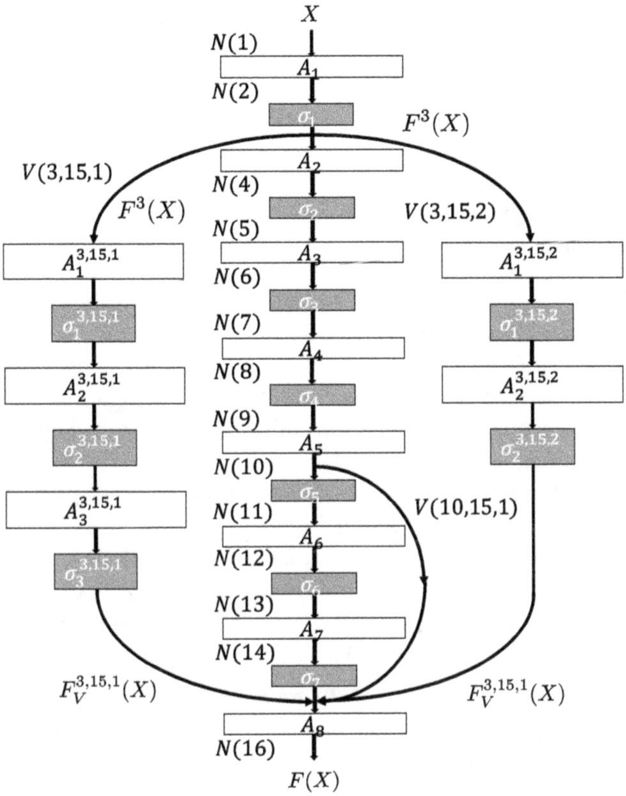

Fig. 5.1 A deep neural network with Residual Connections under the Stem-Vine Framework

ral networks, represents a more sophisticated model. Empirical findings demonstrate that these residual connections effectively reduce training error and expedite training, while also maintaining strong generalization capabilities. However, the absence of theoretical underpinnings to explain or support these empirical results remains a notable gap in the understanding of such architectures.

Our analysis on covering bounds reveals that the complexity of hypothesis spaces computed by deep neural networks remains consistent, regardless of the distribution of weight matrices-whether they reside in the stem, the vines, or even without any vines at all-when the total number of weight matrices is fixed. This observation suggests that the presence or absence of residual connections does not inherently alter the overall hypothesis complexity. By integrating established results from statistical learning theories (Lemmas 3.1 and 3.2), our findings suggest that the generalization capability of deep neural networks with residual connections can be comparably effective to those without, particularly in challenging scenarios where model complexity plays a critical role. This theoretical insight sheds light on why deep neural networks incorporating residual connections exhibit strong generalization capabilities comparable to traditional chain-like architectures, while maintaining competitive training performance.

5.3.2.1 Covering Bound for Deep Neural Networks with Residuals

In this subsection, we give a covering bound generally for any deep neural network with residual connections.

Theorem 5.6 (Covering Bound for Deep Neural Network) *Suppose a deep neural network is constituted by a stem and a series of vines.*

For the stem, let $(\varepsilon_1, \dots, \varepsilon_L)$ *be given, along with* L_N *fixed nonlinearities* $(\sigma_1, \dots, \sigma_{L_N})$. *Suppose the* L *weight matrices* $(A_1, \dots, A_L)$ *lies in* $\mathcal{B}_1 \times \dots \times \mathcal{B}_L$, *where* $\mathcal{B}_i$ *is a ball centered at* 0 *with radius of* s_i, *i.e.,* $\|A_i\| \le s_i$. *Suppose the vertex that directly follows the weight matrix* A_i *is* $N(M(i))$ ($M(i)$ *is the index of the vertex). All* $M(i)$ *constitute an index set* I_M. *When the output* $F_{M(j-1)}(X)$ *of the weight matrix* A_{j-1} *is fixed, suppose all output hypotheses* $F_{M(j)}(X)$ *of the weight matrix* A_j *constitute a hypothesis space* $\mathcal{H}_{M(j)}$ *with an* $\varepsilon_{M(j)}$*-cover* $\mathcal{W}_{M(j)}$ *with covering number* $\mathcal{N}_{M(j)}$. *Specifically, we define* $M(0) = 0$ *and* $F_0(X) = X$.

Each vine $V(u, v, i)$, $(u, v, i) \in I_V$ *is also a chain-like neural network that constructed by multiple weight matrices* $A_j^{u,v,i}$, $j \in \{1, \dots, L^{u,v,i}\}$, *and nonlinearities* $\sigma_j^{u,v,i}$, $j \in \{1, \dots, L_N^{u,v,i}\}$. *Suppose for any weight matrix* $A_j^{u,v,i}$, *there is a* $s_j^{u,v,i} > 0$ *such that* $\|A_j^{u,v,i}\|_\sigma \le s_j^{u,v,i}$. *Also, all nonlinearities* $\sigma_j^{u,v,i}$ *are Lipschitz continuous. Similar to the stem, when the input of the vine* $F_u(X)$ *is fixed, suppose the vine* $V(u, v, i)$ *computes a hypothesis space* $\mathcal{H}_V^{u,v,i}$, *constituted by all hypotheses* $F_V^{u,v,i}(X)$, *has an* $\varepsilon_{u,v,i}$*-cover* $\mathcal{W}_V^{u,v,i}$ *with covering number* $\mathcal{N}_V^{u,v,i}$.

Eventually, we denote the hypothesis space computed by the neural network is $\mathcal{H}$. *Then there exists an* ε *in terms of* ε_i, $i = \{1, \dots, L\}$ *and* $\varepsilon_{u,v,i}$, $(u, v, i) \in I_V$, *such that the following inequality holds:*

$$\mathcal{N}(\mathcal{H}, \varepsilon, \|\cdot\|) \le \prod_{j=1}^{L} \sup_{F_{M(j)}} \mathcal{N}_{M(j+1)} \prod_{(u,v,i)\in I_V} \sup_{F_u} \mathcal{N}_V^{u,v,i}. \tag{5.15}$$

A detailed proof will be given in Sect. 5.3.3.3.

As vines are chain-like neural networks, we can further obtain an upper bound for $\sup_{F_u} \mathcal{N}_V^{u,v,i}$ via a lemma slightly modified from Bartlett et al. (2017). The lemma is summarised as follows.

Lemma 5.1 *(Covering Bound for Chain-like Deep Neural Network; cf. Bartlett et al. (2017), Lemma A.7) Suppose there are* L *weight matrices in a chain-like neural network. Let* $(\varepsilon_1, \dots, \varepsilon_L)$ *be given. Suppose the* L *weight matrices* $(A_1, \dots, A_L)$ *lies in* $\mathcal{B}_1 \times \dots \times \mathcal{B}_L$, *where* $\mathcal{B}_i$ *is a ball centered at* 0 *with the radius of* s_i, *i.e.,* $\mathcal{B}_i = \{A_i : \|A_i\| \le s_i\}$. *Furthermore, suppose the input data matrix* X *is restricted in a ball centred at* 0 *with the radius of* B, *i.e.,* $\|X\| \le B$. *Suppose* F *is a hypothesis* *function computed by the neural network. If we define:*

$$\mathcal{H} = \{F(X) : A_i \in \mathcal{B}_i\}, \tag{5.16}$$

where $i = 1, \ldots, L$ *and* $t \in \{1, \ldots, L^{u,v,s}\}$*. Let* $\varepsilon = \sum_{j=1}^{L} \varepsilon_j \rho_j \prod_{l=j+1}^{L} \rho_l s_l$*. Then we have the following inequality:*

$$\mathcal{N}(\mathcal{H}, \varepsilon, \|\cdot\|) \leq \prod_{i=1}^{L} \sup_{\mathbf{A}_{i-1} \in \boldsymbol{B}_{i-1}} \mathcal{N}_i, \tag{5.17}$$

where $\mathbf{A}_{i-1} = (A_1, \ldots, A_{i-1})$*,* $\boldsymbol{B}_{i-1} = \mathcal{B}_1 \times \ldots \times \mathcal{B}_{i-1}$*, and*

$$\mathcal{N}_i = \mathcal{N}\left(\left\{A_i F_{\mathbf{A}_{i-1}}(X) : A_i \in \mathcal{B}_i\right\}, \varepsilon_i, \|\cdot\|\right). \tag{5.18}$$

Remark 5.1 The mapping induced by a chain-like neural network involves composing a sequence of affine and nonlinear transformations. According to Lemma 5.1, the covering bound for a chain-like network can be decomposed into the product of covering bounds for each layer, as demonstrated in Bartlett et al. (2017). However, when residual connections are introduced, parallel structures emerge within the network, complicating the representation of the overall mapping as a series of transformations. Instead, calculating a covering bound for the entire network requires considering many additions of function spaces (as seen in Eq. (5.13)), which presents challenges for applying previous methods directly. To address this, we introduce a new proof in Sect. 5.3.3.3 to analyze the covering bound under the presence of residual connections and parallel structures.

Contrary to the different proofs, the result for deep neural networks with residual connections share similarities with the one for the chain-like network (see, respectively, Eqs. (5.15) and (5.17)). The similarities lead to the property summarised as follows.

> *The influences on the hypothesis complexity of weight matrices are in the same way, no matter whether they are in the stem or the vines. Specifically, adding an identity vine could not affect the hypothesis complexity of the deep neural network.*

As indicated by Eq. (5.17) in Lemma 5.1, the covering number of the hypothesis computed by a chain-like neural network (including the stem and all the vines) is upper bounded by the product of the covering number of all single layers. Specifically, the contribution of the stem on the covering bound is the product of a series of covering numbers, i.e., $\prod_{j=1}^{L} \sup_{F_{M(j)}} \mathcal{N}_{M(j+1)}$. In the meantime, applying Eq. (5.17) in Lemma 5.1, the contribution $\sup_{F_u} \mathcal{N}_V^{u,v,i}$ of the vine $V(u, v, i)$ can also be decomposed as the product of a series of covering numbers. Apparently, the contributions respectively by the weight matrices in the stem and the ones in the vines have similar formulations. This result gives an insight that residuals would not undermine the generalization capability of deep neural networks. Also, if a vine $V(u, v, i)$ is an identity mapping, the term in Eq. (5.15) that relates to it is definitely 1, i.e., $\mathcal{N}_V^{u,v,i} = 1$. This is because there is no parameter to tune in an identity vine. This result gives an insight that adding an identity vine to a neural network would not affect the hypothesis complexity. However, it is worth noting that the vines could influence the part of the stem in the covering bound, i.e., $\mathcal{N}_{M(j+1)}$ in Eq. (5.15). The mechanism of the cross-influence between the stem and the vines is an open problem.

Intuitively, introducing residual connections to a neural network may not change the hypothesis space. Here, we discuss the following case as an example. Consider a residual connection that links nodes i and j. Suppose the hypothesis functions computed in the nodes i and j are F_i and F_j, respectively. Also, we denote the hypothesis functions computed by the bone and residual connection between nodes i and j as $F_{i,j}$ and $F'_{i,j}$, respectively. See an illumination in Fig. 5.2. Then, we have the following equation

$$F_j = F_{i,j} \circ F_i + F'_{i,j} \circ F_i.$$

Usually, the residual connection is a simpler sub-network of the bone part. Therefore, the hypothesis space constituted by all $F'_{i,j}$ is a subspace of the one of $F_{i,j}$. Thus, the hypothesis space constituted by all $F'_{i,j} \circ F_i$ is a subspace of the one of $F_{i,j} \circ F_i$. In other words, the hypothesis space computed by a residual connection is only a sub-space computed by the bone part. Introducing a residual connection is merging the two spaces by addition operation, which obtains the larger space. This property guarantees that introducing residual connections may not change the hypothesis space.

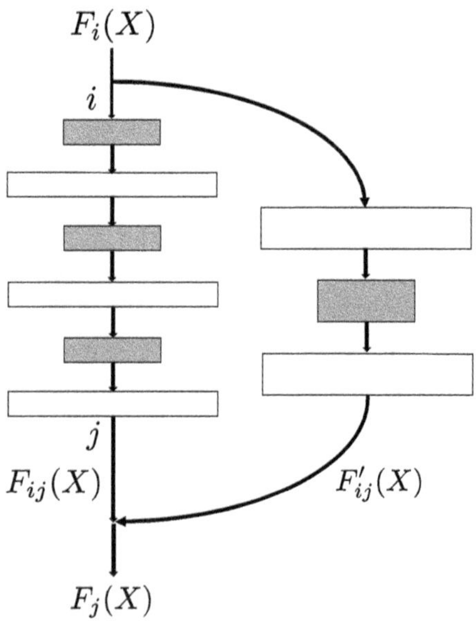

Fig. 5.2 An illustration of influence of a residual connection to the hypothesis space

5.3.2.2 Covering Bound for ResNet

As an example, we analyze the generalization capability of the 34-layer ResNet. Analysis of other deep neural networks under the stem-vine framework is similar. For the convenience, we give a detailed illustration of the 34-layer ResNet under the stem-vine framework in Fig. 5.3.

There are one 34-layer stem and 16 vines in the 34-layer ResNet. Each layer in the stem contains one weight matrix and several Lipschitz-continuous nonlinearities. For most layers with over one nonlinearity, the multiple nonlinearities are connected one by one directly; we merge the nonlinearities as one single nonlinearity.[3] However, the vine links the stem at a vertex between two nonlinearities after the 33-th weight matrix, and thus we cannot merge the two nonlinearities. Hence, the stem of ResNet can be expressed as follows:

$$S_{res} = (A_1, \sigma_1, \ldots, A_{33}, \sigma_{33}, \sigma_{34}, A_{34}, \sigma_{35}) . \tag{5.19}$$

From the vertex that receives the input data to the vertex that outputs classification functions, there are $34 + 35 + 1 = 70$ vertexes (34 is the number of weight matrices and 35 is the number of nonlinearities). We denote them as $N(1)$ to $N(70)$. Additionally, we assume the norm of the the weight matrix A_i has an upper bound s_i, i.e., $\|A_i\|_\sigma \leq s_i$, while the Lipschitz constant of the nonlinearity σ_i is denoted as b_i.

Under the stem-vine framework, the 16 vines in ResNet are respectively denoted as $V(3, 7, 1), V(7, 11, 1), \ldots, V(63, 67, 1)$. Among these 16 vines, there are 3 vines, $V(15, 19, 1)$, $V(31, 35, 1)$, and $V(55, 59, 1)$, that respectively contains one weight matrix, while all others are identity mappings. Let's denote the weight matrices in the vines $V(15, 19, 1)$, $V(31, 35, 1)$, and $V(55, 59, 1)$ respectively as $A_1^{15,19,1}$, $A_1^{31,35,1}$, and $A_1^{55,59,1}$. Suppose the norms of $A_1^{15,19,1}$, $A_1^{31,35,1}$, and $A_1^{55,59,1}$ are respectively upper bounded by $s_1^{15,19,1}$, $s_1^{31,35,1}$, and $s_1^{55,59,1}$. Denote the reference matrices that correspond to weight matrices $(A_1, \ldots, A_{34})$ as $(M_1, \ldots, M_{34})$. Suppose the distance between each weight matrix A_i and the corresponding reference matrix M_i is upper bounded by b_i, i.e., $\|A_i^T - M_i^T\| \leq b_i$. Similarly, suppose there are reference matrices $M_1^{s,t,1}$, $(s, t) \in \{(15, 19), (31, 35), (55, 59)\}$ respectively for weight matrices $A_1^{s,t,1}$, and the distance between $A_1^{s,t}$ and $M_1^{s,t,1}$ is upper bounded by $b_1^{s,t,1}$, i.e., $\|(A_i^{s,t,1})^T - (M_i^{s,t,1})^T\| \leq b_1^{s,t,1}$. We then have the following lemma.

[3] Specifically, if there are two Lipschitz-continuous nonlinearities connected directly somewhere in the stem, such as one max-pooling and one ReLU, we compose the two nonlinearities as one single nonlinearity. The composition is well-defined, as the composition of two Lipschitz-continuous nonlinearities is still a Lipschitz-continuous nonlinearity. The Lipschitz constant of the composition function is the product of the Lipschitz constants respectively of the two nonlinearities. Additionally, the composition would not limit any generality, as in our theory, different nonlinearities with the same Lipschitz constant have the same influence on the generalization bound (this argument is supported by Eq. (5.20) of Lemma 5.2, Theorem 5.7, etc.).

Fig. 5.3 The 34-layer ResNet under the Stem-Vine Framework

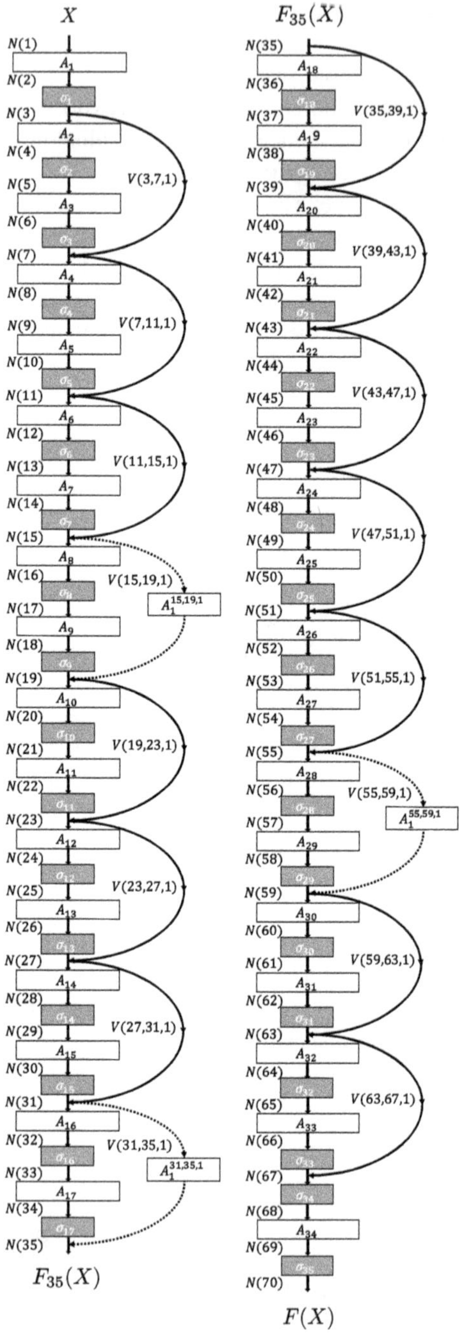

Lemma 5.2 *(Covering Number Bound for ResNet) For a ResNet R satisfies all conditions above, suppose the hypothesis space is $\mathcal{H}_R$. Then, we have*

$$\begin{aligned}\log \mathcal{N}(\mathcal{H}_R, \varepsilon, \|\cdot\|) \leq & \sum_{u\in\{15,31,55\}} \frac{(b_1^{u,u+4,1})^2 \| F_u(X^T)^T \|_2^2}{\varepsilon_{u,u+4,1}^2} \log(2W^2) \\ & + \sum_{j=1}^{34} \frac{b_j^2 \| F_{2j-1}(X^T)^T \|_2^2}{\varepsilon_{2j+1}^2} \log(2W^2) \\ & + \frac{b_{34}^2 \| F_{68}(X^T)^T \|_2^2}{\varepsilon_{70}^2} \log(2W^2)\,, \end{aligned} \tag{5.20}$$

where $\mathcal{N}(\mathcal{H}_R, \varepsilon, \|\cdot\|)$ is the ε-covering number of $\mathcal{H}_R$. When $j = 1, \ldots, 16$,

$$\begin{aligned}\| F_{4j+1}(X)\|_2^2 \leq & \|X\|^2 \rho_1^2 s_1^2 \rho_{2j}^2 s_{2j}^2 \prod_{\substack{1\leq i\leq j-1 \\ i\notin\{4,8,14\}}} \left(\rho_{2i}^2 s_{2i}^2 \rho_{2i+1}^2 s_{2i+1}^2 + 1\right) \\ & \prod_{\substack{1\leq i\leq j-1 \\ i\in\{4,8,14\}}} \left[\rho_{2i}^2 s_{2i}^2 \rho_{2i+1}^2 s_{2i+1}^2 + (s_1^{4i-1,4i+3,1})^2\right], \end{aligned} \tag{5.21}$$

and

$$\begin{aligned}\| F_{4j+3}(X)\|_2^2 \leq & \|X\|^2 \rho_1^2 s_1^2 \prod_{\substack{1\leq i\leq j \\ i\notin\{4,8,14\}}} \left(\rho_{2i}^2 s_{2i}^2 \rho_{2i+1}^2 s_{2i+1}^2 + 1\right) \\ & \prod_{\substack{1\leq i\leq j \\ i\in\{4,8,14\}}} \left[\rho_{2i}^2 s_{2i}^2 \rho_{2i+1}^2 s_{2i+1}^2 + (s_1^{4i-1,4i+3,1})^2\right], \end{aligned} \tag{5.22}$$

and specifically,

$$\begin{aligned}\| F_{68}(X^T)^T\|_2^2 \leq & \|X\|^2 \rho_1^2 s_1^2 \rho_{34}^2 \prod_{\substack{1\leq i\leq 16 \\ i\notin\{4,8,14\}}} \left(\rho_{2i}^2 s_{2i}^2 \rho_{2i+1}^2 s_{2i+1}^2 + 1\right) \\ & \prod_{\substack{1\leq i\leq 16 \\ i\in\{4,8,14\}}} \left[\rho_{2i}^2 s_{2i}^2 \rho_{2i+1}^2 s_{2i+1}^2 + (s_1^{4i-1,4i+3,1})^2\right]. \end{aligned} \tag{5.23}$$

Also, when $j = 1, \ldots, 16$,

$$\varepsilon_{4j+1} = (1+s_1)\rho_1(1+s_{2j})\rho_{2j} \prod_{\substack{1\leq i\leq j-1 \\ i\notin\{4,8,14\}}} [(*)+1] \prod_{\substack{1\leq i\leq j-1 \\ i\in\{4,8,14\}}} \left[(*)+1+s_1^{4i-1,4i+3,1}\right], \tag{5.24}$$

and

$$\varepsilon_{4j+3} =(1+s_1)\rho_1 \prod_{\substack{1\le i\le j\\ i\notin\{4,8,14\}}} [(*)+1] \prod_{\substack{1\le i\le j\\ i\in\{4,8,14\}}} \left[(*)+1+s_1^{4i-1,4i+3,1}\right], \tag{5.25}$$

and for $u = 15, 31, 55$,

$$\varepsilon_{u,u+4,1} = \varepsilon_u \left(1+s_1^{u,u+4,1}\right). \tag{5.26}$$

In above equations/inequalities,

$$\bar{\alpha} =(s_1+1)\rho_1\rho_{34}(s_{34}+1)\rho_{35} \prod_{\substack{1\le i\le 16\\ i\notin\{4,8,14\}}} [(*)+1] \prod_{i\in\{4,8,14\}} \left[(*)+s_1^{4i-1,4i+3,1}+1\right], \tag{5.27}$$

and

$$(*) = \rho_{2i}(s_{2i}+1)\rho_{2i+1}(s_{2i+1}+1). \tag{5.28}$$

A detailed proof is omitted and will be given in Sect. 5.3.3.3.

5.3.2.3 Generalization Bound for ResNet

Lemma 3.2 guarantees that when the covering number of a hypothesis space is upper bounded, the corresponding generalization error is also upper bounded. This lemma provides a theoretical foundation for establishing generalization bounds based on the covering number of the hypothesis space. By leveraging Lemma 5.2, which presents a specific covering bound for ResNet, we can derive a concrete generalization bound for this architecture. In this subsection, we summarize and formalize the generalization bound as Theorem 5.7, providing a clear theoretical link between the complexity of the hypothesis space and the expected generalization performance of ResNet.

For the brevity, we rewrite the radius ε_{2j+1} and $\varepsilon_{u,u+4,1}$ as follows:

$$\varepsilon_{2j+1} := \hat{\varepsilon}_{2j+1}, \tag{5.29}$$

$$\varepsilon_{u,u+4,1} := \hat{\varepsilon}_{u,u+4,1}\varepsilon. \tag{5.30}$$

Additionally, we rewrite Eq. (5.20) of Lemma 5.2 as the following inequality:

$$\log \mathcal{N}(\mathcal{H}, \varepsilon, \|\cdot\|) \le \frac{R}{\varepsilon^2}, \tag{5.31}$$

where

$$R = \sum_{u \in \{15,31,55\}} \frac{(b_1^{u,u+4,1})^2 \| F_u(X^T)^T \|_2^2}{\hat{\varepsilon}_{u,u+4,1}^2} \log(2W^2)$$
$$+ \sum_{j=1}^{33} \frac{b_j^2 \| F_{2j-1}(X^T)^T \|_2^2}{\hat{\varepsilon}_{2j+1}^2} \log(2W^2) + \frac{b_{34}^2 \| F_{68}(X^T)^T \|_2^2}{\hat{\varepsilon}_{70}^2} \log(2W^2) , \quad (5.32)$$

Then, we can obtain the following theorem.

We first define the margin operator $\mathcal{M}$ for the k-class classification task as

$$\mathcal{M} : \mathbb{R}^k \times \{1, ..., k\} \to \mathbb{R}, (v, y) \to v_y - \max_{i \neq y} v_i .$$

Then, ramp loss l_λ is defined as

$$l_\lambda(x) = \begin{cases} 0, & x < -\lambda \\ 1 + x/\lambda, & -\lambda \le x < 0 \\ 1, & x > 0 \end{cases}$$

Furthermore, the empirical ramp risk is defined as

$$\hat{R}_S^\lambda = \frac{1}{m} \sum_{i=1}^{m} l_\lambda(-\mathcal{M}(F(x_i), y_i)).$$

Theorem 5.7 (Generalization Bound for ResNet) *Suppose a ResNet satisfies all conditions in Lemma 5.2. Suppose a given series of examples* $(x_1, y_1), \ldots, (x_m, y_m)$ *are arbitrary i.i.d. variables. Suppose hypothesis function* $F_{\mathcal{A}} : \mathbb{R}^{n_0} \to \mathbb{R}^{n_L}$ *is computed by a ResNet with weight matrices* $\mathcal{A} = (A_1, \ldots, A_{34}, A_1^{15,19,1}, A_1^{31,35,1}, A_1^{55,59,1})$*. Then for any margin* $\lambda > 0$ *and any real* $\delta \in (0, 1)$*, with probability at least* $1 - \delta$*, we have the following inequality:*

$$\mathbb{P}\{\arg\max_i F(x)_i \neq y\} \le \hat{R}_S^\lambda(F) + \frac{8}{m^{\frac{3}{2}}} + \frac{36}{m}\sqrt{R}\log m + 3\sqrt{\frac{\log(1/\delta)}{2m}} , \quad (5.33)$$

where R is defined as Eq. (5.32).

This generalization bound is established via the hypothesis complexity which is measured by the covering number of the hypothesis space. This relationship is characterised by Lemma 3.1. Intuitively, Lemma 3.1 follows the principle of Occam's razor, which demonstrates a negative correlation between the generalization ability of an algorithm and its hypothesis complexity. Directly applying Lemma 3.1 with the covering number bound (Theorem 5.6), we can obtain Theorem 5.7. A proof is omitted here and will be given in Sect. 5.3.3.5.

Indicated by Theorem 5.7, the generalization bound of ResNet relies on its covering bound. Specifically, when the sample size n and the probability δ are fixed, the generalization error satisfies that

$$\mathbb{P}\{\arg\max_i F(x)_i \neq y\} - \hat{R}_S^{\lambda}(F) = O\left(\sqrt{R}\right), \tag{5.34}$$

where R expresses the magnitude of the covering number (R/ε^2 is an ε-covering bound).

Incorporating the characteristics applicable to any neural network within the stem-vine framework, Eq. (5.34) provides two insights into the impact of residual connections on neural network generalization capability: (1) The influence of weight matrices on generalization remains invariant regardless of their location (stem or vines); (2) The addition of an identity vine does not affect generalization. These findings offer a theoretical rationale for ResNet's equivalent generalization capability compared to chain-like neural networks. This implies that the placement of weight matrices within the network structure, whether in the stem or the vines, does not fundamentally alter the network's capability to generalize. Additionally, the inclusion of identity vines, which do not introduce additional transformations, maintains this generalization capability without detriment.

As indicated in Eq. (5.33), the expected risk of ResNet, which corresponds to the expectation of the test error, is composed of the sum of the empirical risk (training error) and the generalization error. Notably, residual connections have demonstrated a remarkable capability to reduce the training error of neural networks across various tasks. Our findings provide a theoretical basis for understanding why ResNet exhibits substantially lower test errors in these scenarios. This elucidates the underlying mechanism by which ResNet achieves superior performance by effectively mitigating training error, leading to enhanced generalization ability in practical applications.

5.3.2.4 Practical Implementation

In addition to the sample size m, our generalization bound (Eq. (5.33)) highlights a positive correlation with the norms of all weight matrices in the neural network. Specifically, weight matrices with higher norms contribute to a higher generalization bound, suggesting a potential trade-off with generalization performance. This observation validates the importance of techniques such as weight decay, which aims to regulate the norms of weight matrices to optimize generalization performance in deep learning models. By controlling the norms of weight matrices, practitioners can mitigate the risk of overfitting and enhance the model's ability to generalize to unseen data.

Weight decay, original introduced by Krogh and Hertz (1992), has become a widely adopted regularization technique in the training of deep neural networks. In the context of learning theory, weight decay involves augmenting the typical training objective with an additional term that penalizes large weights. It involves adding the L_2 norm of all weights as a regularization term to the loss function during training. This term penalizes large weight values, aiming to prevent overfitting by discouraging complex model configurations that fit the training data too closely. By constraining the magnitude of weight norms, weight decay promotes smoother and more stable training dynamics, which often leads to improved generalization performance on unseen data. This regularization method is particularly effective in controlling model complexity and mitigating the risk of overfitting, contributing to the robustness and stability of deep learning models.

Remark 5.2 The technique of weight decay can improve the generalization ability of deep neural networks. It refers to adding the L_2 norm of the weights $W = (W_1, \ldots, W_D)$ to the objective function as a regularization term:

$$\mathcal{L}'(W) = \mathcal{L}(W) + \frac{1}{2}\lambda \sum_{i=1}^{D} W_i^2 ,$$

where λ is a tuneable parameter, $\mathcal{L}(W)$ is the original objective function, and $\mathcal{L}'(w)$ is the objective function with weight decay.

The term $\frac{1}{2}\lambda \sum_{i=1}^{D} W_i^2$ can be easily re-expressed by the L_2 norms of all the weight matrices. Therefore, using weight decay can control the magnitude of the norms of all the weights matrices not to increase too much. Also, our generalization bound (Eq. (5.33)) provides a positive correlation between the generalization bound and the norms of all the weight matrices. Thus, our work gives a justification for why weight decay leads to a better generalization ability.

A recent systematic experiment conducted by Li et al. (2018c) studies the influence of weight decay on the loss surface of the deep neural networks (Li et al. 2018c). It trains a 9-layer VGGNet (Chen et al. 2017) on the dataset CIFAR-10 (Krizhevsky and Hinton 2009) by employing stochastic gradient descent with batch sizes of 128 (0.26% of the training set of CIFAR-10) and 8192 (16.28% of the training set of CIFAR-10). The results demonstrate that by employing weight decay, SGD can find flatter minima[4] of the loss surface with lower test errors as shown in Fig. 5.4 (original presented as Li et al. (2018c), p. 6, Fig. 3). Other technical advances and empirical analysis include (Galloway et al. 2018; Zhang et al. 2018b; Chen et al. 2018a; Park et al. 2019).

5.3.3 *Proofs of Section 5.3*

This appendix compiles several proofs that were omitted from Sect. 5.3.2. First, we present a proof establishing the covering bound for an affine transformation induced

[4] The flatness (or equivalently sharpness) of the loss surface around the minima is considered as an important index expressing the generalization ability. However, the mechanism remains elusive. For more details, please refers to (Keskar et al. 2017) and (Dinh et al. 2017).

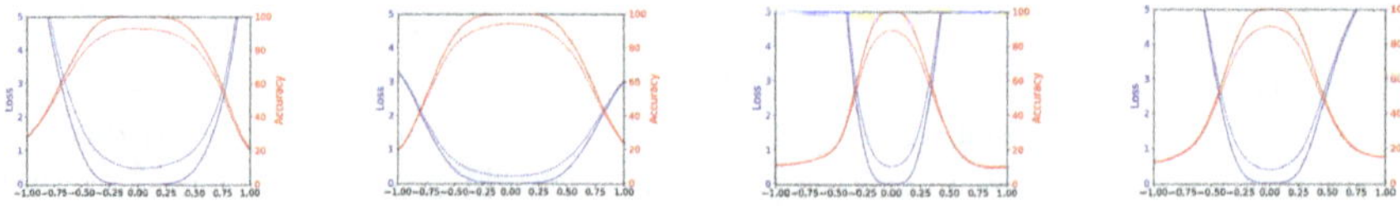

(a) 0, 128, 7.37% (b) 5×10^{-4}, 128, 6.00% (c) 0, 8192, 11.07% (d) 5×10^{-4}, 8192, 10.19%

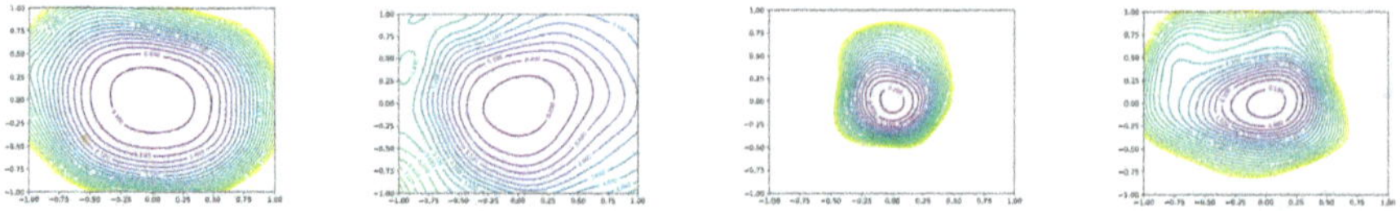

(e) 0, 128, 7.37% (f) 5×10^{-4}, 128, 6.00% (g) 0, 8192, 11.07% (h) 5×10^{-4}, 8192, 10.19%

Fig. 5.4 Illustrations of the 1D and 2D visualization of the loss surface around the solutions obtained with different weight decay and batch size. The numbers in the title of each subfigure is respectively the parameter of weight decay, batch size, and test error. The data and figures are originally presented in (Li et al. 2018c)

by a single weight matrix, which serves as a foundation for subsequent analyses. Next, we provide a detailed proof of the covering bound for deep neural networks within the stem-vine framework (Theorem 5.6). Additionally, we include a proof demonstrating the covering bound specific to ResNet architectures (Lemma 5.2). Finally, we provide a comprehensive proof of the generalization bound for ResNet (Theorem 5.7). These proofs collectively contribute to a deeper understanding of the theoretical underpinnings discussed in the main section.

5.3.3.1 Proof of the Covering Bound for the Hypothesis Space of a Single Weight Matrix

In this subsection, we provide an upper bound for the covering number of the hypothesis space induced by a single weight matrix A. This covering bound relies on *Maurey sparsification lemma* (Pisier 1981) and has been introduced in machine learning by previous works (see, e.g.,(Zhang 2002; Bartlett et al. 2017)).

Suppose a data matrix X is the input of a weight matrix A. All possible values of the output XA constitute a space. We use the following lemma to express the complexity of all XA via the covering number.

Lemma 5.3 *(Bartlett et al.; see (Bartlett et al. 2017), Lemma 3.2) Let conjugate exponents (p, q) and (r, s) be given with $p \leq 2$, as well as positive reals (a, b, ε) and positive integer m. Let matrix $X \in \mathbb{R}^{n \times d}$ be given with $\|X\|_p \leq b$. Let $\mathcal{H}_A$ denote the family of matrices obtained by evaluating X with all choices of matrix A:*

$$\mathcal{H}_A \triangleq \left\{XA | A \in \mathbb{R}^{d \times m}, \|A\|_{q,s} \leq a\right\} . \tag{5.35}$$

Then

$$\log \mathcal{N}(\mathcal{H}_A, \varepsilon, \|\cdot\|_2) \leq \lceil * \rceil \frac{a^2 b^2 m^{2/r}}{\varepsilon^2} \log(2dm) . \tag{5.36}$$

5.3.3.2 Covering Bound for the Hypothesis Space of Chain-like Neural Network

This subsection considers the upper bound for the covering number of the hypothesis space induced by the stem of a deep neural network. Intuitively, following the stem from the first vertex $N(1)$ to the last one $N(L)$, every weight matrices and nonlinearities increase the complexity of the hypothesis space that could be computed by the stem. Following this intuition, we use an induction method to approach the upper bound. The result is summarized as Lemma 5.1. This lemma is originally given in the work by Bartlett et al. (2017). Here to make this work complete, we recall the main part of the proof but omit the part for ε.

Proof of Lemma 5.1 We use an induction procedure to prove the lemma.

(1) The covering number of the hypothesis space computed by the first weight matrix A_1 can be straightly upper bounded by Lemma 5.3.

(2) The vertex after the j-th nonlinearity is $N(2j+1)$. Suppose $\mathcal{W}_{2j+1}$ is an ε-cover of the hypothesis space $\mathcal{H}_{2j+1}$ induced by the output hypotheses in the vertex $N(2j+1)$. Suppose there is a weight matrix A_{j+1} directly follows the vertex $N(2j+1)$. We then analyze the contribution of the weight matrix A_{j+1}. Assume that there exists an upper bound s_{j+1} of the norm of A_{j+1}. For any $F_{2j+1}(X) \in \mathcal{H}_{2j+1}$, there exists a $W(X) \in \mathcal{W}_{2j+1}$ such that

$$\|F_{2j+1}(X) - W(X)\| \leq \varepsilon_{2j+1} . \tag{5.37}$$

Lemma 5.3 guarantees that for any $W(X) \in \mathcal{W}_{2j+1}$ there exists an ε_{2j+1}-cover $\mathcal{W}_{2j+2}(W)$ for the function space $\{W(X)A_{j+1} : W(X) \in \mathcal{W}_{2j+1}, \|A_{j+1}\| \leq s_{j+1}\}$, i.e., for any $W'(X) \in \hat{\mathcal{H}}_{2j+1}$, there exists a $V(X) \in \{W(X)A_{j+1} : W(X) \in \mathcal{W}_{2j+1}, \|A_{j+1}\| \leq s_{j+1}\}$ such that

$$\|W'(X) - V(X)\| \leq \varepsilon_{2j+1} . \tag{5.38}$$

As for any $F'_{2j+1}(X) \in \mathcal{H}_{2j+2} \triangleq \{F_{2j+1}(X)A_{j+1} : F_{2j+1}(X) \in \mathcal{H}_{2j+1}, \|A_{j+1}\| \leq c\}$, there is a $F_{2j+1}(X) \in \mathcal{H}_{2j+1}$ such that

$$F'_{2j+1}(X) = F_{2j+1}(X)A_{j+1} . \tag{5.39}$$

Thus, applying Eqs. (5.37), (5.38), and (5.39), we prove the following inequality

$$\begin{aligned}\|F'_{2j+1}(X) - V(X)\| =&\|F_{2j+1}(X)A_{j+1} - V(X)\| \\ \leq&\|F_{2j+1}(X)A_{j+1} - W(X)A_{j+1}\| + \|W(X)A_{j+1} - V(X)\| \\ \leq&\|F_{2j+1}(X) - W(X)\|\|A_{j+1}\| + \varepsilon_{2j+1} \\ \leq&s_{j+1}\varepsilon_{2j+1} + \varepsilon_{2j+1} \\ =&(s_{j+1} + 1)\varepsilon_{2j+1} .\end{aligned} \tag{5.40}$$

Therefore, $\bigcup_{W \in \mathcal{W}_{2j+1}} \mathcal{W}_{2j+2}(W)$ is a $(s_{j+1} + 1)\varepsilon_{2j+1}$-cover of $\mathcal{H}_{2j+2}$. Let's denote $(s_{j+1} + 1)\varepsilon_{2j+1}$ as ε_{2j+2}. Apparently,

$$\begin{aligned}\mathcal{N}(\mathcal{H}_{2j+2}, \varepsilon_{2j+2}, \|\cdot\|) \leq& \left|\bigcup_{W \in \mathcal{W}_{2j+1}} \mathcal{W}_{2j+2}(W)\right| \leq \left|\mathcal{W}_{2j+1}\right| \cdot \sup_{W \in \mathcal{W}_{2j+1}} \left|\mathcal{W}_{2j+2}(W)\right| \\ \leq&\mathcal{N}(\mathcal{H}_{2j+1}, \varepsilon_{2j+1}, \|\cdot\|) \sup_{\substack{(A_1,\ldots,A_j) \\ \forall j \leq j,\ A_i \in \mathcal{B}_i}} \mathcal{N}\left((**), \varepsilon_{2j+1}, \|\cdot\|_{2j+1}\right) ,\end{aligned} \tag{5.41}$$

where

$$(**) = \left\{A_{j+1}F_{2j+1}(X) : A_{j+1} \in \mathcal{B}_{j+1}\right\} .$$

Thus, $\mathcal{N}(\mathcal{W}_{2j+1}, \varepsilon_{2j+1}, \|\cdot\|) \cdot \mathcal{N}(\mathcal{W}_{2j+2}, \varepsilon_{2j+2}, \|\cdot\|)$ is an upper bound for the ε_{2j+2}-covering number of the hypotheses space $\mathcal{H}_{i+1}$.

(3) The vertex after the j-th weight matrix is $N(2j - 1)$. Suppose $\mathcal{W}_{2j-1}$ is an ε_{2j-1}-cover of the hypothesis space $\mathcal{H}_{2j-1}$ induced by the output hypotheses in the vertex $N(2j - 1)$. Suppose there is a nonlinearity σ_j directly follows the vertex $N(2j - 1)$. We then analyze the contribution of the nonlinearity σ_j. Assume that the nonlinearity σ_j is ρ_j-Lipschitz continuous. Apparently, $\sigma_j(\mathcal{W}_{2j-1})$ is a $\rho\varepsilon_{2j-1}$-cover of the hypothesis space $\sigma_j(\mathcal{H}_{2j-1})$. Specifically, for any $F' \in \sigma(\mathcal{H}_{2j-1})$, there exits a $F \in \mathcal{H}_{2j-1}$ that $F' = \sigma_j(F)$. Since $\mathcal{W}_{2j-1}$ is an ε_{2j-1}-cover of the hypothesis space $\mathcal{H}_{2j-1}$, there exists a $W \in \mathcal{W}_{2j-1}$ such that

$$\|F - W_{2j-1}\| \leq \varepsilon_{2j-1} . \tag{5.42}$$

Therefore, we have the following equation

$$\|F' - \sigma_j(W_{2j-1})\| = \|\sigma_j(F) - \sigma_j(W_{2j-1})\| \leq \rho_j\|F - W_{2j-1}\| = \rho_j\varepsilon_{2j-1} . \tag{5.43}$$

We thus prove that $\mathcal{W}_{2j} \triangleq \sigma_j(\mathcal{W}_{2j-1})$ is a $\rho_j\varepsilon_{2j-1}$-cover of the hypothesis space $\sigma_j(\mathcal{H}_{2j-1})$. Additionally, the covering number remains the same while applying a nonlinearity to the neural network.

By analyzing the influence of weight matrices and nonlinearities one by one, we can prove Eq. (5.17). As for ε, the above part indeed gives an constructive method to obtain ε from all ε_i and $\varepsilon_{u,v,j}$. Here we omit the explicit formulation of ε in terms of ε_i and $\varepsilon_{u,v,j}$, since it could not benefit our theory. This completes the proof.

5.3.3.3 Covering Bound for the Hypothesis Space of Deep Neural Networks with Residual Connections

In Sect. 5.3.2.1, we give a covering bound generally for all deep neural networks with residual connections. The result is summarised as Theorem 5.6. In this subsection, we give a detailed proof of Theorem 5.6.

Proof of Theorem 5.6. To approach the covering bound for the deep neural networks with residuals, we first analyze the influence of adding a vine to a deep neural network, and then use an induction method to obtain a covering bound for the whole network.

All vines are connected with the stem at two points that is respectively after a nonlinearity and before a weight matrix. When the input $F_u(X)$ of the vine $V(u, v, i)$ is fixed, suppose all the hypothesis functions $F_V^{u,v,i}(X)$ computed by the vine $V(u, v, i)$ constitute a hypothesis space $\mathcal{H}_V^{u,v,i}$. As a vine is also a chain-like neural network constructed by stacking a series of weight matrices and nonlinearities, we can straightly apply Lemma 5.1 to approach an upper bound for the covering number of the hypothesis space $\mathcal{H}_V^{u,v,i}$. It is worth noting that vines could be identity mappings. This situation is normal in ResNet–there are 13 out of all the 16 vines are identities. For the circumstances that the vines are identities, the hypothesis space computed by the vine only contains one element–an identity mapping. The covering number of the hypothesis space for the identities are apparently 1.

Applying Lemmas 5.3 and 5.1, there exists an ε_v-cover $\mathcal{W}_v$ for the hypothesis space $\mathcal{H}_v$ with a covering number $\mathcal{N}(\mathcal{H}_v, \varepsilon_i, \|\cdot\|)$, as well as an $\varepsilon_V^{u,v,i}$-cover $\mathcal{W}_V^{u,v,i}$ for the hypothesis space $\mathcal{H}_V^{u,v,i}$ with a covering number $\mathcal{N}(\mathcal{H}_V^{u,v,i}, \varepsilon_i, \|\cdot\|)$.

The hypotheses computed by the vine $V(u, v, i)$ and the deep neural network without $V(u, v, i)$, i.e., respectively, $F_v(X)$ and $F_V^{u,v,i}$, are added element-wisely at the vertex $V(v)$. We denote the space constituted by all $F' \triangleq F_v(X) + F_V^{u,v,i}(X)$ as $\mathcal{H}'_v$.

Let's define a function space as $\mathcal{W}'_v \triangleq \{W_S + W_V : W_S \in \mathcal{W}_v, W_V \in \mathcal{W}_V^{u,v,i}\}$. For any hypothesis $F' \in \mathcal{H}'_v$, there must exist an $F_S \in \mathcal{H}_v$ and $F_V \in \mathcal{H}_V^{u,v,i}$ such that

$$F'(X) = F_S(X) + F_V(X) \,. \tag{5.44}$$

Because $\mathcal{W}_v$ is an ε_v-cover of the hypothesis space $\mathcal{H}_v$. For any $F_S \in \mathcal{H}_v$, there exists an element $W_{F_S}(X) \in \mathcal{W}_v$, such that

$$\|F_S(X) - W_{F_S}(X)\| \leq \varepsilon_v \,. \tag{5.45}$$

Similarly, as $\mathcal{W}_V^{u,v,i}$ is an $\varepsilon_V^{u,v,i}$-cover of $\mathcal{H}_V^{u,v,i}$, we can obtain a similar result. For any $F_V(X) \in \mathcal{H}_V^{u,v,i}$, there exists an element $W_{F_V}(X) \in \mathcal{W}_V^{u,v,i}$, such that

$$\|F_V(X) - W_{F_V}(X)\| \leq \varepsilon_V^{u,v,i} \,. \tag{5.46}$$

Therefore, For any hypothesis $F'(X) \in \mathcal{H}'_v$, there exists an element $W(X) \in \mathcal{W}'$, such that $W(X) = W_{F_S}(X) + W_{F_V}(X)$ satisfying Eqs. (5.45) and (5.46), and

$$\begin{aligned}\|F'(X) - W(X)\| =& \|F_V(X) + F_S(X) - W_{F_V}(X) - W_{F_S}(X)\| \\ \leq& \|F_V(X) - W_{F_V}(X)\| + \|F_S(X) - W_{F_S}(X)\| \\ \leq& \varepsilon_V^{u,v,i} + \varepsilon_v \,. \end{aligned} \tag{5.47}$$

Therefore, the function space $\mathcal{W}'_v$ is an $(\varepsilon_V^{u,v,i} + \varepsilon_v)$-cover of the hypothesis space $\mathcal{H}'_v$. An upper bound for the cardinality of the function space $\mathcal{W}'_v$ is given as below (it is also an $\varepsilon_V^{u,v,i} + \varepsilon_v$-covering number of the hypothesis space $\mathcal{H}'_v$):

$$\begin{aligned}\mathcal{N}(\mathcal{H}'_v, \varepsilon_V^{u,v,i} + \varepsilon_v, \|\cdot\|) \leq& |\mathcal{W}'_v| \leq |\mathcal{W}_v| \cdot |\mathcal{W}_V^{u,v,i}| \\ \leq& \sup_{F_{v-2}} \mathcal{N}(\mathcal{H}_v, \varepsilon_i, \|\cdot\|) \cdot \sup_{F_u} \mathcal{N}(\mathcal{H}_V^{u,v,i}, \varepsilon_V^{u,v,i}, \|\cdot\|) \\ \leq& \sup_{F_{v-2}} \mathcal{N}_v \cdot \sup_{F_u} \mathcal{N}_V^{u,v,i} \,, \end{aligned} \tag{5.48}$$

where $\mathcal{N}_v$ and $\mathcal{N}_V^{u,v,i}$ can be obtained from Eq. (5.17) in Lemma 5.1, as the stem and all the vines are chain-like neural networks.

By adding vines to the stem one by one, we can construct the whole deep neural network. Combining Lemma 5.1 for the covering number of $F_{v-1}(X)$ and $F_u(X)$, we further prove the following inequality:

$$\mathcal{N}(\mathcal{H}, \varepsilon, \|\cdot\|) \leq \prod_{j=1}^{L} \sup_{F_{M(j)}} \mathcal{N}_{M(j+1)} \prod_{(u,v,i)\in I_V} \sup_{F_u} \mathcal{N}_V^{u,v,i} \,. \tag{5.49}$$

Thus, we prove Eq. (5.15) of Theorem 5.6. As for ε, the above part indeed gives an constructive method to obtain ε from all ε_i and $\varepsilon_{u,v,j}$. Here we omit the explicit formulation of ε in terms of ε_i and $\varepsilon_{u,v,j}$, since it could be extremely complex and does not benefit our theory. The proof is completed.

5.3.3.4 Covering Bound for the Hypothesis Space of ResNet

In Sect. 5.3.2.2, we give a covering bound for ResNet. The result is summarized as Lemma 5.2. In this subsection, we give a detailed proof of Lemma 5.2.

Proof of Lemma 5.2. There are 34 weight matrices and 35 nonlinearities in the stem of the 34-ResNet. Let's denote the weight matrices respectively as $A_1, \ldots, A_{34}$ and denote the nonlinearities respectively as $\sigma_1, \ldots, \sigma_{35}$. Apparently, there are $34 + 35 + 1 = 70$ vertexes in the network, where 34 is the number of weight matrices and 35 is the number of nonlinearities. We denote them respectively as $N(1), \ldots, N(70)$. Additionally, there are 16 vines which are respectively denoted as $V(4i - 1, 4i +$

$3, 1), i = \{1, \ldots, 16\}$, where $4i - 1$ and $4i + 3$ are the indexes of the vertexes that the vine connected. Among all the 16 vines, there are 3, $V(15, 19, 1)$, $V(31, 35, 1)$, and $V(55, 59, 1)$, respectively contain one weight matrix, while all others are identities mappings. For the vine $V(4i - 1, 4i + 3, 1)$, $i = 4, 8, 14$, we denote the weight matrix in the vine as $A_1^{4i-1,4i+3,1}$.

Applying Theorem 5.6, we straightly prove the following inequality:

$$\log \mathcal{N}(\mathcal{H}, \varepsilon, \|\cdot\|) \leq \sum_{j=1}^{34} \sup_{F_{2j-1}(X)} \log \mathcal{N}_{2j+1} + \sum_{(u,v,i)\in I_V} \sup_{F_u(X)} \log \mathcal{N}_V^{u,v,1} , \tag{5.50}$$

where $\mathcal{N}_{2j+1}$ is the covering number of the hypothesis space constituted by all outputs $F_{2j+1}(X)$ at the vertex $N(2j + 1)$ when the input $F_{2j-1}(X)$ of the vertex $N(2j - 1)$ is fixed, $\mathcal{N}_V^{u,v,1}$ is the covering number of the hypothesis space constituted by all outputs $F_V^{u,v,i}(X)$ of the vine $V(u, v, 1)$ when the input $F_v(X)$ is fixed, and I_V is the index set $\{(4i - 1, 4i + 3, 1), i = 1, \ldots, 16\}$.

Applying Lemma 5.3, we can further prove an upper bound for the ε_{2j+1}-covering number $\mathcal{N}_{2j+1}$. The bound is expressed as the following inequality:

$$\log \mathcal{N}_{2j+1} \leq \frac{b_{j+1}^2 \| F_{2j+1}(X^T)^T \|_2^2}{\varepsilon_{2j+1}^2} \log(2W^2) , \tag{5.51}$$

where W is the maximum dimension among all features through the ResNet, i.e., $W = \max_i n_i, i = 0, 1, \ldots, L$. Also, we can decompose $\| F_{2j+1}(X^T)^T \|_2^2$ and utilize an induction method to obtain an upper bound for it.

(1) If there is no vine connected with the stem at the vertex $N(2j - 1)$, we have the following inequality:

$$\begin{aligned} \| F_{2j+1}(X^T)^T \|_2 =& \|\sigma_j(A_j F_{2j-1}(X^T))^T - \sigma_j(0)\|_2 \\ \leq& \rho_j \| A_j F_{2j-1}(X^T)^T - 0\|_2 \\ \leq& \rho_j \| A_j \|_\sigma \cdot \| F_{2j-1}(X^T)^T \|_2 . \end{aligned} \tag{5.52}$$

(2) If there is a vine $V(2j - 3, 2j + 1, 1)$ connected at the vertex $N(2j + 1)$, then we prove the following inequality:

$$\begin{aligned} \| F_{2j+1}(X^T)^T \|_2 =& \|\sigma_j(A_j \sigma_j(A_j F_{2j-3}(X^T)))^T + A_1^{2j-3,2j+1,1} F_{2j-3}(X^T)^T \|_2 \\ \leq& \|\sigma_j(A_j \sigma_j(A_j F_{2j-3}(X^T)))^T \|_2 + \| A_1^{2j-3,2j+1,1} F_{2j-3}(X^T)^T \|_2 \\ \leq& \rho_j \| A_j \|_\sigma \rho_{j-1} \| A_{j-1} \|_\sigma \cdot \| F_{2j-3}(X^T)^T \|_2 \\ &+ \| A_1^{2j-3,2j+1,1} \|_\sigma \cdot \| F_{2j-3}(X^T)^T \|_2 \\ =& \left(\rho_j \rho_{j-1} \| A_j \|_\sigma \cdot \| A_{j-1} \|_\sigma + \| A_1^{2j-3,2j+1,1} \|_\sigma \right) \\ & \| F_{2j-3}(X^T)^T \|_2 . \end{aligned} \tag{5.53}$$

Therefore, based on Eqs. (5.52) and (5.53), we can prove the norm of output of ResNet as in the main text.

Similar with $\mathcal{N}_{2j+1}$, we can obtain an upper bound for the $\varepsilon_{u,v,1}$-covering number $\mathcal{N}_V^{u,v,1}$. Suppose the output computed at the vertex $N(u)$ is $F_u(X^T)$. Then, we can prove the following inequality:

$$\log \mathcal{N}_V^{u,v,1} \leq \frac{(b_1^{u,v,1})^2 \| F_u(X^T)^T \|_2^2}{\varepsilon_{u,v,1}^2} \log(2W^2) . \tag{5.54}$$

Applying Eqs. (5.51) and (5.54) to Eq. (5.50), we thus prove Eq. (5.20).

As for the formulation of the radiuses of the covers, we also employ an induction method.

(1) Suppose the radius of the cover for the hypothesis space computed by the weight matrix A_1 and the nonlinearity σ_1 is ε_3. Then, applying Eqs. (5.40) and (5.43), after the weight matrix A_2 and the nonlinearity σ_2, we prove the following equation:

$$\varepsilon_3 = (s_2 + 1)\rho_2 \varepsilon_1 . \tag{5.55}$$

(2) Suppose the radius of the cover for the hypothesis space computed by the weight matrix A_{j-1} and the nonlinearity σ_{j-1} is ε_{2j-1}. Assume there is no vine connected around. Then, similarly, after the weight matrix A_2 and the nonlinearity σ_j, we prove the following equation:

$$\varepsilon_{2j+1} = \rho_j (s_j + 1)\varepsilon_{2j-1} . \tag{5.56}$$

(3) Suppose the radius of the cover at the vertex $N(i)$ is ε_i. Assume there is a vine $V(u, u+4, 1)$ links the stem at the vertex $N(u)$ and $N(u+4)$. Then, similarly, after the weight matrix A_2 and the nonlinearity σ_j, we prove the following equation:

$$\begin{aligned} \varepsilon_{2j+1} =& \varepsilon_{u+2} \left(s_{\frac{u-1}{2}} + 1 \right) \rho_{\frac{u-1}{2}} + \varepsilon_u \left(s_{u,u+4,1} + 1 \right) \\ =& \varepsilon_u \left(s_{\frac{u-1}{2}} + 1 \right) \rho_{\frac{u-1}{2}} \left(s_{\frac{u-3}{2}} + 1 \right) \rho_{\frac{u-3}{2}} + \varepsilon_u \left(s_{u,u+4,1} + 1 \right) \\ =& \varepsilon_u \left(s_{\frac{u-1}{2}} + 1 \right) \left(s_{\frac{u-3}{2}} + 1 \right) \rho_{\frac{u-1}{2}} \rho_{\frac{u-3}{2}} + \varepsilon_u \left(s_{u,u+4,1} + 1 \right) . \end{aligned} \tag{5.57}$$

From Eqs. (5.55), (5.56), and (5.57), we can obtain the following equation

$$\begin{aligned} \varepsilon =& \varepsilon_1 \rho_1 (s_1 + 1) \rho_{34} (s_{34} + 1) \rho_{35} \prod_{\substack{1 \leq i \leq 16 \\ i \notin \{4,8,14\}}} [(* * *) + 1] \\ & \prod_{i \in \{4,8,14\}} \left[(* * *) + s_1^{4i-1,4i+3,1} + 1 \right] , \end{aligned} \tag{5.58}$$

where

$$(***) = \rho_{2i}(s_{2i}+1)\rho_{2i+1}(s_{2i+1}+1)\ . \tag{5.59}$$

Combining the definition of $\bar{\alpha}$:

$$\begin{aligned}\bar{\alpha} =&\rho_1(s_1+1)\rho_{34}(s_{34}+1)\rho_{35}\prod_{\substack{\leq i\leq 16\\ i\notin\{4,8,14\}}}[(***)+1]\\ &\prod_{i\in\{4,8,14\}}\left[(***)+s_1^{4i-1,4i+3,1}+1\right]\ ,\end{aligned} \tag{5.60}$$

we can obtain that

$$\varepsilon_1 = \frac{\varepsilon}{\bar{\alpha}}\ . \tag{5.61}$$

Applying Eqs. (5.55), (5.56), and (5.57), we can prove all ε_{2j+1} and $\varepsilon^{u,u+4,1}$.

The proof is completed.

5.3.3.5 Generalization Bound for ResNet

Proof of Theorem 5.7 We prove this theorem in 2 steps: (1) We first apply Lemma 3.2 to Lemma 5.2 in order to prove an upper bound on the Rademacher complexity of the hypothesis space computed by ResNet; and (2) We then apply the result of (1) to Lemma 3.1 in order to prove a generalization bound.

(1) *Upper bound on the Rademacher complexity.*

Applying Eq. (3.12) of Lemma 3.2 to Eq. (5.31) of Lemma 5.2, we can prove the following inequality:

$$\begin{aligned}\mathfrak{R}(\mathcal{H}_\lambda|_D) \leq& \inf_{\alpha>0}\left(\frac{4\alpha}{\sqrt{m}}+\frac{12}{m}\int_\alpha^{\sqrt{m}}\sqrt{\log\mathcal{N}(\mathcal{H}_\lambda|_D,\varepsilon,\|\cdot|_2)}\mathrm{d}\varepsilon\right)\\ \leq& \inf_{\alpha>0}\left(\frac{4\alpha}{\sqrt{m}}+\frac{12}{m}\int_\alpha^{\sqrt{m}}\frac{\sqrt{R}}{\varepsilon}\mathrm{d}\varepsilon\right)\\ \leq& \inf_{\alpha>0}\left(\frac{4\alpha}{\sqrt{m}}+\frac{12}{m}\sqrt{R}\log\frac{\sqrt{m}}{\alpha}\right)\ .\end{aligned} \tag{5.62}$$

Apparently, the infinimum is reached uniquely at $\alpha = 3\sqrt{\frac{R}{m}}$. Here, we use a simpler and also widely used choice $\alpha = \frac{1}{m}$, and prove the following inequality:

$$\mathfrak{R}(\mathcal{H}_\lambda|_D) \leq \frac{4}{n^{\frac{3}{2}}}+\frac{18}{m}\sqrt{R}\log m\ . \tag{5.63}$$

(2) *Upper bound on the generalization error.*
Combining with Lemma 3.1, we prove the following inequality:

$$\mathbb{P}(\arg\max_i F(x)_i \neq y) \leq \hat{R}_S^{\lambda}(F) + \frac{8}{m^{\frac{3}{2}}} + \frac{36}{m}\sqrt{R}\log m + 3\sqrt{\frac{\log(1/\delta)}{2m}}. \quad (5.64)$$

The proof is completed.

5.4 Vacuous Generalization Guarantee in Deep Learning

Conventional statistical learning theory suggests that the hypotheses learned from datasets of different sizes constitute a Glivenko-Cantelli class (Talagrand 1987; Dudley et al. 1991): the learned hypothesis converges to the target hypothesis, *i.e.*, the generalization bound decreases, as the size of the training dataset increases. Surprisingly, however, Nagarajan and Kolter (2019) reported that many classical uniform-convergence generalization bounds may in fact increase with the training dataset size for interpolators. This behavior is also echoed by the phenomenon of benign overfitting. In response to this work, Negrea et al. (2020) defended uniform convergence by defining a new notion called the structural Glivenko-Cantelli property for learned hypothesis sequences. They proved that (1) unregularized and overparameterized linear regression has a structural Glivenko-Cantelli surrogate hypothesis class and (2) removing a few bits of information from a nonstructural Glivenko-Cantelli hypothesis sequence can yield a sequence with the structural Glivenko-Cantelli property, whose generalization bounds exhibit double descent. Zhou et al. (2020) proved that consistency cannot be shown for any set by uniformly bounding the generalization error in a norm ball. To address this, they proved that zero-error predictors exhibit uniform convergence in a norm ball. Yang et al. (2021) presented an exact comparison between the generalization error and the uniform-convergence generalization bound in random feature models.

References

Alex, Krizhevsky, Ilya Sutskever, and Geoffrey E Hinton. 2012. Imagenet classification with deep convolutional neural networks. In *Advances in Neural Information Processing Systems*, 1097–1105.

Anders, Krogh and John A Hertz. 1992. A simple weight decay can improve generalization. In *Advances in Neural Information Processing Systems*, 950–957.

Andre, Esteva, Brett Kuprel, Roberto A Novoa, Justin Ko, Susan M Swetter, Helen M Blau, and Sebastian Thrun. 2017. Dermatologist-level classification of skin cancer with deep neural networks. *nature*, 542 (7639): 115–118.

Angus, Galloway, Thomas Tanay, and Graham W Taylor. 2018. Adversarial training versus weight decay. *arXiv preprint* arXiv:1804.03308.

Behnam, Neyshabur, Ryota Tomioka, and Nathan Srebro. 2015. Norm-based capacity control in neural networks. In *Conference on Learning Theory*, 1376–1401.

Behnam, Neyshabur, Srinadh Bhojanapalli, and Nathan Srebro. 2017. A PAC-Bayesian approach to spectrally-normalized margin bounds for neural networks. *arXiv preprint* arXiv:1707.09564

Ben, Taskar, Carlos Guestrin, and Daphne Koller. 2004. Max-margin Markov networks. In *Advances in Neural Information Processing Systems*, 25–32.

Chenxi, Liu, Liang-Chieh Chen, Florian Schroff, Hartwig Adam, Wei Hua, Alan L Yuille, and Li Fei-Fei. 2019. Auto-deeplab: Hierarchical neural architecture search for semantic image segmentation. In *IEEE/CVF conference on computer vision and pattern recognition*, 82–92.

David, Silver, Aja Huang, Chris J. Maddison, Arthur Guez, Laurent Sifre, George Van Den Driessche, Julian Schrittwieser, Ioannis Antonoglou, Veda Panneershelvam, and Marc Lanctot. 2016. Mastering the game of go with deep neural networks and tree search. *Nature* 529 (7587): 484–489.

Eunchun, Park, B. Wade Brorsen, and Ardian Harri. 2019. Using bayesian kriging for spatial smoothing in crop insurance rating. *American Journal of Agricultural Economics* 101 (1): 330–351.

G Pisier. 1981. Remarques sur un résultat non publié de b. maurey. *Séminaire Analyse fonctionnelle (dit" Maurey-Schwartz")*, 1–12.

Gao, Huang, Zhuang Liu, Laurens Van Der Maaten, and Kilian Q Weinberger. 2017. Densely connected convolutional networks. In *IEEE Conference on Computer Vision and Pattern Recognition*, 4700–4708.

Geert, Litjens, Thijs Kooi, Babak Ehteshami Bejnordi, Arnaud Arindra Adiyoso. Setio, Francesco Ciompi, Mohsen Ghafoorian, Jeroen Awm Van Der. Laak, Bram Van Ginneken, and Clara I. Sánchez. 2017. A survey on deep learning in medical image analysis. *Medical Image Analysis* 42: 60–88.

Guodong, Zhang, Chaoqi Wang, Bowen Xu, and Roger Grosse. 2018b. Three mechanisms of weight decay regularization. *arXiv preprint* arXiv:1810.12281.

Hao, Li, Zheng Xu, Gavin Taylor, Christoph Studer, and Tom Goldstein. 2018c. Visualizing the loss landscape of neural nets. In *Advances in Neural Information Processing Systems*.

Ian, H Witten, Eibe Frank, Mark A Hall, and Christopher J Pal. 2016. *Data Mining: Practical machine learning tools and techniques*. Morgan Kaufmann.

J, Janai, F Güney, A Behl, and A Geiger. 2017. Computer vision for autonomous vehicles: problems, datasets and state of the art. arxiv e-prints. *arXiv preprint* arXiv:1704.05519.

Jeffrey, Negrea, Gintare Karolina Dziugaite, and Daniel Roy. 2020. In defense of uniform convergence: Generalization via derandomization with an application to interpolating predictors. In *International Conference on Machine Learning*, 7263–7272.

Jinghui, Chen, Dongruo Zhou, Yiqi Tang, Ziyan Yang, Yuan Cao, and Quanquan Gu. 2018a. Closing the generalization gap of adaptive gradient methods in training deep neural networks. *arXiv preprint* arXiv:1806.06763.

Jonas, Gehring, Michael Auli, David Grangier, Denis Yarats, and Yann N Dauphin. 2017. Convolutional sequence to sequence learning. In *International Conference on Machine learning*, 1243–1252.

Kaiming, He, Georgia Gkioxari, Piotr Dollár, and Ross Girshick. 2017. Mask r-cnn. In *IEEE international conference on computer vision*, 2961–2969.

Kaiming, He, Xiangyu Zhang, Shaoqing Ren, and Jian Sun. 2016a. Deep residual learning for image recognition. In *IEEE Conference on Computer Vision and Pattern Recognition*.

Klaus, Greff, Rupesh K. Srivastava, Jan Koutník, Bas R. Steunebrink, and Jürgen. Schmidhuber. 2016. Lstm: A search space odyssey. *IEEE transactions on neural networks and learning systems* 28 (10): 2222–2232.

Koltchinskii, Vladimir, and Dmitry Panchenko. 2002. Empirical margin distributions and bounding the generalization error of combined classifiers. *The Annals of Statistics* 30 (1): 1–50.

Krizhevsky, Alex, and Geoffrey Hinton. 2009. *Learning multiple layers of features from tiny images*. Citeseer: Technical report.

Laurent, Dinh, Razvan Pascanu, Samy Bengio, and Yoshua Bengio. 2017. Sharp minima can generalize for deep nets. In *International Conference on Machine learning*

LeCun, Yann, Yoshua Bengio, and Geoffrey Hinton. 2015. Deep learning. *Nature* 521 (7553): 436.

Liang-Chieh, Chen, George Papandreou, Iasonas Kokkinos, Kevin Murphy, and Alan L. Yuille. 2017. Deeplab: Semantic image segmentation with deep convolutional nets, atrous convolution, and fully connected crfs. *IEEE transactions on pattern analysis and machine intelligence* 40 (4): 834–848.

Lijia, Zhou, Danica J Sutherland, and Nati Srebro. 2020. On uniform convergence and low-norm interpolation learning. In *Advances in Neural Information Processing Systems*, 6867–6877.

Matthew, E Peters, Mark Neumann, Mohit Iyyer, Matt Gardner, Christopher Clark, Kenton Lee, and Luke Zettlemoyer. 2018. Deep contextualized word representations. In *Annual Conference of the North American Chapter of the Association for Computational Linguistics*, 2227–2237.

Michael, B Chang, Abhishek Gupta, Sergey Levine, and Thomas L Griffiths. 2018. Automatically composing representation transformations as a means for generalization. *arXiv preprint* arXiv:1807.04640.

Michel Talagrand. 1987. The glivenko-cantelli problem. *The Annals of Probability*, 837–870.

Nick, Harvey, Christopher Liaw, and Abbas Mehrabian. 2017. Nearly-tight VC-dimension bounds for piecewise linear neural networks. In *Annual Conference on Learning Theory*, 1064–1068.

Nitish Shirish, Keskar, Dheevatsa Mudigere, Jorge Nocedal, Mikhail Smelyanskiy, and Ping Tak Peter Tang. 2017. On large-batch training for deep learning: Generalization gap and sharp minima. In *International Conference on Learning Representations*.

Noah, Golowich, Alexander Rakhlin, and Ohad Shamir. 2018. Size-independent sample complexity of neural networks. In *Annual Conference on Learning Theory*, 297–299.

Paul, Goldberg, and Mark Jerrum. 1993. Bounding the vapnik-chervonenkis dimension of concept classes parameterized by real numbers. In *Proceedings of the sixth annual conference on Computational learning theory*, 361–369.

Paul, W Goldberg, and Mark R. Jerrum. 1995. Bounding the Vapnik-Chervonenkis dimension of concept classes parameterized by real numbers. *Machine Learning* 18 (2–3): 131–148.

Peter, Bartlett and John Shawe-Taylor. 1999. Generalization performance of support vector machines and other pattern classifiers. *Advances in Kernel methods-Support Vector Learning*, 43–54.

Peter, L Bartlett, Dylan J Foster, and Matus J Telgarsky. 2017. Spectrally-normalized margin bounds for neural networks. In *Advances in Neural Information Processing Systems*, 6240–6249.

Peter, L Bartlett, Vitaly Maiorov, and Ron Meir. 1999. Almost linear VC dimension bounds for piecewise polynomial networks. In *Advances in Neural Information Processing Systems*, 190–196.

Richard, M Dudley, Evarist Giné, and Joel Zinn. 1991. Uniform and universal glivenko-cantelli classes. *Journal of Theoretical Probability* 4 (3): 485–510.

Robert, E Schapire, Yoav Freund, Peter Bartlett, and Wee Sun Lee. 1998. Boosting the margin: A new explanation for the effectiveness of voting methods. *The Annals of Statistics* 26 (5): 1651–1686.

Roman, Novak, Yasaman Bahri, Daniel A Abolafia, Jeffrey Pennington, and Jascha Sohl-Dickstein. 2018. Sensitivity and generalization in neural networks: an empirical study. In *International Conference on Learning Representations*.

Saining, Xie, Ross Girshick, Piotr Dollár, Zhuowen Tu, and Kaiming He. 2017. Aggregated residual transformations for deep neural networks. In *IEEE Conference on Computer Vision and Pattern Recognition*, 1492–1500.

Stephen, Merity, Nitish Shirish Keskar, and Richard Socher. 2018. Regularizing and optimizing lstm language models. In *International Conference on Learning Representations*.

Takeru, Miyato, Toshiki Kataoka, Masanori Koyama, and Yuichi Yoshida. 2018. Spectral normalization for generative adversarial networks. In *International Conference on Learning Representations*.

Tong Zhang. 2002. Covering number bounds of certain regularized linear function classes. *Journal of Machine Learning Research*, 2 (Mar): 527–550.

Tsung-Yi, Lin, Piotr Dollár, Ross Girshick, Kaiming He, Bharath Hariharan, and Serge Belongie. 2017. Feature pyramid networks for object detection. In *IEEE conference on computer vision and pattern recognition*, 2117–2125.

Vaishnavh, Nagarajan and J Zico Kolter. 2019. Uniform convergence may be unable to explain generalization in deep learning. In *Advances in Neural Information Processing Systems*.

Vladimir Vapnik. 2006. *Estimation of Dependences based on Empirical Data*. Springer Science & Business Media.

Yichun, Shi, Charles Otto, and Anil K. Jain. 2018. Face clustering: representation and pairwise constraints. *IEEE Transactions on Information Forensics and Security* 13 (7): 1626–1640.

Zhuozhuo, Tu, Fengxiang He, and Dacheng Tao. 2020. Understanding generalization in recurrent neural networks. In *International Conference on Learning Representations*.

Zitong, Yang, Yu Bai, and Song Mei. 2021. Exact gap between generalization error and uniform convergence in random feature models. In *International Conference on Machine Learning*, 11704–11715.

Chapter 6
Stochastic Gradient Descent as an Implicit Regularization

Overparameterization creates a colossal hypothesis space. Although Golowich et al. (2018) have proved generalization bounds that have explicit independence on the network size, some implicit dependence may be observed.

Stochastic gradient methods (SGMs) have been widely used for training deep learning models. However, a key limitation of SGMs is their tendency to explore only a limited portion of the entire hypothesis space. This limitation can be seen as a form of implicit regularization, which constrains the effective capacity of the learned model. Importantly, this constraint is influenced by both the specific SGM algorithm and the training data.

By leveraging this inherent regularization of SGMs, researchers are exploring the possibility of establishing algorithm-dependent and data-dependent generalization bounds. This approach holds promise for surpassing the limitations of conventional statistical learning theory, which often relies on bounds based on the Vapnik-Chervonenkis (VC) dimension or Rademacher complexity. In essence, these new bounds aim to capture the interplay between the chosen SGM, the training data, and the model's generalization ability.

6.1 Stochastic Gradient Methods (SGMs)

A natural tool for optimizing the expected risk is gradient descent (GD) methods. Denote an estimator with parameter θ by F_θ. Specifically, the gradient of the empirical risk in terms of the t-th iteration parameters $\theta(t)$ can be expressed as

$$g(\theta(t)) = \nabla_{\theta(t)} R(\theta(t)) = \nabla_{\theta(t)} \mathbb{E}[l(F_{\theta(t)}(X), Y)],$$

F. He and D. Tao, *Foundations of Deep Learning*, Machine Learning: Foundations, Methodologies, and Applications, https://doi.org/10.1007/978-981-16-8233-9_6

and the corresponding update equation is defined as follows:

$$\theta(t+1) = \theta(t) - \eta g(\theta(t)),$$

where $\theta(t)$ represents the parameters in iteration t and $\eta > 0$ is the learning rate.

In SGD, the gradient $g(\theta)$ is estimated from minibatches of the training sample set. Let S be the index set of a minibatch, in which all indices are drawn in an i.i.d. manner from $\{1, 2, \ldots, m\}$, where m is the number of training samples. Then, the iterative process of SGD on minibatch S can be defined as follows:

$$\hat{g}_S(\theta(t)) = \nabla_{\theta(t)} \hat{R}_S(\theta(t)) = \frac{1}{|S|} \sum_{n \in S} \nabla_{\theta(t)} l(F_{\theta(t)}(x_n), y_n) \tag{6.1}$$

and

$$\theta(t+1) = \theta(t) - \eta \hat{g}(\theta(t)), \tag{6.2}$$

where

$$\hat{R}_S(\theta) = \frac{1}{|S|} \sum_{n \in S} l(F_\theta(x_n), y_n)$$

is the empirical risk on the minibatch and $|S|$ is the cardinality of the set S. For brevity, we adopt the notation

$$l(F_\theta(X_n), Y_n) = l_n(\theta)$$

in the rest of this section.

Additionally, suppose that in step i, we represent the distribution of a parameter by Q_i, with the initial distribution being Q_0 and the convergent distribution being Q. Then, SGD is the process of finding Q by starting from Q_0 and proceeding through a series of Q_i.

6.2 Generalization Bounds on Convex Loss Surfaces

Intensive studies have been conducted on the generalizability of SGMs under both continuous and convex losses.

Ying and Pontil (2008) proved a generalization bound of the following order for a fixed learning rate of $\eta_t \simeq m^{-\frac{2\zeta}{2\zeta+1}}$:

$$O\left(m^{-\frac{2\zeta}{2\zeta+1}} \log m\right),$$

where ζ is a constant.

Later, Dieuleveut and Bach (2016) improved this result to the following average order when $2\zeta + \gamma > 1$:

$$O\left(m^{-\frac{2\min\{\zeta,1\}}{2\min\{\zeta,1\}+\gamma}}\right).$$

Other related works include (Lin et al. 2016; Lin and Rosasco 2016; Chen et al. 2016; Wei et al. 2017). However, in deep learning, the loss surface is usually highly nonconvex, which invalidates the previous results.

6.3 Generalization Bounds on Nonconvex Loss Surfaces

Results for nonconvex loss surfaces also exist in the literature. Related works include analyses obtaining generalization bounds based on algorithmic stability and Probably Approximately Correct (PAC)-Bayesian theory.

Algorithmic stability. Bousquet and Elisseeff (2002) proposed employing *algorithmic stability* to measure the stability of the output hypothesis when the training sample set is disturbed. Many versions of algorithmic stability have been proposed; a popular one is presented as follows.

Definition 6.1 (Uniform stability; cf. Bousquet and Elisseeff (2002), Definition 6) A machine learning algorithm $\mathcal{A}$ is uniformly stable if for any neighbouring sample pair S and S' that differ by only one example, we have the following inequality:

$$\left|\mathbb{E}_{\mathcal{A}(S)} l(\mathcal{A}(S), z) - \mathbb{E}_{\mathcal{A}(S')} l(\mathcal{A}(S'), z)\right| \leq \beta,$$

where z is an arbitrary example; $\mathcal{A}(S)$ and $\mathcal{A}(S')$ are the output hypotheses learned on the training sets S and S', respectively; and β is a positive real constant. The constant β is called the uniform stability of algorithm $\mathcal{A}$.

It is natural for an algorithm that is insensitive to disturbances in the training data to have good generalizability. Following this intuition, several generalization bounds have been proven based on algorithmic stability (Bousquet and Elisseeff 2002; Xu et al. 2011). For example, based on uniform stability, (Bousquet and Elisseeff 2002) proved the following theorem in relation to *generalization in expectation*.

Theorem 6.1 *Let algorithm $\mathcal{A}$ be β-uniformly stable. Then,*

$$|\mathbb{E}_{S,\mathcal{A}}[\hat{R}_S[\mathcal{A}(S)] - R[\mathcal{A}(S)]]| \leq \beta.$$

In recent works, SGMs have commonly been modelled in terms of stochastic differential equations (SDEs). Minibatch stochastic gradients introduce randomness into an SGM (*i.e.*, a stochastic gradient can be decomposed into the stochastic estimate of the full gradient plus noise). Therefore, the parameter updating in an SGM can be formulated as a stochastic process. Thus, the output hypothesis is indeed drawn from the steady distribution of this stochastic process. We can ultimately analyse the generalization and optimization of deep learning on the basis of this steady distribution.

An SGM is usually formulated as shown in Eqs. (6.1) and (6.2). The stochastic gradient introduces extra noise into the parameter updating. When the gradient noise can be modelled as a Gaussian distribution, the SGM reduces to stochastic gradient Langevin dynamics (SGLD), expressed as follows (Mandt et al. 2017):

$$\Delta\theta(t) = \theta(t+1) - \theta(t) = -\eta \hat{g}_S(\theta(t)) = -\eta g(\theta) + \sqrt{\frac{2\eta}{\beta}}\Delta W, \Delta W \sim N(0, I),$$

where η is the learning rate and $\beta > 0$ is an inverse temperature parameter.

Hardt et al. (2016b) proved the following upper bound for the uniform stability, which is exponentially correlated with the aggregated step size and the smoothness parameter.

Theorem 6.2 *Suppose that the loss function is ϵ-smooth, L-Lipschitz, and no larger than* 1 *for all data. If we run an SGM with monotonically increasing learning rates $\alpha_t \leq c/t$ for T steps, then the SGM is uniformly stable with*

$$\beta \leq \frac{1 + 1/\epsilon c}{m - 1}(2cL^2)^{\frac{1}{\epsilon c+1}} T^{\frac{1}{\epsilon c+1}}.$$

The proof of this theorem mimics a proof of convergence. Suppose that the loss function l is L-Lipschitz constant with respect to the weights for any example, *i.e.*,

$$|l(w, z) - l(w', z)| \leq L\|w - w'\|.$$

In other words, the stability can be measured in terms of the stability of the weights. Therefore, one can simply analyse how the weights diverge as a function of time t. Generalization bounds depending on the learning rates and the number of iterations can then be presented. Furthermore, we can easily drive a generalization bound based on this theorem.

Under five additional assumptions, Raginsky et al. (2017) proved the following upper bound for the expected excess risk:

$$\tilde{O}\left(\frac{\beta(\beta + U)^2}{\lambda_*}\delta^{1/4}\log\left(\frac{1}{\epsilon} + \epsilon\right)\right) + \tilde{O}\left(\frac{(\beta + U)^2}{\lambda_* + m} + \frac{U\log(\beta + 1)}{\beta}\right),$$

provided that

$$k = \tilde{\Omega}\left(\frac{\beta(\beta + U)}{\lambda_*\epsilon^4}\log^5\left(\frac{1}{\epsilon}\right)\right)$$

and

$$\eta \leq \left(\frac{\epsilon}{\log(1/\epsilon)}\right)^4,$$

where U is the number of parameters, m is the number of training samples, the inverse temperature parameter satisfies $\beta > 1 \vee 2/m$, λ_* is a factor that controls the exponential rate of convergence of the SGM dynamics to the stationary distribution, and k is the number of iteration; *i.e.*, for any datum z and any parameter w, for some $m > 0$ and $b > 0$,

$$< w, \nabla h(w, z) >\leq m\|w\|^2 - b$$

and

$$\epsilon \in \left(0, \frac{m}{4M^2} \wedge e^{-\tilde{\Omega}\left(\frac{\lambda_*}{\beta(\beta+U)}\right)}\right).$$

The proof proceeds through three important steps:

(1) Convergence to Langevin Dynamics: Initially, we establish that with a sufficiently small learning rate, the behavior of SGLD resembles that of a continuous-time Langevin dynamic process under the 2-Wasserstein distance. This convergence conceptually links discrete updates in optimization with continuous-time diffusion processes, offering insights into the stability of parameter updates.

(2) Transition to Gibbs Distribution: As training progresses across numerous epochs, the Langevin dynamics are shown to converge to a Gibbs distribution. This transition signifies the establishment of a probabilistic equilibrium, providing a probabilistic view of the model's parameter space.

(3) Gibbs Algorithm Construction: Building upon these convergences, we construct a Gibbs sampling algorithm designed to approximate an empirical risk minimizer. This algorithmic approach encapsulates the theoretical insights into a practical framework for model optimization.

Understanding these dynamics and their convergence to probabilistic distributions is essential for grasping the optimization and generalization aspects of deep learning. However, it's important to note that achieving stable convergence, particularly towards stationary distributions, can be challenging and dependent on factors like model complexity and model dimensionality (at the exponential rate).

A canonical mutual-information-based generalization bound has been given in Xu and Raginsky (2017):

$$\sqrt{\frac{2\sigma^2}{m} I(S; \theta)}, \tag{6.3}$$

where S is the training sample, θ represents the model parameters, and the loss is σ-sub-Gaussian under the data distribution. This result arises from the following result on the informational theoretical "stability" of a two-input function $f(X, Y)$:

$$|f(X, Y) - f(\bar{X}, \bar{Y})| \leq \sqrt{2\sigma^2 I(X; Y)},$$

if $f(\bar{X}, \bar{Y})$ is sub-Gaussian under distribution $P_{\bar{X},\bar{Y}} = P_X \times P_Y$, where P_X and P_Y are the distributions of X and Y, respectively. Especially, noting that the mutual information is calculated between the learned weights and all training data; it is an "on-average" estimate of how much information has been transformed into the model learned from the data.

Based on this result, Pensia et al. (2018) gave a generalization error bound for SGLD,

$$\sqrt{\frac{R^2}{m}\sum_{t=1}^{T}\frac{\eta_t^2 L^2}{\sigma_t^2}}, \tag{6.4}$$

where the hypothesis is assumed to be R-sub-Gaussian, η_t is the learning rate in the t-th iteration, L is the upper bound on the weight update $\Delta\theta_t$, and the gradient noise is assumed to be $N(0, \sigma_t^2 I)$. This result was later extended by Bu et al. (2020):

- Suppose that the loss $l(\theta, z)$ is R-sub-Gaussian under $z \sim \mu$, where θ represents the model parameters, and μ is the data-generating distribution. Then, the generalization error has the following upper bound:

$$\frac{1}{m}\sum_{i=1}^{m}\sqrt{2R^2 I(\theta; z_i)}, \tag{6.5}$$

  where z_i is the i-th datum and m is the sample size.
- Suppose that the loss $l(\tilde{\theta}, \tilde{z})$ is R-sub-Gaussian under the following distribution:

$$P_{\tilde{\theta},\tilde{z}} = P_\theta \otimes P_z.$$

  Then, the generalization error has the same upper bound expressed in Eq. (6.5).

In contrast with Xu and Raginsky (2017), Bu et al. (2020) proposed an "individual" version of a metric to measure the information transformed into a model learned from data; see Eq. (6.5), where the whole training sample S is replaced by individual data $I(\theta; z_i)$.

Mou et al. (2018a) modelled SGLD as Langevin diffusion dynamics and then proved an $O(1/m)$ generalization bound and an $O\left(1/\sqrt{m}\right)$ generalization bound for SGLD via algorithmic stability and PAC-Bayesian theory, respectively:

- **Algorithmic stability:** Assume that C is an upper bound of the loss function, the hypothesis is L-Lipschitz continuous, and the learning rate satisfies $\eta_t \le \frac{\log 2}{\beta L^2}$ for $\forall t$; then, the generalization error has the following upper bound on average:

$$O\left(\frac{2LC}{m}\left(\beta\sum_{t=1}^{N}\eta_t\right)^2\right), \tag{6.6}$$

  where N is the number of iterations.

- **PAC-Bayesian theory:** Suppose that the empirical risk has a l_2-norm regularization on the weights, *i.e.*, $\frac{\lambda}{2}\|\theta\|^2$, and that the loss function is sub-Gaussian. Then, there is the following high-probability generalization bound:

$$O\left(\sqrt{\frac{\beta}{m}\sum_{k=1}^{N}\eta_k e^{-\frac{\lambda}{3}(T_N-T_k)}\mathbb{E}[\|g_k(\theta_k)\|^2]}\right),$$

  where $T_k=\sum_{i=1}^{k}\eta_i$, β is the temporal parameter, and $g_k(\theta_k)$ is the gradient at iteration k.

He and Su (2020) empirically showed that neural networks are *locally elastic*: the prediction for an instance x' given by predictor learned by SGD is not sensitive to that of a dissimilar example x. This phenomenon inspired Deng et al. (2020) to propose a localized version of algorithmic stability.

Definition 6.2 (Locally elastic stability; cf. Deng et al. (2020), Definition 1) A machine learning algorithm $\mathcal{A}$ has locally elastic stability $\beta_m(\cdot,\cdot)$ with respect to the loss function l if, for any sample $S\in\mathcal{Z}^m$, $z_i\in S$, and $z\in\mathcal{Z}$, the following inequality holds:

$$\left|l(\mathcal{A}(S),z)-l(\mathcal{A}(S^{\backslash i}),z)\right|\leq\beta_m(z_i,z).$$

This localized algorithmic stability leads to a tighter generalization bound.

Theorem 6.3 *Let algorithm $\mathcal{A}$ have $\beta_m(\cdot,\cdot)$-locally elastic stability. Suppose that $\|z\|\leq 1$ for any $z\in\mathcal{Z}$ and that $\beta_m(\cdot,\cdot)=\frac{\beta(\cdot,\cdot)}{m}$, where $\beta(\cdot,\cdot)$ is independent of the sample size m and $\|\beta(\cdot,\cdot)\|\leq M_\beta$. Then, for any $0<\delta<1$ and $\eta>0$, with probability at least $1-\delta$, we have*

$$\begin{aligned}&\left|R(\mathcal{A}(S))-\hat{R}_S(\mathcal{A}(S))\right|\\ &\leq\frac{2\sup_{z'\in\mathcal{Z}}\mathbb{E}\beta(z',z)}{m}+2\left(2\sup_{z'\in\mathcal{Z}}\mathbb{E}\beta(z',z)+\eta+M_l\right)\sqrt{\frac{2\log(2/\delta)}{m}}.\end{aligned}$$

Chen et al. (2018b) proved the existence of a trade-off between stability and convergence for all iterative algorithms under either the convex smooth or the strong convex smooth assumption, leading to an $O(1/m)$ generalization bound. Other advances include the works of (He et al. 2019; Tzen et al. 2018; Zhou et al. 2018; Wen et al. 2019; Li et al. 2019).

6.4 Generalization Bounds Relying on Data-Dependent Priors

In Bayesian statistics, prior distributions play a crucial role in encoding our beliefs or expectations about model parameters before observing any data. Unlike conventional machine learning algorithms that assume full understanding of the data, Bayesian methods require us to specify priors independently of the training data. This ensures that our models start with a neutral or uninformative stance regarding the underlying data characteristics.

Common choices for priors in Bayesian inference include uniform distributions (assigning equal probability to all possible values) or Gaussian distributions (assuming a normal distribution of parameter values). These priors are often selected to reflect minimal assumptions about the data, allowing the data itself to shape the posterior distribution through Bayesian updating.

Importantly, Bayesian priors are designed to have diminishing influence as more data is incorporated into the model during training. This property helps ensure that the model's performance improves based primarily on the observed data rather than the initial prior assumptions.

Recent advances in probabilistic learning, such as PAC-Bayesian theory, aim to extend traditional Bayesian principles to more complex scenarios, relaxing strict assumptions about data-independent priors while preserving the theoretical underpinnings of Bayesian inference. This approach allows for a more flexible and nuanced integration of prior knowledge and observed data in learning algorithms.

Distribution-dependent priors. In machine learning theory, distribution-dependent priors are designed to leverage knowledge about the underlying data distribution without directly relying on the specific training dataset used. This concept is rooted in the notion that the data-generating process and its distribution exist independently of the actual training data collection (Lever et al. 2013). By incorporating such priors, researchers aim to refine generalization bounds, which are fundamental in assessing the theoretical performance and predictive ability of machine learning models. Specifically, the utilization of distribution-dependent priors has been shown to be able to considerably tighten the generalization bounds.

Following this line of reasoning, Lever et al. (2013) tightened the PAC-Bayesian bound (see Theorem 2.7) to the following bound. Given any positive constant C, there is that with probability at least $1 - \delta > 0$, with respect to the drawn sample, the generalization error is upper bounded as follows:

$$R(Q(h)) - C^* \hat{R}_S(Q(h)) \le \frac{C^*}{Cm}\left(\lambda\sqrt{\frac{2}{m}\log\frac{2\xi(m)}{\delta}} + \frac{\gamma^2}{2m} + \log\frac{2}{\delta}\right),$$

where $Q(h)$ is the posterior of the learned model, with density

$$q(h) \propto e^{-\gamma \hat{R}_S(Q(h))},\ \xi(m) = O(\sqrt{m}),\ C^* = \frac{C}{1 - e^{-C}},$$

and γ is a positive constant.

Li et al. (2019) further designed a *Bayes-stability* framework to derive generalization bounds based on distribution-dependent priors. They tightened the result of Mou et al. (2018a) (Eq. 6.6) to the following formula:

$$O\left(\frac{C}{m}\sqrt{\mathbb{E}_S\left[\sum_{t=1}^{T}\frac{\eta_t^2}{\sigma_t^2}g_e(t)\right]}\right),$$

where $\frac{\sigma_t}{\sqrt{2}}I$ is assumed as the gradient noise, $C > 0$ is the upper bound of the loss function, T is the number of iterations, η_t is the step size at iteration t, and

$$g_e(t) = \mathbb{E}_{W_{t-1}}\left[\frac{1}{m}\sum_{i=1}^{m}\|\nabla F(W_{t-1}, z_i)\|^2\right].$$

Data-dependent priors. Negrea et al. (2019) further pushed the frontier of generalization bound analysis by constructing priors that are not independent of the data. Suppose that set $S_J \subset S$ has a size of $n < m$. It is natural to design a prior exploiting S_J to derive a data-dependent forecast of the posterior Q. We denote this subset by $S_J = \{z_{j_1}, \ldots, z_{j_n}\}$, where all indices constitute a set J. The index set J is randomly drawn from $\{1, \ldots, m\}$. Suppose that we have the following generalization bound:

$$\mathbb{E}^{\mathcal{F}}\left[R(\theta) - \hat{R}_{S_J}(\theta)\right] \leq B,$$

where $\hat{R}_{S_J}$ is the empirical risk on the set S_J, $B > 0$ is $\mathcal{F}$ measurable, and $\mathcal{F}$ is a σ-field such that $\sigma(S_J) \subset \mathcal{F} \perp \sigma(S_J^c)$. Additionally, we denote by U the uncertainty introduced by the machine learning algorithm. Negrea et al. (2019) proved the following theorem, which shows considerable improvement over the one given by Xu and Raginsky (2017) (Eq. 6.3).

Theorem 6.4 *Let $J \subset [m]$, $|J| = n$ be uniformly distributed and independent of the training sample set S and the model parameters θ. Suppose that the loss is σ-sub-Gaussian under the data distribution. When the posterior $Q = \mathbb{P}(\theta|S)$ and the prior P are $\sigma(S_J)$-measurable data-dependent distributions on the data space,*

$$\mathbb{E}\left[R(\theta) - \hat{R}_S(\theta)\right] \leq \sqrt{\frac{2\sigma^2}{m-n}I(\theta; S_J^c)} \leq \sqrt{\frac{2\sigma^2}{m-n}\mathbb{E}[KL(Q||P)]}.$$

Moreover, let U be independent of S and θ. When the posterior $Q = \mathbb{P}(\theta|S, U)$ and the prior P are $\sigma(S_J, U)$-measurable data-dependent distributions on the data space,

$$\mathbb{E}\left[R(\theta) - \hat{R}_S(\theta)\right] \leq \sqrt{\frac{2\sigma^2}{m-n}I^{S_J,U}(\theta; S_J^c)} \leq \sqrt{\frac{2\sigma^2}{m-n}\mathbb{E}^{S_J,U}[KL(Q||P)]}.$$

Further works following this approach include (Haghifam et al. 2020) in cooperation with (Steinke and Zakynthinou 2020; Dziugaite et al. 2020; Hellström and Durisi 2020) and their applications (Cherian et al. 2020).

6.5 The Role of Learning Rate and Batch Size in Shaping Generalizability

The last decade has seen the dramatic success of deep neural networks (Goodfellow et al. 2016) based on the stochastic gradient descent (SGD) optimization method (Bottou et al. 1998; Sutskever et al. 2013). The task of fine tuning the hyper-parameters of SGD to make neural networks generalize well is both critical and challenging. Some works have addressed the strategies of tuning hyper-parameters (Dinh et al. 2017; Goyal et al. 2017; Keskar et al. 2017) and the generalization ability of SGD (Chen et al. 2018b; Hardt et al. 2016b; Lin et al. 2016; Mou et al. 2018a; Pensia et al. 2018). However, there is still a lack of solid evidence to validate the effectiveness of training strategies for tuning the hyper-parameters of neural networks.

In this section, we present both theoretical and empirical evidence for a more effective training strategy for deep neural networks:

> When employing SGD to train deep neural networks, we should ensure that the batch size is not too large and the learning rate is not too low, to make the networks generalize well.

This strategy gives a guide to tune the hyper-parameters that helps neural networks achieve good test performance when the training error is small. It is derived from the following property:

> The generalizability of deep neural networks has a negative correlation with the ratio of batch size to learning rate.

Regarding theoretical evidence, we prove a novel PAC-Bayes McAllester (1999a, b) upper bound for the generalization error of deep neural networks trained by SGD. The positive correlation between the proposed generalization bound with the ratio between the batch size and the learning rate indicates that the generalizability of neural networks is poor. This result builds the theoretical foundation of the training strategy.

From the empirical aspect, we conduct extensive systematic experiments while strictly controlling unrelated variables to investigate the influence of batch size and learning rate on generalizability. Specifically, we trained 1,600 neural networks based on two popular architectures, ResNet-110 (He et al. 2016a, b) and VGG-19 (Simonyan and Zisserman 2015), on two standard datasets, CIFAR-10 and CIFAR-100 (Krizhevsky and Hinton 2009). The accuracies on the test set of all the networks are collected for analysis. Since the training error is almost the same across all the networks (it is almost 0), the test accuracy is an informative index to express the model generalizability. Evaluation is then performed on 164 groups of the collected data. The Spearman's rank-order correlation coefficients and the corresponding p

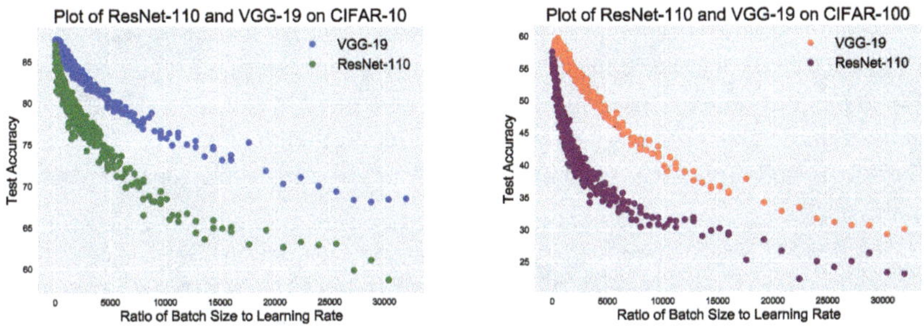

Fig. 6.1 Scatter plots of accuracy on test set to ratio of batch size to learning rate. Each point represents a model. Totally 1,600 points are plotted

values (Spearman 1987) demonstrate that the correlation is statistically significant (the probability that the correlation is wrong is smaller than 0.005), which fully supports the training strategy (Fig. 6.1).

6.5.1 Theoretical Evidence

In this section, we explore and develop the theoretical foundations for the training strategy. The main ingredient is a PAC-Bayes generalization bound of deep neural networks based on the SGD optimization method. The generalization bound has a positive correlation with the ratio of batch size to learning rate. This correlation validates the effectiveness of the presented training strategy.

6.5.1.1 A Generalization Bound for SGD

Let $l_n(\theta) = l(\theta, x_n)$ be the contribution to the overall loss from a single observation $x_n, n = 1, \ldots, m$. Apparently, both $l_n(\theta)$ and $\hat{R}_S(\theta)$ are un-biased estimations of the expected risk $R(\theta)$, while $\nabla_\theta l_n(\theta)$ and $\hat{g}_S(\theta)$ are both un-biased estimations of the gradient $g(\theta) = \nabla_\theta R(\theta)$:

$$\mathbb{E}\left[l_n(\theta)\right] = \mathbb{E}\left[\hat{R}_S(\theta)\right] = R(\theta), \tag{6.7}$$

$$\mathbb{E}\left[\nabla_\theta l_n(\theta)\right] = \mathbb{E}\left[\hat{g}_S(\theta)\right] = g(\theta) = \nabla_\theta R(\theta), \tag{6.8}$$

where the expectations are in terms of the corresponding examples (X, Y).

An assumption (see, e.g., Weinan (2017); Mandt et al. (2017)) for the estimations is that all the gradients $\{\nabla_\theta l_n(\theta)\}$ calculated from individual data points are i.i.d. and drawn from a Gaussian distribution centered at $g(\theta) = \nabla_\theta R(\theta)$:

$$\nabla_\theta l_n(\theta) \sim N(g(\theta), C). \tag{6.9}$$

where C is the covariance matrix and is a constant matrix for all θ. As covariance matrices are (semi) positive-definite, for brevity, we suppose that C can be factorized as $C = BB^\top$ for brevity. This assumption can be justified by the central limit theorem when the sample size m is large enough compared to the batch size S.

Therefore, the stochastic gradient is also drawn from a Gaussian distribution centered at $g(\theta)$:

$$\hat{g}_B(\theta) = \frac{1}{|G|}\sum_{n\in G} \nabla_\theta l_n(\theta) \sim N\left(g(\theta), \frac{1}{|G|}C\right). \tag{6.10}$$

SGD uses the stochastic gradient $\hat{g}_G(\theta)$ to iteratively update the parameter θ to minimize the function $R(\theta)$:

$$\Delta\theta(t) = \theta(t+1) - \theta(t) = -\eta\hat{g}_G(\theta(t)) = -\eta g(\theta) + \frac{\eta}{\sqrt{|G|}}B\Delta W,\ \Delta W \sim N(0, I). \tag{6.11}$$

Equation (6.11) expresses a well-known stochastic process called the Ornstein-Uhlenbeck process (Uhlenbeck and Ornstein 1930).

Furthermore, we assume that the loss function in the local region around the minimum is convex and 2-order differentiable:

$$R(\theta) = \frac{1}{2}\theta^\top A\theta, \tag{6.12}$$

where A is the Hessian matrix around the minimum and is a (semi) positive-definite matrix. This assumption has been primarily demonstrated by empirical works (see Li et al. (2018c, p. 1, Figs. 1(a) and 1(b) and p. 6, Figs. 4(a) and 4(b))). Without loss of generality, we assume that the global minimum of the objective function $R(\theta)$ is 0 and achieves at $\theta = 0$. General cases can be obtained by translation operations, which would not change the geometry of objective function and the corresponding generalization ability.

From the results of Ornstein-Uhlenbeck process, Eq. (6.11) has an analytic stationary distribution:

$$q(\theta) = M\exp\left\{-\frac{1}{2}\theta^\top\Sigma^{-1}\theta\right\}, \tag{6.13}$$

where M is the normalizer (Gardiner 1985).

This assumption that the local minima are convex and 2-order differentiable has been primarily proved by empirical works (see Fig. 6.2 which is originally presented in a recent work by Li et al. (2018c, pp. 1, Figs. 1(a) and 1(b) and pp. 6, Figs. 4(a) and 4(b))).

Estimating SGD as a continuous-time stochastic process dates back to works by Kushner and Yin (2003); Ljung et al. (2012). For a detailed justification, please refer to a recent work [see Mandt et al. 2017, pp. 6–8, Sect. 3.2].

The generation bound we obtained is based on the PAC-Bayesian framework that dates back to works by McAllester (1999a, b). From the PAC-Bayesian perspec-

Fig. 6.2 The risk surfaces with/without skips (Li et al. 2018c)

(a) ResNet-56 with skips (b) DenseNet (112 layers) with skips

tive, the hypothesis function learned by a stochastic machine learning algorithm is drawn randomly (but still governed by several "laws") from the hypothesis class. The generalizability of an algorithm refers to its ability to perform well on unseen data beyond the training set. In the context of machine learning, this property is negatively impacted by the divergence (measured by Kullback-Leibler divergence) between the distribution of the output hypothesis and the prior distribution used in Bayesian methods. Essentially, when the output distribution diverges significantly from the prior distribution (which is often assumed to be Gaussian or uniform), the algorithm may struggle to generalize effectively to new, unseen data. This trade-off highlights the challenge of balancing empirical risk minimization with the need to explore diverse areas of the hypothesis space defined by the prior distribution.

Suppose the prior distribution over the parameter space Θ is P. Let Q represent the distribution on the parameter space Θ expressing the learned hypothesis function. We then define the expected risk with respect to the distribution Q is shown as follows

$$R(Q) = \mathbb{E}_{\theta \sim Q} R(\theta).$$

Similarly, the empirical risk with respect to the distribution Q is defined as

$$\hat{R}_S(Q) = \mathbb{E}_{\theta \sim Q} \hat{R}_S(\theta).$$

Then, a classic result uniformly bounding the expected risk $R(Q)$ in terms of the empirical risk $\hat{R}_S(Q)$ and the KL divergence $KL(Q||P)$ is as follows.

Lemma 6.1 (see (*McAllester 1999a*)**, Theorem 1)** *For any positive real $\delta \in (0, 1)$, with probability at least $1 - \delta$ over a sample of size m, we have the following inequality for all distributions Q:*

$$R(Q) \leq \hat{R}_S(Q) + \sqrt{\frac{KL(Q||P) + \log \frac{1}{\delta} + \log m + 2}{2m - 1}}, \tag{6.14}$$

where $KL(Q||P)$ is the KL divergence between the distributions Q and P and is

$$KL(Q||P) = \mathbb{E}_{\theta\sim Q}\left(\log\frac{Q(\theta)}{P(\theta)}\right). \tag{6.15}$$

We derive a generalization bound for SGD with considerations to simplify technical aspects for clarity. In this context, we avoid delving into complex issues related to measurability and integrability to keep the discussion succinct and accessible. We also rely on Fubini's theorem to facilitate integration across multiple variables, assuming that the order of integration can be interchanged. Additionally, we assume the existence and uniqueness of stable (stationary) solutions for all stochastic differential equations involved.

Theorem 6.5 *For any positive real $\delta \in (0, 1)$, with probability at least $1-\delta$ over a training sample set of size m, we have the following inequality for the distribution Q of the output hypothesis function of SGD:*

$$R(Q) \leq \hat{R}_S(Q) + \sqrt{\frac{\frac{\eta}{2|G|}tr(CA^{-1}) - \log(\det(\Sigma)) - d + 2\log\left(\frac{1}{\delta}\right) + 2\log m + 4}{4m-2}}. \tag{6.16}$$

and

$$\Sigma A + A\Sigma = \frac{\eta}{|G|}C, \tag{6.17}$$

where A is the Hessian matrix of the loss function around the local minimum, B is from the covariance matrix of the gradients calculated by single sample points , and d is the dimension of the parameter θ (network size).

The proof of this generalization bound involves two key steps that draw upon concepts from SDEs and the PAC-Bayes framework. (1) We use SDE results to identify the stationary solution of the latent Ornstein-Uhlenbeck process described by Eq. (6.11), which captures the iterative update process of SGD. This step helps establish the behavior and stability of SGD over time. (2) We adapt the PAC-Bayes framework, which is a theoretical framework for generalization analysis in machine learning, to derive the generalization bound based on insights gained from the stationary distribution identified in the first step. The PAC-Bayes framework allows us to reason about the performance and predictive ability of SGD in terms of its convergence to a stable solution and its ability to generalize well.

To prove 6.5, we first present the following lemma.

Lemma 6.2 (cf. (*Mandt et al. 2017*), pp. 27-18, Appendix B) *Under the 2-order differentiable assumption (Eq. 6.12), the Ornstein-Uhlenbeck process (Eq. 6.11)'s stationary distribution,*

$$q(\theta) = M\exp\left\{-\frac{1}{2}\theta^\top\Sigma^{-1}\theta\right\}, \tag{6.18}$$

has the following property,

$$A\Sigma + \Sigma A = \frac{\eta}{|G|}C. \tag{6.19}$$

Here, we recall the proof to make this section complete.

Proof Form a result in Ornstein-Uhlenbeck process (Gardiner 1985), we know that the parameter θ has the following analytic solution,

$$\theta(t) = \theta(0)e^{-At} + \sqrt{\frac{\eta}{|G|}}\int_0^t e^{-A(t-t')}B\mathrm{d}W(t'), \tag{6.20}$$

where $W(t')$ is a white noise and follows $N(0, I)$. From Eq. (6.18), we know that

$$\Sigma = \mathbb{E}_{\theta\sim Q}\left[\theta\theta^\top\right]. \tag{6.21}$$

Therefore, we complete the proof by the following equation,

$$\begin{aligned} A\Sigma + \Sigma A =& \frac{\eta}{|G|}\int_{-\infty}^t Ae^{-A(t-t_0)}Ce^{-A(t-t_0)}\mathrm{d}t' + \frac{\eta}{|G|}\int_{-\infty}^t e^{-A(t-t_0)}Ce^{-A(t-t_0)}\mathrm{d}t'A \\ =& \frac{\eta}{|G|}\int_{-\infty}^t \frac{\mathrm{d}}{\mathrm{d}t'}Ae^{-A(t-t_0)}Ce^{-A(t-t_0)} = \frac{\eta}{|G|}C. \end{aligned} \tag{6.22}$$

Proof of Theorem 6.5 In the PAC-Bayesian framework (Lemma 6.1), an essential part is the KL divergence between the distribution of the learned hypothesis and the priori on the hypothesis space. The prior distribution can be interpreted as the distribution of the initial parameters, which are usually settled according to Gaussian distributions or uniform distributions.[1] Here, we use a standard Gaussian distribution $N(0, I)$ as the priori. Suppose the densities of the stationary distribution Q and the prior distribution P are respectively $p(\theta)$ and $q(\theta)$ in terms of the parameter θ in the following equations:

---

[1] In scenarios where there is limited prior knowledge about latent model parameters, it is common practice to set priors as distributions that convey minimal information, such as Gaussian or uniform distributions. This choice is driven by the following two primary considerations. (1) Algorithms based on Bayesian statistics are expected to converge to stationary distributions given sufficient time and data. The existence and uniqueness of these stationary solutions are assumed based on the latent stochastic differential equation governing the iterative process. This assumption provides a theoretical foundation for the convergence of Bayesian algorithms, ensuring that the learned model stabilizes over time. (2) Setting priors requires caution because we cannot assume prior knowledge about the target hypothesis function before initiating model training. Therefore, the choice of non-informative priors ensures that the learning process remains unbiased and adapts to the available data without undue influence from assumed prior beliefs. This approach is fundamental in statistical learning theory to avoid overfitting and to facilitate robust generalization to new, unseen data.

$$p(\theta) = \frac{1}{\sqrt{2\pi \det(I)}} \exp\left\{-\frac{1}{2}\theta^\top I \theta\right\}, \tag{6.23}$$

$$q(\theta) = \frac{1}{\sqrt{2\pi \det(\Sigma)}} \exp\left\{-\frac{1}{2}\theta^\top \Sigma^{-1} \theta\right\}, \tag{6.24}$$

where Eq. (6.24) comes from Eq. (6.18) by calculating the normalizer M. Therefore,

$$\log\left(\frac{q(\theta)}{p(\theta)}\right) = \frac{1}{2}\log\left(\frac{1}{\det(\Sigma)}\right) + \frac{1}{2}\left(\theta^\top I \theta - \theta^\top \Sigma^{-1}\theta\right). \tag{6.25}$$

Applying Eqs. (6.25) to (6.15), we can calculate the KL divergence between the distributions Q and P (we assume $\Theta = \mathbb{R}^d$):

$$\begin{aligned}
KL(Q||P) &= \int_{\theta\in\Theta} \log\left(\frac{q(\theta)}{p(\theta)}\right) q(\theta)\mathrm{d}\theta \\
&= \int_{\theta\in\Theta} \left[\frac{1}{2}\log\left(\frac{1}{\det(\Sigma)}\right) + \frac{1}{2}\left(\theta^\top I \theta - \theta^\top \Sigma^{-1}\theta\right)\right] q(\theta)\mathrm{d}\theta \\
&= \frac{1}{2}\log\left(\frac{1}{\det(\Sigma)}\right) + \frac{1}{2}\int_{\theta\in\Theta} \theta^\top I \theta p(\theta)\mathrm{d}\theta - \frac{1}{2}\int_{\mathbb{R}^{|S|}} \theta^\top \Sigma^{-1}\theta q(\theta)\mathrm{d}\theta \\
&= \frac{1}{2}\log\left(\frac{1}{\det(\Sigma)}\right) + \frac{1}{2}\mathbb{E}_{\theta\sim\mathcal{N}(0,\Sigma)}\theta^\top I \theta - \frac{1}{2}\mathbb{E}_{\theta\sim\mathcal{N}(0,\Sigma)}\theta^\top \Sigma^{-1}\theta \\
&= \frac{1}{2}\log\left(\frac{1}{\det(\Sigma)}\right) + \frac{1}{2}\mathrm{tr}(\Sigma - I).
\end{aligned} \tag{6.26}$$

From Eq. (6.19), we have that

$$A\Sigma + \Sigma A = \frac{\eta}{|G|}C. \tag{6.27}$$

Therefore,

$$A\Sigma A^{-1} + \Sigma = \frac{\eta}{|G|}CA^{-1}. \tag{6.28}$$

After calculating the trace of the both sides, we have the following equation,

$$\mathrm{tr}\left(A\Sigma A^{-1} + \Sigma\right) = \mathrm{tr}\left(\frac{\eta}{|S|}CA^{-1}\right). \tag{6.29}$$

The left-hand side (LHS) is as follows,

$$\mathrm{LHS} = \mathrm{tr}\left(A\Sigma A^{-1} + \Sigma\right) = \mathrm{tr}\left(A\Sigma A^{-1}\right) + \mathrm{tr}\left(\Sigma\right) = \mathrm{tr}\left(\Sigma\right) + \mathrm{tr}\left(\Sigma\right) = 2\mathrm{tr}\left(\Sigma\right). \tag{6.30}$$

Therefore,

$$\mathrm{tr}\,(\Sigma) = \frac{1}{2}\mathrm{tr}\left(\frac{\eta}{|G|}CA^{-1}\right) = \frac{1}{2}\frac{\eta}{|G|}\mathrm{tr}\left(CA^{-1}\right). \tag{6.31}$$

At the same time, we can easily calculate that

$$\mathrm{tr}(I) = d, \tag{6.32}$$

as $I \in \mathbb{R}^{d\times d}$, where d is the dimension of the parameter θ.

Insert Eqs. (6.31) and (6.32) to Eq. (6.26), we can prove the following inequality,

$$KL(Q||P) \leq \frac{1}{4}\frac{\eta}{|G|}tr(CA^{-1}) - \frac{1}{2}\log(\det(\Sigma)) - \frac{1}{2}d. \tag{6.33}$$

Equation (6.33) provides an upper bound on the KL divergence between the stationary distribution of SGD weights and the prior distribution over the hypothesis space. This divergence quantifies how far the learned distribution is away from the prior. By leveraging the monotonous nature of the generalization bound with respect to the KL divergence, we can extend this insight to derive a PAC-Bayesian generalization bound for SGD. This process involves incorporating the KL divergence bound (Eq. (6.33)) into the PAC-Bayesian framework outlined in Eq. (6.14) of Lemma 6.1, enabling us to quantify the trade-off between model complexity and generalization performance in a probabilistic context.

6.5.1.2 A Special Case of the Generalization Bound

In this subsection, we study a special case with two more assumptions for further understating the influence of the gradient fluctuation on our proposed generalization bound.

Assumption 6.6 The matrices A and Σ are symmetric. □

Assumption 6.6 can be translated as that both the local geometry around the global minima and the stationary distribution are homogeneous to every dimension of the parameter space. This assumption indicates that the product ΣA of matrices A and Σ is also symmetric.

Based on Assumptions 6.6, we can further prove the following theorem.

Theorem 6.7 *When Assumptions 6.6 holds, under all the conditions of Theorem 6.5, the stationary distribution of SGD has the following generalization bound,*

$$\begin{aligned} R(Q) \leq\ & \hat{R}_S(Q) \\ & + \sqrt{\frac{\frac{\eta}{2|G|}tr(CA^{-1}) + d\log\left(\frac{2|G|}{\eta}\right) - \log(\det(CA^{-1})) - d + 2\log\left(\frac{1}{\delta}\right) + 2\log m + 4}{4m-2}}. \end{aligned} \tag{6.34}$$

To prove Theorem 6.7, we first present a Lemma.

Lemma 6.3 *When Assumptions 6.6, the KL divergence between the stationary distribution Q of SGD and the prior distribution P is satisfies the following inequality*

$$KL(Q||P) \leq \frac{\eta}{4|G|} tr(CA^{-1}) + \frac{1}{2} d \log\left(\frac{2|G|}{\eta}\right) - \frac{1}{2}\log(\det(CA^{-1})) - \frac{1}{2}d. \tag{6.35}$$

Lemma 6.3 gives an upper bound for the distance between the distribution of the output hypothesis by SGD and the prior distribution of the hypothesis space. It measures how far SGD can explore in the hypothesis space. Based on it, we can further prove the following theorem that controls the generalization error of the special case under Assumptions 6.6.

Proof Apply Assumptions 6.6 to Eq. (6.19), we can prove the following three equations:

$$\Sigma A + A\Sigma = \frac{\eta}{|S|}C, \;\; 2\Sigma A = \frac{\eta}{|S|}C, \;\; \text{and } \Sigma = \frac{\eta}{2|S|}CA^{-1}. \tag{6.36}$$

Therefore,

$$\det(\Sigma) = \det\left(\frac{\eta}{2|S|}CA^{-1}\right) = \left(\frac{\eta}{2|S|}\right)^d \det\left(CA^{-1}\right). \tag{6.37}$$

Thus,

$$\begin{aligned}\log\left(\det(\Sigma)\right) &= \log\left[\left(\frac{\eta}{2|S|}\right)^d \det\left(CA^{-1}\right)\right] \\ &= -\, d\log\left(\frac{2|S|}{\eta}\right) + \log\left[\det\left(CA^{-1}\right)\right].\end{aligned} \tag{6.38}$$

Applying Eqs. (6.36) and (6.32) to Eq. (6.26), we can obtain Eq. (6.35).
The proof is completed. □

Then, we can directly obtain Theorem 6.7.

Proof of Theorem 6.7 Apply Eq. (6.35) of Lemma 6.3 to Eq. (6.14) of Lemma 6.1 of Lemma 6.1, we can directly obtain Eq. (6.34). The proof is completed.

Intuitively, our generalization bound links the generalization ability of the deep neural networks trained by SGD with three factors:

Local geometry around minimum. The determinant of the Hessian matrix A provides insights into the curvature and shape of the objective function's surface around a local minimum. A larger determinant $det(A)$ indicates a sharper or narrower minimum, suggesting more abrupt changes in the function's values around that point. Research (Keskar et al. 2017; Goyal et al. 2017) indicates that sharp minima, characterized by high values of $det(A)$, often correspond to models that overfit the training data and exhibit poorer generalization to unseen data. Understanding these local geometric properties is crucial for assessing the behavior and performance of optimization algorithms in deep learning.

Gradient fluctuation. The covariance matrix C (or equivalently the matrix B) characterizes how much the gradient estimates vary across different data points during the training process. This variation represents the inherent noise in SGD. By introducing this noise into the gradient computation, SGD can explore a broader range of solutions during optimization. This stochastic behavior allows SGD to navigate away from sharp or narrow local minima, potentially leading to solutions with better generalization performance on unseen data. The notion of injecting noise through SGD has been influential in understanding its robustness and ability to escape poor solutions during training.

Hyper-parameters. The relationship between batch size $|S|$ and learning rate η affects how gradients are computed and utilized during the training process. A larger batch size typically provides a more accurate estimate of the gradient, reducing its variance. On the other hand, a higher learning rate can lead to larger updates in parameter space based on these gradients. Specifically, under the following assumption, our generalization bound has a positive correlation with the ratio of batch size to learning rate. The interplay between these two factors influences the stability and convergence of the training process and can impact the generalization performance of the model. By analyzing the generalization bound in relation to the ratio of batch size to learning rate, we gain insights into how to optimize these hyperparameters for better model performance and generalization.

Specifically, under the following assumption, our generalization bound has a positive correlation with the ratio of batch size to learning rate.

Assumption 6.8 The network size is large enough:

$$d > \frac{\mathrm{tr}(CA^{-1})\eta}{2|S|}, \tag{6.39}$$

where d is the number of the parameters, C expresses the magnitude of individual gradient noise, A is the Hessian matrix around the global minima, η is the leaning rate, and $|S|$ is the batch size. □

This assumption can be justified that the network sizes of neural networks are usually extremely large. This property is also called *overparametrization* (Du et al. 2019b; Brutzkus et al. 2018; Allen-Zhu et al. 2019b). We can obtain the following corollary by combining Theorem 6.7 and Assumption 6.8.

Corollary 6.1 *When all conditions of Theorem 6.7 and Assumption 6.8 hold, the generalization bound of the network has a positive correlation with the ratio of batch size to learning rate.*

Proof We first define

$$I = \frac{\eta}{2|S|}tr(CA^{-1}) + d\log\left(\frac{2|S|}{\eta}\right) - \log(\det(CA^{-1})) - d + 2\log\left(\frac{1}{\delta}\right) + 2\log m + 4. \tag{6.40}$$

Then the generalization Eq. (6.34) becomes,

$$R(Q) \leq \hat{R}_S(Q) + \sqrt{\frac{I}{4m-2}}, \tag{6.41}$$

We thus calculate the derivative of I with respect to the ratio $\frac{|S|}{\eta}$ in order to check whether the generalization bound has a positive correlation withe the ratio. For the brevity, we define $k = |S|/\eta$.

$$\begin{aligned}\frac{\partial I}{\partial k} =& \frac{\partial}{\partial k}\left[\frac{1}{2k}tr(CA^{-1}) + d\log(2k) - \log(\det(CA^{-1})) - d + 2\log\left(\frac{1}{\delta}\right) + 2\log m + 4\right] \\ =& \frac{\partial}{\partial k}\left[\frac{1}{2k}tr(CA^{-1}) + d\log(2k) - \log(\det(CA^{-1})) - d + 2\log\left(\frac{1}{\delta}\right) + 2\log m + 4\right] \\ =& -\frac{1}{2k^2}tr(CA^{-1}) + \frac{d}{k}\end{aligned} \tag{6.42}$$

Therefore, when Assumption 6.8 holds, we have

$$d > \frac{tr(CA^{-1})\eta}{2|S|} = \frac{1}{2k}tr(CA^{-1}). \tag{6.43}$$

Thus,

$$\frac{\partial I}{\partial k} > 0. \tag{6.44}$$

So, I and further the generalization bound has a positive correlation with the ratio of batch size to learning rate.

The proof is completed. □

It reveals the negative correlation between the generalizability and the ratio. This property further derives the training strategy, which requires us to control the ratio and ensure that it is not too large in order to achieve a good generalization when training deep neural networks using SGD.

6.5.2 Empirical Evidence

To evaluate the training strategy from the empirical aspect, we conduct extensive systematic experiments to investigate the influence of the batch size and learning rate on the generalizability of deep neural networks trained by SGD. To deliver rigorous results, our experiments strictly control all unrelated variables. The empirical results show that there is a statistically significant negative correlation between the generalizability of the networks and the ratio of the batch size to the learning rate, which builds a solid empirical foundation for the training strategy.

6.5.2.1 Implementation Details

To guarantee that the empirical results generally apply to any case, our experiments are conducted based on two popular architectures, ResNet-110 (He et al. 2016a, b) and VGG-19 (Simonyan and Zisserman 2015), on two standard datasets, CIFAR-10 and CIFAR-100 (Krizhevsky and Hinton 2009), which can be downloaded from https://www.cs.toronto.edu/~kriz/cifar.html. The separation of the training and test sets we used are the same as the official version.

We trained 1,600 models with 20 batch sizes, $S_{BS} = \{16, 32, 48, 64, 80, 96, 112, 128, 144, 160, 176, 192, 208, 224, 240, 256, 272, 288, 304, 320\}$, and 20 learning rates, $S_{LR} = \{0.01, 0.02, 0.03, 0.04, 0.05, 0.06, 0.07, 0.08, 0.09, 0.10, 0.11, 0.12, 0.13, 0.14, 0.15, 0.16, 0.17, 0.18, 0.19, 0.20\}$. All SGD training techniques, such as momentum, are disabled. Also, both batch size and learning rate are constant in our experiments. Every model with a specific pair of batch size and learning rate is trained for 200 epochs. The test accuracies of all 200 epochs are collected for analysis. We select the highest accuracy on the test set to express the generalizability of each model, since the training error is almost the same across all models (they are all nearly 0).

The collected data is then utilized to investigate three correlations: (1) the correlation between network generalizability and the batch size, (2) the correlation between network generalizability and the learning rate, and (3) the correlation between network generalizability and the ratio of batch size to learning rate, where the first two are preparations for the final one. Specifically, we calculate the Spearman's rank-order correlation coefficients(SCCs) and the corresponding p groups of the collected data to investigate the statistical significance of the correlations. Almost all results indicate that the correlations are statistically significant ($p < 0.005$).[2] The p values of the correlation between the test accuracy and the ratio are all lower than 10^{-180} (see Table 6.3).

The architectures of our models are similar to a popular implementation of ResNet-110 and VGG-19.[3] Additionally, our experiments are conducted on a computing cluster with GPUs of NVIDIA® Tesla™ V100 16GB and CPUs of Intel® Xeon® Gold 6140 CPU @ 2.30GHz.

---

[2] The definition of "statistically significant" has various versions, such as $p < 0.05$ and $p < 0.01$. This section uses a more rigorous one ($p < 0.005$).

[3] See Wei Yang, https://github.com/bearpaw/pytorch-classification, 2017.

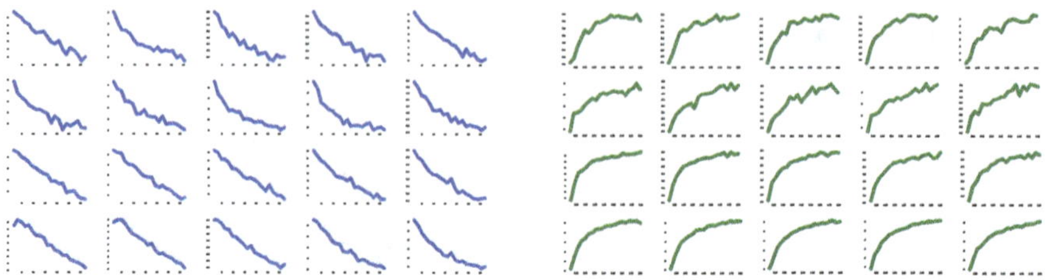

(a) Test accuracy-batch size (b) Test accuracy-learning rate

Fig. 6.3 Curves of test accuracy to batch size and learning rate. The four rows are respectively for (1) ResNet-110 trained on CIFAR-10, (2) ResNet-110 trained on CIFAR-100, (3) VGG-19 trained on CIFAR-10, and (4) VGG-19 trained on CIFAR-10. Each curve is based on 20 networks

6.5.2.2 Empirical Results on the Correlation

Correlation between generalization ability and batch size. When the learning rate is fixed as an element of S_{LR}, we train ResNet-110 and VGG-19 on CIFAR10 and CIFAR100 with 20 batch sizes of S_{BS}. The plots of test accuracy to batch size are illustrated in Fig. 6.3a. We list $1/4$ of all plots due to space limitation. The rest of the plots are in the supplementary materials. We then calculate the SCCs and the p values as Table 6.1, where bold p values refer to the statistically significant observations, while underlined ones refer to those that are not significant (This convention applies to Table 6.2). The results clearly show that there is a statistically significant negative correlation between generalization ability and batch size.

Correlation between generalization ability and learning rate. When the batch size is fixed as an element of S_{BS}, we train ResNet-110 and VGG-19 on CIFAR10 and CIFAR100 respectively with 20 learning rates S_{LR}. The plot of the test accuracy to the learning rate is illustrated in Fig. 6.3b, which include $1/4$ of all plots due to space limitation. The rest of the plots are in the supplementary materials. We then calculate the SCC and the p values as Table 6.2 shows. The results clearly show that there is a statistically significant positive correlation between the learning rate and the generalization ability of SGD.

Correlation between generalization ability and ratio of batch size to learning rate. We plot the test accuracies of ResNet-110 and VGG-19 on CIFAR-10 and CIFAR-100 to the rate of batch size to learning rate in Fig. 6.1. Totally over 1,600 points are plotted. Additionally, we perform Spearman's rank-order correlation test on all the accuracies of ResNet-110 and VGG-19 on CIFAR-10 and CIFAR-100. The SCC and p values show that the correlation between the ratio and the generalization ability is statistically significant as Table 6.3 demonstrate. Each test is performed on 400 models. The results strongly support the training strategy.

Table 6.1 SCC and p values of batch size to test accuracy for different learning rate (LR)

LR	ResNet-110 on CIFAR-10		ResNet-110 on CIFAR-100		VGG-19 on CIFAR-10		VGG-19 on CIFAR-100	
	SCC	p	SCC	p	SCC	p	SCC	p
0.01	−0.96	**2.6×10^{-11}**	−0.92	**5.6×10^{-8}**	−0.98	**3.7×10^{-14}**	−0.99	**7.1×10^{-18}**
0.02	−0.96	**1.2×10^{-11}**	−0.94	**1.5×10^{-9}**	−0.99	**3.6×10^{-17}**	−0.99	**7.1×10^{-18}**
0.03	−0.96	**3.4×10^{-11}**	−0.99	**1.5×10^{-16}**	−0.99	**7.1×10^{-18}**	−1.00	**1.9×10^{-21}**
0.04	−0.98	**1.8×10^{-14}**	−0.98	**7.1×10^{-14}**	−0.99	**9.6×10^{-19}**	−0.99	**3.6×10^{-17}**
0.05	−0.98	**3.7×10^{-14}**	−0.98	**1.3×10^{-13}**	−0.99	**7.1×10^{-18}**	−0.99	**1.4×10^{-15}**
0.06	−0.96	**1.8×10^{-11}**	−0.97	**6.7×10^{-13}**	−1.00	**1.9×10^{-21}**	−0.99	**1.4×10^{-15}**
0.07	−0.98	**5.9×10^{-15}**	−0.94	**5.0×10^{-10}**	−0.98	**8.3×10^{-15}**	−0.97	**1.7×10^{-12}**
0.08	−0.97	**1.7×10^{-12}**	−0.97	**1.7×10^{-12}**	−0.98	**2.4×10^{-13}**	−0.97	**1.7×10^{-12}**
0.09	−0.97	**4.0×10^{-13}**	−0.98	**3.7×10^{-14}**	−0.98	**1.8×10^{-14}**	−0.96	**1.2×10^{-11}**
0.10	−0.97	**1.9×10^{-12}**	−0.96	**8.7×10^{-12}**	−0.98	**8.3×10^{-15}**	−0.93	**2.2×10^{-9}**
0.11	−0.97	**1.1×10^{-12}**	−0.98	**1.3×10^{-13}**	−0.99	**2.2×10^{-16}**	−0.93	**2.7×10^{-9}**
0.12	−0.97	**4.4×10^{-12}**	−0.96	**2.5×10^{-11}**	−0.98	**7.1×10^{-13}**	−0.90	**7.0×10^{-8}**
0.13	−0.94	**1.5×10^{-9}**	−0.98	**1.3×10^{-13}**	−0.97	**1.7×10^{-12}**	−0.89	**1.2×10^{-7}**
0.14	−0.97	**2.6×10^{-12}**	−0.91	**3.1×10^{-8}**	−0.97	**6.7×10^{-13}**	−0.86	**1.1×10^{-6}**
0.15	−0.96	**4.6×10^{-11}**	−0.98	**1.3×10^{-13}**	−0.95	**8.3×10^{-11}**	−0.79	**3.1×10^{-5}**
0.16	−0.95	**3.1×10^{-10}**	−0.96	**8.7×10^{-12}**	−0.95	**1.4×10^{-10}**	−0.77	**6.1×10^{-5}**
0.17	−0.95	**2.4×10^{-10}**	−0.95	**2.6×10^{-10}**	−0.91	**2.3×10^{-8}**	−0.68	**1.3×10^{-3}**
0.18	−0.97	**4.0×10^{-12}**	−0.97	**1.1×10^{-12}**	−0.93	**2.6×10^{-9}**	−0.66	**2.8×10^{-3}**
0.19	−0.94	**6.3×10^{-10}**	−0.95	**8.3×10^{-11}**	−0.90	**8.0×10^{-8}**	−0.75	**3.4×10^{-4}**
0.20	−0.91	**3.6×10^{-8}**	−0.98	**1.3×10^{-13}**	−0.95	**6.2×10^{-11}**	−0.57	1.4×10^{-2}

6.6 Interplay of Optimization and Bayesian Inference

An interesting interaction can be found between optimization and Bayesian inference (Xiang 2020): (1)generally, Bayesian inference (or sampling) can be viewed as optimization in the probability space (*e.g.*, in stochastic gradient Markov chain Monte Carlo, stochastic gradient MCMC) and can be scaled through optimization (*e.g.*, in variational inference), and (2) stochastic gradient MCMC also involves performing Bayesian approximation.

Bayesian inference. Bayesian learning for optimizing parametric machine learning models is to obtain the posterior of the model parameters and then seek the best parameter settings. However, the analytical expression for the posterior is unknown in most real-world applications. To address this problem, classical methods, such as the Metropolis-Hastings algorithm (Hastings 1970; Geman and Geman 1984) and hybrid Monte Carlo (HMC; (Duane et al. 1987)), adopts MCMC methods to infer the posterior. Due to the promising performance, Bayesian inference has been applied to many applications, including topic models (Larochelle and Lauly 2012; Zhang et al. 2020), Bayesian neural networks (Louizos and Welling 2017; Roth and Pernkopf 2018), and generative models (Kingma and Welling 2014; Goodfellow et al. 2014a; Kobyzev et al. 2020).

Table 6.2 SCC and p values of learning rate to test accuracy for different batch size (BS)

BS	ResNet-110 on CIFAR-10		ResNet-110 on CIFAR-100		VGG-19 on CIFAR-10		VGG-19 on CIFAR-100	
	SCC	p	SCC	p	SCC	p	SCC	p
16	0.60	5.3×10^{-3}	0.84	**3.2×10^{-6}**	0.62	**3.4×10^{-3}**	−0.80	2.6×10^{-5}
32	0.60	5.0×10^{-3}	0.90	**9.9×10^{-8}**	0.78	**4.9×10^{-5}**	−0.14	5.5×10^{-1}
48	0.84	**3.2×10^{-6}**	0.89	**1.8×10^{-7}**	0.87	**4.9×10^{-7}**	0.37	1.1×10^{-1}
64	0.67	**1.2×10^{-3}**	0.89	**1.0×10^{-7}**	0.91	**2.0×10^{-8}**	0.91	**1.1×10^{-6}**
80	0.80	**2.0×10^{-5}**	0.99	**4.8×10^{-16}**	0.95	**2.4×10^{-10}**	0.87	**4.5×10^{-6}**
96	0.79	**3.3×10^{-5}**	0.89	**2.4×10^{-7}**	0.94	**5.2×10^{-9}**	0.94	**1.5×10^{-9}**
112	0.90	**8.8×10^{-8}**	0.91	**2.7×10^{-8}**	0.97	**2.6×10^{-12}**	0.95	**1.2×10^{-10}**
128	0.95	**8.3×10^{-11}**	0.92	**1.1×10^{-8}**	0.98	**2.2×10^{-14}**	0.99	**4.8×10^{-16}**
144	0.85	**2.1×10^{-6}**	0.98	**7.7×10^{-14}**	0.90	**6.2×10^{-8}**	0.98	**3.5×10^{-15}**
160	0.90	**4.3×10^{-8}**	0.94	**5.0×10^{-10}**	0.95	**3.3×10^{-10}**	0.99	**7.1×10^{-18}**
176	0.94	**5.0×10^{-10}**	0.99	**3.6×10^{-17}**	0.91	**2.3×10^{-8}**	0.98	**1.8×10^{-14}**
192	0.94	**6.7×10^{-10}**	0.94	**5.0×10^{-10}**	0.95	**6.2×10^{-11}**	0.97	**2.6×10^{-12}**
208	0.91	**3.6×10^{-8}**	0.97	**6.7×10^{-12}**	0.98	**6.1×10^{-14}**	0.99	**2.5×10^{-17}**
224	0.90	**9.0×10^{-8}**	0.98	**3.7×10^{-14}**	0.93	**2.2×10^{-9}**	0.98	**1.3×10^{-13}**
240	0.78	**4.6×10^{-5}**	0.95	**2.4×10^{-10}**	0.98	**8.3×10^{-15}**	0.99	**9.6×10^{-19}**
256	0.83	**4.8×10^{-6}**	0.94	**5.0×10^{-10}**	0.99	**4.8×10^{-16}**	0.97	**5.4×10^{-12}**
272	0.95	**2.4×10^{-10}**	0.96	**2.5×10^{-11}**	0.97	**4.0×10^{-13}**	0.99	**1.5×10^{-16}**
288	0.94	**9.8×10^{-10}**	0.92	**1.5×10^{-18}**	0.95	**8.3×10^{-11}**	0.99	**1.5×10^{-16}**
304	0.81	**1.5×10^{-5}**	0.97	**4.0×10^{-13}**	0.95	**6.2×10^{-11}**	1.00	**3.7×10^{-24}**
320	0.94	**1.4×10^{-9}**	0.95	**8.3×10^{-11}**	0.97	**2.6×10^{-12}**	1.00	**7.2×10^{-20}**

Table 6.3 SCC and p values of the ratio of batch size to learning rate and test accuracy

ResNet-110 on CIFAR-10		ResNet-110 on CIFAR-100		VGG-19 on CIFAR-10		VGG-19 on CIFAR-100	
SCC	p	SCC	p	SCC	p	SCC	p
−0.97	**3.3×10^{-235}**	−0.98	**5.3×10^{-293}**	−0.98	**6.2×10^{-291}**	−0.94	**6.1×10^{-180}**

Scaling Bayesian inference via optimization. Bayesian inference has a prohibitive computation complexity and thus is time consuming on large-scale data. Several approaches have been proposed to address this challenge:

(1) *Stochastic gradient Markov chain Monte Carlo* (SGMCMC) (Ma et al. 2015) introduces stochastic gradient estimation (Robbins and Monro 1951) into MCMC. The family of SGMCMC algorithms includes SGLD (Welling and Teh 2011), stochastic gradient Riemannian Langevin dynamics (SGRLD) (Patterson and Teh 2013), stochastic gradient Fisher scoring (SGFS) (Ahn et al. 2012), stochastic gradient Hamiltonian Monte Carlo (SGHMC) (Chen et al. 2014), and the stochastic gradient Nosé-Hoover thermostat (SGNHT) (Ding et al. 2014). This section analyses SGLD as an example of an SGMCMC scheme.

(2) *Variational inference* (Hoffman et al. 2013; Blei et al. 2017; Zhang et al. 2018a) employs a two-step process to infer the posterior:

1. A family of distributions is defined as

$$Q = \{q_\lambda | \lambda \in \Lambda\},$$

where λ represents the distribution parameter(s).

2. The member of the distribution family Q that is the closest to the posterior $p(\theta | S)$ under the Kullback-Leibler (KL) divergence is sought, *i.e.*,

$$\min_\lambda \mathrm{KL}(q_\lambda(\theta) \| p(\theta | S)).$$

Minimizing the KL divergence is equivalent to maximizing the evidence lower bound (ELBO), defined as follows:

$$\mathrm{ELBO}(\lambda, S) = \mathbb{E}_{q_\lambda} \log p(\theta, S) - \mathbb{E}_{q_\lambda} \log q_\lambda(\theta).$$

Understanding stochastic-gradient-based optimization via Bayesian inference. Weinan (2017) suggested seeking to understand demystify the process of optimization in deep learning via SDEs. Mandt et al. (2017) suggested gaining an understanding of SGMs via approximate Bayesian inference. Following this line of reasoning, extensive works on SGMs has been produced, as discussed in the previous subsections.

References

Alon, Brutzkus, Amir Globerson, Eran Malach, and Shai Shalev-Shwartz. 2018. SGD learns over-parameterized networks that provably generalize on linearly separable data. In *International Conference on Learning Representations*.

Ankit, Pensia, Varun Jog, and Po-Ling Loh. 2018. Generalization error bounds for noisy, iterative algorithms. In *2018 IEEE International Symposium on Information Theory (ISIT)*, 546–550.

Aolin, Xu and Maxim Raginsky. 2017. Information-theoretic analysis of generalization capability of learning algorithms. In *Advances in Neural Information Processing Systems*, 2524–2533.

Belinda, Tzen, Tengyuan Liang, and Maxim Raginsky. 2018. Local optimality and generalization guarantees for the Langevin algorithm via empirical metastability. In *Conference On Learning Theory*, 857–875.

Bottou, Léon., et al. 1998. Online learning and stochastic approximations. *On-line learning in neural networks* 17 (9): 142.

Cheng, Xiang. 2020. *The Interplay between Sampling and Optimization*. PhD thesis, EECS Department, University of California, Berkeley.

Christos, Louizos and Max Welling. 2017. Multiplicative normalizing flows for variational Bayesian neural networks. In *International Conference on Machine learning*, 2218–2227.

Crispin, W Gardiner. 1985. *Handbook of stochastic methods*, vol. 3. Berlin: Springer.

David, A McAllester. 1999. PAC-Bayesian model averaging. *In Annual Conference of Learning Theory* 99: 164–170.

David, A McAllester. 1999. Some PAC-Bayesian theorems. *Machine Learning* 37 (3): 355–363.

David, M Blei, Alp Kucukelbir, and Jon D. McAuliffe. 2017. Variational inference: A review for statisticians. *Journal of the American statistical Association* 112 (518): 859–877.

Diederik, P Kingma and Max Welling. 2014. Auto-encoding variational Bayes. In *International Conference on Learning Representations.*

Dieuleveut, Aymeric, and Francis Bach. 2016. Nonparametric stochastic approximation with large step-sizes. *The Annals of Statistics* 44 (4): 1363–1399.

Duane, Simon. 1987. Anthony D Kennedy, Brian J Pendleton, and Duncan Roweth. *Hybrid Monte Carlo. Physics Letters B* 195 (2): 216–222.

Fredrik, Hellström and Giuseppe Durisi. 2020. Generalization bounds via information density and conditional information density. *arXiv preprint* arXiv:2005.08044.

Geman, Stuart, and Donald Geman. 1984. Stochastic relaxation, Gibbs distributions, and the Bayesian restoration of images. *IEEE Transactions on Pattern Analysis and Machine Intelligence* 6: 721–741.

George, E Uhlenbeck, and Leonard S. Ornstein. 1930. On the theory of the brownian motion. *Physical Review* 36 (5): 823.

Gintare, Karolina Dziugaite, Kyle Hsu, Waseem Gharbieh, and Daniel M Roy. 2020. On the role of data in pac-bayes bounds. *arXiv preprint* arXiv:2006.10929.

Hangfeng, He and Weijie Su. 2020. The local elasticity of neural networks. In *International Conference on Learning Representations.*

Hao, Li, Zheng Xu, Gavin Taylor, Christoph Studer, and Tom Goldstein. 2018c. Visualizing the loss landscape of neural nets. In *Advances in Neural Information Processing Systems.*

Hao, Zhang, Bo Chen, Yulai Cong, Dandan Guo, Hongwei Liu, and Mingyuan Zhou. 2020. Deep autoencoding topic model with scalable hybrid Bayesian inference. *IEEE Transactions on Pattern Analysis and Machine Intelligence.*

Harold Kushner and G George Yin. *Stochastic approximation and recursive algorithms and applications*, volume 35. Springer Science & Business Media, 2003.

He, Fengxiang, and Tongliang Liu. 2019. *and Dacheng Tao.* Control batch size and learning rate to generalize well: Theoretical and empirical evidence. In Advances in Neural Information Processing Systems.

Herbert, Robbins and Sutton Monro. 1951. A stochastic approximation method. *The Annals of Mathematical Statistics*, 400–407.

Huan, Xu., Constantine Caramanis, and Shie Mannor. 2011. Sparse algorithms are not stable: A no-free-lunch theorem. *IEEE Transactions on Pattern Analysis and Machine Intelligence* 34 (1): 187–193.

Hugo, Larochelle and Stanislas Lauly. 2012. A neural autoregressive topic model. In *Advances in Neural Information Processing Systems*, 2708–2716.

Ian, Goodfellow, Jean Pouget-Abadie, Mehdi Mirza, Bing Xu, David Warde-Farley, Sherjil Ozair, Aaron Courville, and Yoshua Bengio. 2014a. Generative adversarial nets. In *Advances in Neural Information Processing Systems*, 2672–2680.

Ian, Goodfellow, Yoshua Bengio, Aaron Courville, and Yoshua Bengio. 2016. *Deep learning,* vol. 1. MIT Press.

Ilya, Sutskever, James Martens, George Dahl, and Geoffrey Hinton. 2013. On the importance of initialization and momentum in deep learning. In *International conference on machine learning*, 1139–1147.

Ivan, Kobyzev, Simon Prince, and Marcus Brubaker. 2020. Normalizing flows: An introduction and review of current methods. *IEEE Transactions on Pattern Analysis and Machine Intelligence.*

Jeffrey, Negrea, Mahdi Haghifam, Gintare Karolina Dziugaite, Ashish Khisti, and Daniel M Roy. 2019. Information-theoretic generalization bounds for sgld via data-dependent estimates. In *Advances in Neural Information Processing Systems*, 11015–11025.

Jian, Li, Xuanyuan Luo, and Mingda Qiao. 2019. On generalization error bounds of noisy gradient methods for non-convex learning. *arXiv preprint* arXiv:1902.00621.

John, J Cherian, Andrew G Taube, Robert T McGibbon, Panagiotis Angelikopoulos, Guy Blanc, Michael Snarski, Daniel D Richman, John L Klepeis, and David E Shaw. 2020. Efficient hyperparameter optimization by way of pac-bayes bound minimization. *arXiv preprint* arXiv:2008.06431.

Junhong, Lin and Lorenzo Rosasco. 2016. Optimal learning for multi-pass stochastic gradient methods. In *Advances in Neural Information Processing Systems*, 4556–4564.

Junhong, Lin, Raffaello Camoriano, and Lorenzo Rosasco. 2016. Generalization properties and implicit regularization for multiple passes SGM. In *International Conference on Machine learning*, 2340–2348.

Kaiming, He, Xiangyu Zhang, Shaoqing Ren, and Jian Sun. 2016a. Deep residual learning for image recognition. In *IEEE Conference on Computer Vision and Pattern Recognition*.

Kaiming, He, Xiangyu Zhang, Shaoqing Ren, and Jian Sun. 2016b. Identity mappings in deep residual networks. In *European Conference on Computer Vision*.

Karen, Simonyan and Andrew Zisserman. 2015. Very deep convolutional networks for large-scale image recognition. In *International Conference on Learning Representations*.

Krizhevsky, Alex, and Geoffrey Hinton. 2009. *Learning multiple layers of features from tiny images*. Citeseer: Technical report.

Laurent, Dinh, Razvan Pascanu, Samy Bengio, and Yoshua Bengio. 2017. Sharp minima can generalize for deep nets. In .

Lennart, Ljung, Georg Pflug, and Harro Walk. 2012. *Stochastic approximation and optimization of random systems*, volume 17. Birkhäuser.

Lever, Guy, François Laviolette, and John Shawe-Taylor. 2013. Tighter pac-bayes bounds through distribution-dependent priors. *Theoretical Computer Science* 473: 4–28.

Mahdi, Haghifam, Jeffrey Negrea, Ashish Khisti, Daniel M Roy, and Gintare Karolina Dziugaite. 2020. Sharpened generalization bounds based on conditional mutual information and an application to noisy, iterative algorithms. *arXiv preprint* arXiv:2004.12983.

Matthew, D Hoffman, David M. Blei, Chong Wang, and John Paisley. 2013. Stochastic variational inference. *Journal of Machine Learning Research* 14 (1): 1303–1347.

Max, Welling and Yee W Teh. 2011. Bayesian learning via stochastic gradient Langevin dynamics. In *International Conference on Machine learning*, 681–688.

Maxim, Raginsky, Alexander Rakhlin, and Matus Telgarsky. 2017. Non-convex learning via stochastic gradient Langevin dynamics: A nonasymptotic analysis. In *Conference on Learning Theory*, 1674–1703.

Moritz, Hardt, Ben Recht, and Yoram Singer. 2016b. Train faster, generalize better: Stability of stochastic gradient descent. In *International Conference on Machine learning*, 1225–1234.

Nan, Ding, Youhan Fang, Ryan Babbush, Changyou Chen, Robert D Skeel, and Hartmut Neven. 2014. Bayesian sampling using stochastic gradient thermostats. In *Advances in Neural Information Processing Systems*, 3203–3211.

Nitish, Shirish Keskar, Dheevatsa Mudigere, Jorge Nocedal, Mikhail Smelyanskiy, and Ping Tak Peter Tang. 2017. On large-batch training for deep learning: Generalization gap and sharp minima. In *International Conference on Learning Representations*.

Noah, Golowich, Alexander Rakhlin, and Ohad Shamir. 2018. Size-independent sample complexity of neural networks. In *Annual Conference on Learning Theory*, 297–299.

Olivier, Bousquet, and André Elisseeff. 2002. Stability and generalization. *Journal of Machine Learning Research* 2: 499–526.

Priya, Goyal, Piotr Dollár, Ross Girshick, Pieter Noordhuis, Lukasz Wesolowski, Aapo Kyrola, Andrew Tulloch, Yangqing Jia, and Kaiming He. 2017. Accurate, large minibatch sgd: training imagenet in 1 hour. *arXiv preprint* arXiv:1706.02677.

Roth, Wolfgang, and Franz Pernkopf. 2018. Bayesian neural networks with weight sharing using Dirichlet processes. *IEEE Transactions on Pattern Analysis and Machine Intelligence* 42 (1): 246–252.

Sam, Patterson and Yee Whye Teh. 2013. Stochastic gradient Riemannian Langevin dynamics on the probability simplex. In *Advances in Neural Information Processing Systems*, 3102–3110.

Simon, S. Du, Xiyu Zhai, Barnabas Poczos, and Aarti Singh. 2019b. Gradient descent provably optimizes over-parameterized neural networks. In *International Conference on Learning Representations*.

Spearman, Charles. 1987. The proof and measurement of association between two things. *The American journal of psychology* 100 (3/4): 441–471.

Stephan, Mandt, Matthew D. Hoffman, and David M. Blei. 2017. Stochastic gradient descent as approximate Bayesian inference. *Journal of Machine Learning Research* 18 (1): 4873–4907.

Sungjin, Ahn, Anoop Korattikara, and Max Welling. 2012. Bayesian posterior sampling via stochastic gradient Fisher scoring. *arXiv preprint* arXiv:1206.6380.

Thomas, Steinke and Lydia Zakynthinou. 2020. Reasoning about generalization via conditional mutual information. *arXiv preprint* arXiv:2001.09122.

Tianqi, Chen, Emily Fox, and Carlos Guestrin. 2014. Stochastic gradient Hamiltonian Monte Carlo. In *International Conference on Machine learning*, 1683–1691.

W Keith, Hastings. 1970. Monte Carlo sampling methods using Markov chains and their applications.

Weinan, E. 2017. A proposal on machine learning via dynamical systems. *Communications in Mathematics and Statistics* 5 (1): 1–11.

Wenlong, Mou, Liwei Wang, Xiyu Zhai, and Kai Zheng. 2018a. Generalization bounds of sgld for non-convex learning: Two theoretical viewpoints. In *Annual Conference On Learning Theory*.

Xi, Chen, Jason D Lee, Xin T Tong, and Yichen Zhang. 2016. Statistical inference for model parameters in stochastic gradient descent. *arXiv preprint* arXiv:1610.08637.

Yeming, Wen, Kevin Luk, Maxime Gazeau, Guodong Zhang, Harris Chan, and Jimmy Ba. 2019. Interplay between optimization and generalization of stochastic gradient descent with covariance noise. *arXiv preprint* arXiv:1902.08234.

Yi, Zhou, Yingbin Liang, and Huishuai Zhang. 2018. Generalization error bounds with probabilistic guarantee for SGD in nonconvex optimization. *arXiv preprint* arXiv:1802.06903.

Yi-An, Ma, Tianqi Chen, and Emily Fox. 2015. A complete recipe for stochastic gradient mcmc. In *Advances in Neural Information Processing Systems*, 2917–2925.

Ying, Yiming, and Massimiliano Pontil. 2008. Online gradient descent learning algorithms. *Foundations of Computational Mathematics* 8 (5): 561–596.

Yuansi, Chen, Chi Jin, and Bin Yu. 2018b. Stability and convergence trade-off of iterative optimization algorithms. *arXiv preprint* arXiv:1804.01619.

Yuheng, Bu, Shaofeng Zou, and Venugopal V Veeravalli. 2020. Tightening mutual information based bounds on generalization error. *IEEE Journal on Selected Areas in Information Theory*.

Yuting, Wei, Fanny Yang, and Martin J Wainwright. 2017. Early stopping for kernel boosting algorithms: A general analysis with localized complexities. In *Advances in Neural Information Processing Systems*, 6065–6075.

Zeyuan, Allen-Zhu., Yuanzhi Li, and Zhao Song. 2019. A convergence theory for deep learning via over-parameterization. In .

Zhang, Cheng, Judith Bütepage, Hedvig Kjellström, and Stephan Mandt. 2018. Advances in variational inference. *IEEE Transactions on Pattern Analysis and Machine Intelligence* 41 (8): 2008–2026.

Zhun, Deng, Hangfeng He, and Weijie J Su. 2020. Toward better generalization bounds with locally elastic stability. *arXiv preprint* arXiv:2010.13988.

Chapter 7
The Geometry of the Loss Surfaces

A major barrier recognized by the whole community is that the loss surfaces of deep neural networks are extremely nonconvex and nonsmooth. Such nonconvexity and nonsmoothness make the analysis of the optimization and generalization properties of such networks prohibitively difficult. An intuitive approach is to bypass these geometrical properties to seek a theoretical explanation. However, some papers argue that these "intimidating" geometrical properties themselves are the major factors shaping the properties of deep neural networks and the key to explaining deep learning.

7.1 Linear Networks Have No Spurious Local Minima

Linear neural networks are neural networks whose activations are all linear functions. For linear neural networks, the loss surfaces do not have any spurious local minima: all local minima are equally good, i.e., they are all global minima.

Kawaguchi (2016) proved that linear neural networks do not have any spurious local minima under several assumptions: (1) the loss functions are squared losses; (2) XX^T and XY^T are both of full rank, where X is the data matrix and Y is the label matrix; (3) the dimensionality of the input layer is larger than that of the output layer; and (4) all eigenvalues of the matrix $YX^\top \left(XX^T\right)^{-1} XY^T$ are different from each other. Lu and Kawaguchi (2017) replaced these assumptions with a single, more restrictive assumption, namely, that the data matrix X and the label matrix Y are both of full rank. Later, Zhou and Liang (2018) proved that all critical points are global minima when certain conditions hold. The authors proved this based on a new result regarding the analytical formulation of the critical points.

F. He and D. Tao, *Foundations of Deep Learning*, Machine Learning: Foundations, Methodologies, and Applications, https://doi.org/10.1007/978-981-16-8233-9_7

7.2 Nonlinear Activations Bring Infinite Spurious Local Minima

Neural networks have been successfully deployed in many real-world applications (LeCun et al. 2015; Witten et al. 2016; Silver et al. 2016; He et al. 2016; Litjens et al. 2017). Despite this, the theoretical foundations of neural networks are somewhat premature. To cover the many deficiencies in our knowledge of deep learning theory, the investigation into the loss surfaces of neural networks is of fundamental importance. Understanding the loss surface would be helpful in several relevant research areas, such as the ability to estimate data distributions, the optimization of neural networks, and the generalization to unseen data.

This section studies the role of nonlinearities in activation functions which in turn shape the loss surfaces of neural networks. Our results demonstrate that the impact of nonlinearities is profound.

First, we prove that the loss surfaces of nonlinear neural networks are substantially different to those of linear neural networks, in which local minima are created equal, and are all global minima (Kawaguchi 2016; Baldi and Hornik 1989; Lu and Kawaguchi 2017; Freeman and Bruna 2017; Zhou and Liang 2018; Laurent and von Brecht 2018; Yun et al. 2018). By contrast,

> *Neural networks with arbitrary depth and arbitrary piecewise linear activations (excluding linear functions) have infinitely many spurious local minima under arbitrary continuously differentiable loss functions.*

This result only relies on four mild assumptions that cover most practical circumstances: (1) the training sample set is linearly inseparable; (2) all training sample points are distinct; (3) the output layer is narrower than the other hidden layers; and (4) there exists some turning point in the piece-wise linear activations that the sum of the slops on the two sides does not equal to 0.

Our result significantly extends the existing study on the existence of spurious local minimum. For example, Zhou and Liang (2018) prove that one-hidden-layer neural networks with two nodes in the hidden layer and two-piece linear (ReLU-like) activations have spurious local minima; Swirszcz et al. (2016) prove that ReLU networks have spurious local minima under the squared loss when most of the neurons are not activated; Safran and Shamir (2018) present a computer-assisted proof that two-layer ReLU networks have spurious local minima; a recent work Yun et al. (2019) have proven that neural networks with two-piece linear activations have infinite spurious local minima, but the results only apply to the networks with one hidden layer and one-dimensional outputs; and a concurrent work Goldblum et al. (2020) proves that for multi-layer perceptrons of any depth, the performance of every local minimum on the training data equals to a linear model, which is also verified by experiments.

The proposed theorem is proved in three stages: (1) we prove that neural networks with one hidden layer and two-piece linear activations have spurious local minima; (2) we extend the conditions to neural networks with arbitrary hidden layers and two-piece linear activations; and (3) we further extend the conditions to neural networks with arbitrary depth and arbitrary piecewise linear activations. Since some parameters of the constructed spurious local minima are from continuous intervals, we have obtained infinitely many spurious local minima. At each stage, the proof follows a two-step strategy that: (a) constructs an infinite series of local minima; and (b) constructs a point in the parameter space whose empirical risk is lower than the constructed local minimum in Step (a). This strategy is inspired by Yun et al. (2019), upon which we have made significant and non-trivial development.

Second, we draw a "big picture" for the loss surfaces of nonlinear neural networks. Soudry and Hoffer (2018) highlight a *smooth and multilinear partition* of the loss surfaces of neural networks. The nonlinearities in the piecewise linear activations partition the loss surface of any nonlinear neural network into multiple smooth and multilinear open cells. Specifically, every nonlinear point in the activation functions creates a group of the non-differentiable boundaries between the cells, while the linear parts of activations correspond to the smooth and multilinear interiors. Based on the partition, we discover the degenerative nature of the large amounts of local minima from the following aspects:

- **Every local minimum is globally minimal within a cell.** This property demonstrates that the local geometry within every cell is similar to the global geometry of linear networks, although technically, they are substantially different. It applies to any one-hidden-layer neural network with two-piece linear activations for regression under convex loss. We rigorously prove this property in two stages: (1) we prove that within every cell, the empirical risk $\hat{R}_S$ is convex with respect to a variable $\hat{W}$ mapped from the weights W by a mapping Q. Therefore, the local minima with respect to the variable $\hat{W}$ are also the global minima in the cell; and then (2) we prove that the local optimality is maintained under the constructed mapping. Specifically, the local minima of the empirical risk $\hat{R}_S$ with respect to the parameter W are also the local minima with respect to the variable $\hat{W}$. We thereby prove this property by combining the convexity and the correspondence of the minima. This proof is technically novel and non-trivial, despite the natural intuitions.
- **Equivalence classes and quotient space of local minimum valleys.** All local minima in a cell are concentrated as a *local minimum valley*: on a local minimum valley, all local minima are connected with each other by a continuous path, on which the empirical risk is invariant. Further, all these local minima constitute an equivalence class. This local minima valley may have several parallel valleys that are in the same equivalence class but do not appear because of the restraints from cell boundaries. If such constraints are ignored, all the equivalence classes constitute a quotient space. The constructed mapping Q is exactly the quotient map. This result coincides with the property of *mode connectivity* that the minima found by gradient-based methods are connected by a path in the parameter space with almost invariant empirical risk (Garipov et al. 2018; Draxler et al. 2018; Kuditipudi et al. 2019). Additionally, this property suggests that we would need to study every local minimum valley as a whole.

- **Linear collapse.** Linear neural networks are covered by our theories as a simplified case. When all activations are linear, the partitioned loss surface collapses to one single cell, in which all local minima are globally optimal, as suggested by the existing works on linear networks (Kawaguchi 2016; Baldi and Hornik 1989; Lu and Kawaguchi 2017; Freeman and Bruna 2017; Zhou and Liang 2018; Laurent and von Brecht 2018; Yun et al. 2018).

7.2.1 Neural Networks Have Infinite Spurious Local Minima

This section investigates the existence of spurious local minima on the loss surfaces of neural networks. We find that almost all practical neural networks have infinitely many spurious local minima. This result stands for any neural network with arbitrary depth and arbitrary piecewise linear activations excluding linear functions under arbitrary continuously differentiable loss.

7.2.1.1 Preliminaries

Consider a training sample set $\{(x_1, y_1), (x_2, y_2), \ldots, (x_m, y_m)\}$ of size m. Suppose the dimensions of feature X_i and label Y_i are d_X and d_Y, respectively. By aggregating the training sample set, we obtain the feature matrix $X \in \mathbb{R}^{d_X \times m}$ and label matrix $Y \in \mathbb{R}^{d_Y \times m}$.

Suppose a neural network has L layers. Denote the weight matrix, bias, and activation in the j-th layer respectively by $W_j \in \mathbb{R}^{d_j \times d_{j-1}}$, $b_j \in \mathbb{R}^{d_j}$, and $h_j : \mathbb{R}^{d_j \times m} \to \mathbb{R}^{d_j \times m}$, where d_j is the dimension of the output of the j-th layer. Also, for the input matrix X, the output of the j-th layer is denoted as the $Y^{(j)}$ and the output of the j-th layer before the activation is denoted as the $\tilde{Y}^{(j)}$,

$$\tilde{Y}^{(j)} = W_j Y^{(j-1)} + b_j \mathbf{1}_m^T, \tag{7.1}$$

$$Y^{(j)} = h\left(W_j Y^{(j-1)} + b_j \mathbf{1}_m^T\right). \tag{7.2}$$

The output of the network is defined as follows,

$$\hat{Y} = h_L\left(W_L h_{L-1}\left(W_{L-1} h_{L-2}\left(\ldots h_1\left(W_1 X + b_1 \mathbf{1}_m^T\right)\ldots\right) + b_{L-1}\mathbf{1}_m^T\right) + b_L \mathbf{1}_m^T\right). \tag{7.3}$$

Also, we define $Y^{(0)} = X$, $Y^{(L)} = \hat{Y}$, $d_0 = d_X$, and $d_L = d_Y$. In some situations, we use $\hat{Y}\left([W_i]_{i=1}^L, [b_i]_{i=1}^L\right)$ to clarify the parameters, as well as $\tilde{Y}^{(j)}$, $Y^{(j)}$, etc.

This section discusses neural networks with piecewise linear activations. A part of the proof uses two-piece linear activations h_{s_-,s_+} which are defined as follows,

$$h_{s_-,s_+}(x) = \mathbf{I}_{\{x \le 0\}} s_- x + \mathbf{I}_{\{x > 0\}} s_+ x, \tag{7.4}$$

where $|s_+| \neq |s_-|$ and $\mathbf{I}_{\{\cdot\}}$ is the indicator function.

Remark 7.1 Piecewise linear functions are dense in the space of continuous functions. In other words, for any continuous function, we can always find a piecewise linear function to estimate it with arbitrary small distance.

This section uses continuously differentiable loss to evaluate the performance of neural networks. Continuous differentiability is defined as follows.

Definition 7.1 (*Continuously differentiable*) We call a function $f : \mathbb{R}^{d_X} \to \mathbb{R}$ continuously differentiable with respect to the variable X if: (1) the function f is differentiable with respect to X; and (2) the gradient $\nabla f(X)$ of the function f is continuous with respect to the variable X.

7.2.1.2 Main Result

The theorem in this section relies on the following assumptions.

Assumption 7.1 The training data cannot be fit by a linear model. □

Assumption 7.2 All data points are distinct. □

Assumption 7.3 All hidden layers are wider than the output layer. □

Assumption 7.4 For the piece-wise linear activations, there exists some turning point that the sum of the slops on the two sides does not equal to 0. □

These assumptions are respectively justified as follows: (1) most real-world datasets are extremely complex and cannot be simply fit using linear models; (2) it is easy to guarantee that the data points are distinct by employing data cleansing methods; (3) for regression and many classification tasks, the width of output layer is limited and narrower than the hidden layers; and (4) this assumption is invalid only for activations like $f(x) = a|x|$.

Based on these four assumptions, we can prove the following theorem.

Theorem 7.5 *Neural networks with arbitrary depth and arbitrary piecewise linear activations (excluding linear functions) have infinitely many spurious local minima under arbitrary continuously differentiable loss whose derivative can equal* 0 *only when the prediction and label are the same.*

In practice, most loss functions are continuously differentiable and the derivative can equal 0 only when the prediction and label are the same, such as squared loss and cross-entropy loss (see Appendix 7.2.3.2, Lemmas 7.2 and 7.3).

One can also remove Assumption 7.4, if Assumption 7.3 is replaced by the following assumption, which is mildly more restrictive (see a detailed proof in Sect. 7.2.3.5–7.2.3.6).

Assumption 7.6 The dimensions of the layers satisfy that:

$$d_1 \geq d_Y + 2,$$
$$d_i \geq d_Y + 1, i = 2, \ldots, L - 1.$$

The above result demonstrates that introducing nonlinearities into activations substantially reshapes the loss surface: they bring infinitely many spurious local minima into the loss surface. This result highlights the substantial difference from linear neural networks that all local minima of linear neural networks are equally good, and therefore, they are all global minima (Kawaguchi 2016; Baldi and Hornik 1989; Lu and Kawaguchi 2017; Freeman and Bruna 2017; Zhou and Liang 2018; Laurent and von Brecht 2018; Yun et al. 2018).

Some works have noticed the existence of spurious local minima on the loss surfaces of nonlinear neural networks, which however has a limited applicable domain (Choromanska et al. 2015; Swirszcz et al. 2016; Safran and Shamir 2018; Yun et al. 2019). A notable work by Yun et al. (2019) proves that one-hidden-layer neural networks with two-piece linear (ReLU-like) activations for one-dimensional regression have infinitely many spurious local minima under squared loss. This work first constructs a series of local minima and then proves that they are spurious. This idea inspires some of the results in Theorem 7.5. However, the result in Theorem 7.5 makes a significant and non-trivial development that extends the conditions to arbitrary depth, piecewise linear activations excluding linear functions, and continuously differentiable loss.

7.2.1.3 Proof Skeleton

This section presents the skeleton of the proof. Theorem 7.5 is proved in three stages. We first prove a simplified version of Theorem 7.5 and then extend the conditions in the last two stages.

Yun et al. (2019) and the method discussed in this section share a common strategy: (a) creating a sequence of local minima using a linear classifier; and (b) showing the existence of a constructed new point with lower empirical risk to demonstrate that the constructed local minima are spurious. However, the specifics of how these local minima are constructed vary significantly due to differences in the loss function used and the dimensions of the output space. These distinctions highlight the nuanced nature of the strategy and its application across different contexts.

We have extended the work of Yun et al. (2019) in three key ways: (1) Generalizing from one hidden layer to arbitrary depth: Our approach aims to demonstrate that neural networks of arbitrary depth exhibit infinite spurious local minima. We introduced a new strategy that employs transformation operations to direct data flow through the same linear parts of the activations, facilitating the construction of spurious local minima. (2) Extending from squared loss to arbitrary differentiable loss: Yun et al. (2019) obtained the analytic derivatives of the loss function to construct and prove the existence of spurious local minima. However, this method cannot be directly applied to arbitrary differentiable loss functions lacking analytic formulations. To establish that a loss surface under an arbitrary differentiable loss has infinite spurious local minima, we developed a new proof technique based on Taylor series expansions and a new separation lemma. (3) Expanding from one-dimensional to arbitrary-dimensional output: Our work investigates neural networks with arbitrary-dimensional output, dealing with the calculus of functions whose domain and codomain are a matrix space and a vector space, respectively. This extension contrasts with the one-dimensional output scenario, which deals solely with the codomain of real number spaces. Exploring higher-dimensional outputs significantly increases the complexity of the proof process, requiring specialized mathematical techniques to demonstrate the presence of infinite spurious local minima.

Stage (1): Neural networks with one hidden layer and two-piece linear activations.

We first prove that nonlinear neural networks with one hidden layer and two-piece linear activation functions (ReLU-like activations) have spurious local minima. The proof in this stage further follows a two-step strategy:

(a) We first construct local minima of the empirical risk $\hat{R}_S$ (see Appendix 7.2.3.3, Lemma 7.4). These local minimizers are constructed based on a linear neural network which has the same network size (dimension of weight matrices) and evaluated under the same loss. The design of the hidden layer guarantees that the components of the output $\tilde{Y}^{(1)}$ in the hidden layer before the activation are all positive. The activation is thus effectively reduced to a linear function. Therefore, the local geometry around the local minima with respect to the weights W is similar to those of linear neural networks. Further, the design of the output layer guarantees that its output $\hat{Y}$ is the same as the linear neural network. This construction helps to utilize the results of linear neural networks to solve the problems in nonlinear neural networks.

(b) We then prove that all the constructed local minima in Step (a) are spurious (see Appendix 7.2.3.3, Theorem 7.9). Specifically, we assumed by Assumption 7.1 that the dataset cannot be fit by a linear model. Therefore, the gradient $\nabla_{\hat{Y}} \hat{R}_S$ of the empirical risk $\hat{R}_S$ with respect to the prediction $\hat{Y}$ is not zero. Suppose the i-th row of the gradient $\nabla_{\hat{Y}} \hat{R}_S$ is not zero. Then, we use Taylor series and a preparation lemma (see Appendix 7.2.3.6, Lemma 7.7) to construct another point in the parameter space that has smaller empirical risk. Therefore, we prove that the constructed local minima are spurious. Furthermore, the constructions involve some parameters that are randomly picked from a continuous interval. Thus, we constructed infinitely many spurious local minima.

Stage (2) - Neural networks with arbitrary hidden layers and two-piece linear activations.

We extend the condition in Stage (1) to any neural network with arbitrary depth and two-piece linear activations. The proof in this stage follows the same two-step strategy but has different implementations:

(a) We first construct a series of local minima of the empirical risk $\hat{R}_S$ (see Appendix 7.2.3.4, Lemma 7.5). The construction guarantees that every component of the output $\tilde{Y}^{(i)}$ in each layer before the activations is positive, which secure all the input examples flow through the same part of the activations. Thereby, the nonlinear activations are reduced to linear functions. Also, our construction guarantees that the output $\hat{Y}$ of the network is the same as a linear network with the same weight matrix dimensions.

(b) We then prove that the constructed local minima are spurious (see Appendix 7.2.3.4, Theorem 7.10). The idea is to find a point in the parameter space that has the same empirical risk $\hat{R}_S$ with the constructed point in Stage (1), Step (b).

Stage (3) - Neural networks with arbitrary hidden layer and piecewise linear activations.

We further extend the conditions in Stage (2) to any neural network with arbitrary depth and arbitrary piecewise linear activations. We continue to adapt the two-step strategy in this stage:

(a) We first construct a local minimizer of the empirical risk $\hat{R}_S$ based on the results in Stages (1) and (2) (see Appendix 7.2.3.5, Lemma 7.6). This construction is based on Stage (2), Step (a). The difference of the construction in this stage is that every linear part in the activations can be of a finite interval. The constructed weight matrices apply several uniform scaling and translation operations on the outputs of hidden layers in order to guarantee that all the input training sample points flow through the same linear parts of the activations. We thereby effectively reduce the nonlinear activations to linear functions. Also, our construction guarantees that the output $\hat{Y}$ of the neural network equals to that of the corresponding linear neural network.

(b) We then prove that the constructed local minima are spurious (see Appendix 7.2.3.5). We use the same strategy in Stage (2), Step (b). Some adaptations are implemented for the new conditions.

7.2.2 A Big Picture of the Loss Surface

This section draws a big picture for the loss surfaces of neural networks. Based on a recent result by Soudry and Hoffer (2018), we present four profound properties of the loss surface that collectively characterize how the nonlinearities in activations shape the loss surface.

7.2.2.1 Preliminaries

The discussions in this section use the following concepts.

Definition 7.2 (*Open ball and open set*) The open ball in $\mathcal{H}$ centered at $x \in \mathcal{H}$ and of radius $r > 0$ is defined by $B_r(h) = \{x : \|x - h\| < r\}$. A subset $A \subset \mathcal{H}$ of a space $\mathcal{H}$ is called a open set, if for every point $h \in A$, there exists a positive real $r > 0$, such that the open ball $B_r(h)$ with center h and radius r is in the subset A: $B_r(h) \subset A$.

Definition 7.3 (*Interior point and interior*) For a subset $A \subset \mathcal{H}$ of a space $\mathcal{H}$, a point $h \in A$ is called an interior point of A, if there exists a positive real $r > 0$, such that the open ball $B_r(h)$ with center h and radius r is in the subset A: $B_r(h) \subset A$. The set of all the interior points of the set A is called the interior of the set A.

Definition 7.4 (*Limit point, closure, and boundary*) For a subset $A \subset \mathcal{H}$ of a space $\mathcal{H}$, a point $h \in A$ is called a limit point, if for every $r > 0$, the open ball $B_r(h)$ with center h and radius r contains some point of A: $B_r(h) \cap A \neq \emptyset$. The closure $\bar{A}$ of the set A consists of the union of the set A and all its limit points. The boundary ∂A is defined as the set of points which are in the closure of set A but not in the interior of set A.

Definition 7.5 (*Multilinear*) A function f: $\mathcal{X}_1 \times \mathcal{X}_2 \to \mathcal{Y}$ is called multilinear if for arbitrary $x_1^1, x_1^2 \in \mathcal{X}$, $x_2^1, x_2^2 \in \mathcal{X}_2$, and constants λ_1, λ_2, μ_1, and μ_2, we have

$$
\begin{aligned}
&f(\lambda_1 x_1^1 + \lambda_2 x_1^2, \mu_1 x_2^1 + \mu_2 x_2^2) \\
&= \lambda_1\mu_1 f(x_1^1, x_2^1) + \lambda_1\mu_2 f(x_1^1, x_2^2) + \lambda_2\mu_1 f(x_1^2, x_2^1) + \lambda_2\mu_2 f(x_1^2, x_2^2).
\end{aligned}
$$

Remark 7.2 The definition of "multilinear" implies that the domain of any multilinear function f is a connective and convex set, such as the smooth and multilinear cells below.

Definition 7.6 (*Equivalence class, and quotient space*) Suppose X is a linear space. $[x] = \{v \in X : v \sim x\}$ is an equivalence class, if there is an equivalent relation $\sim$ on $[x]$, such that for any $a, b, c \in [x]$, we have: (1) reflexivity: $a \sim a$; (2) symmetry: if $a \sim b$, $b \sim a$; and (3) transitivity: if $a \sim b$ and $b \sim c$, $a \sim c$. The quotient space and quotient map are defined to be $X/\sim = \{\{v \in X : v \sim x\} : x \in X\}$ and $x \to [x]$, respectively.

7.2.2.2 Main Results

In this section, the loss surface is defined under convex loss with respect to the prediction $\hat{Y}$ of the neural network. Convex loss covers many popular loss functions in practice, such as the squared loss for the regression tasks and many others based on norms. The triangle inequality of the norms secures the convexity of the corresponding loss functions. The convexity of the squared loss is checked in the appendix (see Appendix 7.3, Lemma 7.8).

We now present four propositions to express the loss surfaces of nonlinear neural networks. These propositions give four major properties of the loss surface that collectively draw a big picture for the loss surface.

We first recall a lemma by Soudry and Hoffer (2018). It proves that the loss surfaces of neural networks have smooth and multilinear partitions.

Lemma 7.1 (Smooth and multilinear partition; cf. Soudry and Hoffer (2018)) *The loss surfaces of neural networks of arbitrary depth with piecewise linear functions excluding linear functions are partitioned into multiple smooth and multilinear open cells, while the boundaries are nondifferentiable.*

Based on the smooth and multilinear partition, we prove four propositions as follows.

Theorem 7.7 (Analogous convexity) *For one-hidden-layer neural networks with two-piece linear activation for regression under convex loss, within every cell, all local minima are equally good, and also, they are all global minima in the cell.*

Theorem 7.8 (Equivalence classes of local minimum valleys) *Suppose all conditions of Theorem 7.7 hold. Assume the loss function is strictly convex. Then, all local minima in a cell are concentrated as a local minimum valley: they are connected with each other by a continuous path and have the same empirical risk. Additionally, all local minima in a cell constitute an equivalence class.*

Corollary 7.1 (Quotient space of local minimum valleys) *Suppose all conditions of Theorem 7.8 hold. There might exist some "parallel" local minimum valleys in the equivalence class of a local minimum valley. They do not appear because of the constraints from the cell boundaries. If we ignore such constraints, all equivalence classes of local minima valleys constitute a quotient space.*

Corollary 7.2 (Linear collapse) *The partitioned loss surface collapses to one single smooth and multilinear cell, when all activations are linear.*

7.2.2.3 Discussions and Proof Techniques

The four propositions collectively characterize how the nonlinearities in activations shape the loss surfaces of neural networks. This section discusses the results and the structure of the proofs. A detailed proof is omitted here and given in Appendix 7.3.

Smooth and multilinear partition. Intuitively, the nonlinearities in the piecewise linear activation functions partition the surface into multiple smooth and multilinear cells. Zhou and Liang (2018), Soudry and Hoffer (2018) highlight the partition of the loss surface. We restate it here to make the picture self-contained. A similar but also markedly different notions recently proposed by Hanin and Rolnick (2019) demonstrate that the input data space is partitioned into multiple linear regions, while our work focuses on the partition in the parameter space.

Every local minimum is globally minimal within a cell. In convex optimization, convexity guarantees that all the local minima are global minima. This theorem proves that the local minima within a cell are equally good, and also, they are all global minima in the cell. This result is not surprising provided the excellent training performance of deep learning algorithms. However, the proof is technically non-trivial.

Soudry and Hoffer (2018) proved that the local minima in a cell are the same. However, some points near the boundary that have a smaller empirical risk and are not locally minimal may exist. Unfortunately, the proof by Soudry and Hoffer (2018) cannot exclude this possibility. By contrast, our proof completely solves this problem. Furthermore, our proof holds for any convex loss, including squared loss and cross-entropy loss, while the proof by Soudry and Hoffer (2018) only holds for squared loss.

Developing our proof presented notable challenges due to the limitations of applying techniques used for linear networks to nonlinear contexts. Linear networks, characterized by sequential weight matrix products, possess straightforward geometrical properties where each linear activation function scales the output by a constant. In contrast, the loss surface in a cell of a nonlinear neural network lacks this simplicity owing to the intricate behaviors induced by nonlinear activations. Our approach involves addressing these complexities within the proof framework.

We first prove that the empirical risk $\hat{R}_S$ is a convex function within every cell with respect to a variable $\hat{W}$ which is calculated from the weights W. Therefore, all local minima of the empirical risk $\hat{R}_S$ with respect to $\hat{W}$ are also globally optimal in the cell. Every cell corresponds to a specific series of linear parts of the activations. Therefore, in any fixed cell, the activation h_{s_-,s_+} can be expressed by the slopes of the corresponding linear parts as the following equations,

$$\hat{R}_S(W_1, W_2) = \frac{1}{m}\sum_{i-1}^{m} l\left(y_i, W_2 h(W_1 x_i)\right) = \frac{1}{m}\sum_{i=1}^{m} l\left(y_i, W_2 \mathrm{diag}\left(A_{\cdot,i}\right) W_1 x_i\right), \tag{7.5}$$

where $A_{\cdot,i}$ is the i-th column of matrix

$$A = \begin{bmatrix} h'_{s_-,s_+}((W_1)_{1,\cdot}x_1) & \cdots & h'_{s_-,s_+}((W_1)_{1,\cdot}x_m) \\ \vdots & \ddots & \vdots \\ h'_{s_-,s_+}((W_1)_{d_1,\cdot}x_1) & \cdots & h'_{s_-,s_+}((W_1)_{d_1,\cdot}x_m) \end{bmatrix}.$$

Matrix A is constituted by collecting the slopes of the activation h at every point $(W_1)_{i,\cdot}x_j$.

Different elements of the matrix A can be multiplied either one of $\{s_-, s_+\}$. Therefore, we cannot use a single constant to express the effect of this activation, and thus, even within the cell, a nonlinear network cannot be expressed as the product of a sequence of weight matrices. This difference ensures that the proofs of deep linear neural networks cannot be transplanted here.

Then, we prove that (see Sect. 7.3.3).

$$W_2\mathrm{diag}\left(A_{\cdot,i}\right) W_1 x_i = A_{\cdot,i}^T \mathrm{diag}(W_2) W_1 x_i. \tag{7.6}$$

Applying Eq. (7.6) to Eq. (7.5), the empirical risk $\hat{R}_S$ equals to a formulation similar to the linear neural networks,

$$\hat{R}_S = \frac{1}{m}\sum_{i=1}^{m} l\left(y_i - A_{\cdot,i}^T \mathrm{diag}(W_2) W_1 x_i\right). \tag{7.7}$$

Afterwards, define $\hat{W}_1 = \mathrm{diag}(W_2)W_1$ and then straighten the matrix $\hat{W}_1$ to a vector $\hat{W}$,

$$\hat{W} = \left((\hat{W}_1)_{1,\cdot} \cdots (\hat{W}_1)_{d_1,\cdot}\right),$$

Define $Q : (W_1, W_2) \mapsto \hat{W}$, and also define,

$$\hat{X} = \left(A_{\cdot,1} \otimes x_1 \cdots A_{\cdot,m} \otimes x_m\right).$$

We can prove the following equations (see Sect. 7.3.3),

$$\left(A_{\cdot,1}^T \hat{W}_1 x_1 \cdots A_{\cdot,m}^T \hat{W}_1 x_m\right) = \hat{W}\hat{X}.$$

Applying Eq. (7.7), the empirical risk is transferred to a convex function as follows,

$$\hat{R}_S = \frac{1}{m}\sum_{i=1}^{m} l\left(y_i, \left(A_{\cdot,i}\right)^T \hat{W}_1 x_i\right) = \frac{1}{m}\sum_{i=1}^{m} l(y_i, \hat{W}\hat{X}_i).$$

We then prove that the local optimality of the empirical risk $\hat{R}_S$ is maintained when the weights W are mapped to the variable $\hat{W}$. Specifically, the local minima of the empirical risk $\hat{R}_S$ with respect to the weight W are also the local minima with respect to the variable $\hat{W}$. The maintenance of optimality is not surprising but the proof is technically non-trivial (see a detailed proof in Sect. 7.3.3).

Equivalence classes and quotient space of local minimum valleys. The constructed mapping Q is a quotient map. Under the setting in the previous property, all local minima in a cell is an equivalence class; they are concentrated as a local minimum valley. However, there might exist some "parallel" local minimum valley in the equivalence class, which do not appear because of the constraints from the cell boundaries. Further for neural networks of arbitrary depth, we also constructed a local minimum valley (the spurious local minima constructed in Sect. 7.2.1). This result explains the property of *mode connectivity* that the minima found by gradient-based methods are connected by a path in the parameter space with almost constant empirical risk, which is proposed in two empirical works (Garipov et al. 2018; Draxler et al. 2018). A recent theoretical work (Kuditipudi et al. 2019) proves that dropout stability and noise stability guarantee the mode connectivity.

Linear collapse. Our theories also cover the case of linear neural networks. Linear neural networks do not have any nonlinearity in their activations. Correspondingly, the loss surface does not have any non-differentiable boundaries. In our theories, when there is no nonlinearity in the activations, the partitioned loss surface collapses to a single smooth, multilinear cell. All local minima wherein are equally good, and also, they are all global minima as follows. This result unites the existing results on linear neural networks (Kawaguchi 2016; Baldi and Hornik 1989; Lu and Kawaguchi 2017; Freeman and Bruna 2017; Zhou and Liang 2018; Laurent and von Brecht 2018; Yun et al. 2018).

7.2.3 Proofs of Sect. 7.2

7.2.3.1 Proof of Theorem 7.5

This section gives a detailed proof of Theorem 7.5. It follows the skeleton presented in Sect. 7.2.1.3.

7.2.3.2 Squared Loss and Cross-Entropy Loss

We first check whether squared loss and cross-entropy loss are covered by the requirements of Theorem 7.5.

Lemma 7.2 *The squared loss is continuously differentiable with respect to the prediction of the model, whose gradient of loss equal to zero when the prediction and the label are different.*

Proof Apparently, the squared loss is differentiable with respect to $\hat{Y}$. Specifically, the gradient with respect to $\hat{Y}$ is as follows,

$$\nabla_{\hat{Y}} \left\| Y - \hat{Y} \right\|^2 = 2\left(Y - \hat{Y}\right),$$

which is continuous with respect to $\hat{Y}$.

Also, when the prediction $\hat{Y}$ does not equals to the label Y, we have

$$\nabla_{\hat{Y}} \left\| Y - \hat{Y} \right\|^2 \neq 0.$$

The proof is completed. □

Lemma 7.3 *The cross-entropy loss is continuously differentiable with respect to the prediction of the model, whose gradient of loss equal to zero when the prediction and the label are different. Also, we assume that the ground-truth label is a one-hot vector.*

Proof For any $i \in [1:m]$, the cross-entropy loss is differentiable with respect to $\hat{Y}_i$. The j-th component of the gradient with respect to the prediction $\hat{Y}_i$ is as follows,

$$\frac{\partial\left(-\sum_{k=1}^{d_Y} Y_{k,i}\log\left(\frac{e^{\hat{Y}_{k,i}}}{\sum_{k=1}^{d_Y} e^{\hat{Y}_{k,i}}}\right)\right)}{\partial \hat{Y}_{j,i}} = \left(\sum_{k=1}^{d_Y} Y_{k,i}\right)\frac{e^{\hat{Y}_{j,i}}}{\sum_{k=1}^{d_Y} e^{\hat{Y}_{k,i}}} - Y_{j,i}. \tag{7.8}$$

which is continuous with respect to $\hat{Y}_i$. So, the cross-entropy loss is continuously differentiable with respect to $\hat{Y}_i$.

Additionally, if the gradient (Eq. (7.8)) is zero, we have the following equations,

$$\left(\sum_{k=1}^{d_Y} Y_{k,i}\right)e^{\hat{Y}_{j,i}} - Y_{j,i}\sum_{k=1}^{d_Y} e^{\hat{Y}_{k,i}} = 0,\ j = 1, 2, \ldots, m.$$

Rewrite it into the matrix form, we have

$$\begin{bmatrix} \sum_{k=1}^{d_Y} Y_{k,i} - Y_{1,i} & -Y_{1,i} & \cdots & -Y_{1,i} \\ -Y_{2,i} & \sum_{k=1}^{d_Y} Y_{k,i} - Y_{2,i} & \cdots & -Y_{2,i} \\ \vdots & \vdots & \ddots & \vdots \\ -Y_{d_Y,i} & \cdots & -Y_{d_Y,i} & \sum_{i=1}^{d_Y} Y_{k,i} - Y_{d_Y,i} \end{bmatrix} \begin{bmatrix} e^{\hat{Y}_{1,i}} \\ e^{\hat{Y}_{2,i}} \\ \vdots \\ e^{\hat{Y}_{d_Y,i}} \end{bmatrix} = 0.$$

Since $\sum_{k=1}^{d_Y} Y_{k,i} = 1$, we can easily check the rank of the left matrix is $d_Y - 1$. So the dimension of the solution space is one. Meanwhile, we have

$$\begin{bmatrix} \sum_{k=1}^{d_Y} Y_{k,i} - Y_{1,i} & -Y_{1,i} & \cdots & -Y_{1,i} \\ -Y_{2,i} & \sum_{k=1}^{d_Y} Y_{k,i} - Y_{2,i} & \cdots & -Y_{2,i} \\ \vdots & \vdots & \ddots & \vdots \\ -Y_{d_Y,i} & \cdots & -Y_{d_Y,i} & \sum_{i=1}^{d_Y} Y_{k,i} - Y_{d_Y,i} \end{bmatrix} \begin{bmatrix} Y_{1,i} \\ Y_{2,i} \\ \vdots \\ Y_{d_Y,i} \end{bmatrix} = 0.$$

Therefore, $0 \neq e^{\hat{Y}_{k,i}} = \lambda Y_{k,i}$, for some $\lambda \in \mathbb{R}$, which contradicts to the assumption that some of the components of Y is 0 ($Y_{i,\cdot}$ is a one-hot vector).

The proof is completed. □

7.2.3.3 Stage (1)

In Stage (1), we prove that deep neural networks with one hidden layer, two-piece linear activation h_{s_-,s_+}, and multi-dimensional outputs have infinite spurious local minima.

This stage is organized as follows: (a) we construct a local minimizer by Lemma 7.4; and (b) we prove that the local minimizer is spurious in Theorem 7.9 by constructing a set of parameters with smaller empirical risk.

Without loss of generality, we assume that $s_+ \neq 0$. Otherwise, suppose that $s_+ = 0$. From the definition of ReLU-like activation (Eq. (7.4)), we have $s_- \neq 0$. Since

$$h_{s_-,s_+}(x) = h_{-s_+,-s_-}(-x),$$

the output of the neural network with parameters $\left\{[W_i]_{i=1}^L, [b_i]_{i=1}^L\right\}$ and activation h_{s_-,s_+} equals to that of the neural network with parameters $\left\{\left[W_i'\right]_{i=1}^L, \left[b_i'\right]_{i=1}^L\right\}$ and activation $h_{-s_+,-s_-}$ where $W_i' = -W_i, b_i' = -b_i, i = 1, 2, \ldots, L-1$ and $W_L' = W_L$, $b_L' = b_L$. Since $\left\{[W_i]_{i=1}^L, [b_i]_{i=1}^L\right\} \to \left\{\left[W_i'\right]_{i=1}^L, \left[b_i'\right]_{i=1}^L\right\}$ is an one-to-one map, it is equivalent to consider either the two networks, with $h_{-s_+,-s_-}(x)$ has non-zero slope when $x > 0$.

Step (a). Construct local minima of the loss surface.

Lemma 7.4 *Suppose that $\tilde{W}$ is a local minimizer of*

$$f(W) \triangleq \frac{1}{m}\sum_{i=1}^{m} l\left(Y_i, W\begin{bmatrix} x_i \\ 1 \end{bmatrix}\right), \tag{7.9}$$

Under Assumption 7.3, any one-hidden-layer neural network has a local minimum at

$$\hat{W}_1 = \begin{bmatrix} \left[\tilde{W}\right]_{\cdot,[1:d_X]} \\ \mathbf{0}_{(d_1-d_Y)\times d_X} \end{bmatrix}, \hat{b}_1 = \begin{bmatrix} \left[\tilde{W}\right]_{\cdot,d_X+1} - \eta\mathbf{1}_{d_Y} \\ -\eta\mathbf{1}_{d_1-d_Y} \end{bmatrix}, \tag{7.10}$$

and

$$\hat{W}_2 = \begin{bmatrix} \frac{1}{s_+} I_{d_Y} & \mathbf{0}_{d_Y\times(d_1-d_Y)} \end{bmatrix}, \hat{b}_2 = \eta\mathbf{1}_{d_Y}, \tag{7.11}$$

where $\hat{W}_1$ and $\hat{b}_1$ are respectively the weight matrix and the bias of the first layer, $\hat{W}_2$ and $\hat{b}_2$ are respectively the weight matrix and the bias of the second layer, and η is a negative constant with absolute value sufficiently large such that

$$\tilde{W}\tilde{X} - \eta\mathbf{1}_{d_Y}\mathbf{1}_m^T > \mathbf{0}, \tag{7.12}$$

where $>$ is element-wise.

Also, the loss in this lemma is continuously differentiable loss whose gradient does not equals to 0 *when the prediction is not the same as the ground-truth label.*

Proof We show that the empirical risk is higher in the neiborhood of $\left\{\left[\hat{W}_i\right]_{i=1}^2, \left[\hat{b}_i\right]_{i=1}^2\right\}$, in order to prove that $\left\{\left[\hat{W}_i\right]_{i=1}^2, \left[\hat{b}_i\right]_{i=1}^2\right\}$ is a local minimizer.

The output of the first layer before the activation is

$$\tilde{Y}^{(1)} = \hat{W}_1 X + \hat{b}_1 \mathbf{1}_m^T = \begin{bmatrix} \hat{W} X - \eta \mathbf{1}_{d_Y} \mathbf{1}_m^T \\ -\eta \mathbf{1}_{d_1 - d_Y} \mathbf{1}_m^T \end{bmatrix}.$$

Because η is a negative constant with absolute value sufficiently large such that Eq. (7.25)) holds, the output above is positive (element-wise), the output of the neural network with parameters $\{\hat{W}_1, \hat{W}_2, \hat{b}_1, \hat{b}_2\}$ is

$$\begin{aligned}
\hat{Y} &= \hat{W}_2 h_{s_-, s_+}\left(\hat{W}_1 X + \hat{b}_1 \mathbf{1}_m^T\right) + \hat{b}_2 \mathbf{1}_m^T \\
&= s_+ \hat{W}_2 \left(\hat{W}_1 X + \hat{b}_1 \mathbf{1}_m^T\right) + \hat{b}_2 \mathbf{1}_m^T \\
&= s_+ \left[\tfrac{1}{s_+} I_{d_Y} \ \mathbf{0}_{d_Y \times (d_1 - d_Y)}\right] \begin{bmatrix} \hat{W} X - \eta \mathbf{1}_{d_Y} \mathbf{1}_m^T \\ -\eta \mathbf{1}_{d_1 - d_Y} \mathbf{1}_m^T \end{bmatrix} + \eta \mathbf{1}_{d_Y} \mathbf{1}_m^T \\
&= \tilde{W} \tilde{X},
\end{aligned}$$

where $\tilde{X}$ is defined as

$$\tilde{X} = \begin{bmatrix} X \\ \mathbf{1}_m^T \end{bmatrix}. \tag{7.13}$$

Therefore, the empirical risk $\hat{R}_S$ in terms of parameters $\{\hat{W}_1, \hat{W}_2, \hat{b}_1, \hat{b}_2\}$ is

$$\hat{R}_S\left(\hat{W}_1, \hat{W}_2, \hat{b}_1, \hat{b}_2\right) = \frac{1}{m} \sum_{i=1}^m l\left(Y_i, \left(\tilde{W} \tilde{X}\right)_{\cdot,i}\right) = \frac{1}{m} \sum_{i=1}^m l\left(Y_i, \tilde{W} \begin{bmatrix} x_i \\ 1 \end{bmatrix}\right) = f(\tilde{W}).$$

Then, we introduce a sufficiently small disturbance $\left\{[\delta_{Wi}]_{i=1}^2, [\delta_{bi}]_{i=1}^2\right\}$ into the parameters $\left\{\left[\hat{W}_i\right]_{i=1}^2, \left[\hat{b}_i\right]_{i=1}^2\right\}$. When the disturbance is sufficiently small, all components of the output of the first layer remain positive. Therefore, the output after the disturbance is

$$\begin{aligned}
&\hat{Y}\left(\left[\hat{W}_i + \delta_{Wi}\right]_{i=1}^2, \left[\hat{b}_i + \delta_{bi}\right]_{i=1}^2\right) \\
&= \left(\hat{W}_2 + \delta_{W2}\right) h_{s_-, s_+}\left(\left(\hat{W}_1 + \delta_{W1}\right) X + \left(\hat{b}_1 + \delta_{b1}\right) \mathbf{1}_m^T\right) + \left(\hat{b}_2 + \delta_{b2}\right) \mathbf{1}_m^T \\
&\overset{(*)}{=} \left(\hat{W}_2 + \delta_{W2}\right) s_+ \left(\left(\hat{W}_1 + \delta_{W1}\right) X + \left(\hat{b}_1 + \delta_{b1}\right) \mathbf{1}_m^T\right) + \left(\hat{b}_2 + \delta_{b2}\right) \mathbf{1}_m^T
\end{aligned}$$

$$
\begin{aligned}
&= s_+\delta_{W2}\left(\left(\hat{W}_1+\delta_{W1}\right)X+\left(\hat{b}_1+\delta_{b1}\right)\mathbf{1}_m^T\right)+s_+\hat{W}_2\delta_{W1}X+s_+\hat{W}_2\delta_{b1}\mathbf{1}_m^T+\delta_{b2}\mathbf{1}_m^T \\
&\quad+\hat{W}_2 s_+(\hat{W}_1X+\hat{b}_1\mathbf{1}_m^T)+\hat{b}_2\mathbf{1}_m^T \\
&= \left(s_+\delta_{W2}\left(\hat{W}_1+\delta_{W1}\right)+s_+\hat{W}_2\delta_{W1}\right)X+\left(s_+\hat{W}_2\delta_{b1}+\delta_{b2}+s_+\delta_{W2}\left(\hat{b}_1+\delta_{b1}\right)\right)\mathbf{1}_m^T \\
&\quad+\hat{W}_2 h_{s_-,s_+}(\hat{W}_1X+\hat{b}_1\mathbf{1}_m^T)+\hat{b}_2\mathbf{1}_m^T \\
&= (\tilde{W}+\delta)\begin{bmatrix}X\\ \mathbf{1}_m^T\end{bmatrix},
\end{aligned}
$$

where Eq. (∗) is because all components of $\left(\hat{W}_1+\delta_{W1}\right)X+\left(b_1'+\delta_{b1}\right)\mathbf{1}_m^T$ are positive, and δ is defined as the following matrix

$$
\delta=\left[s_+\left(\hat{W}_2\delta_{W1}+\delta_{W2}\hat{W}_1+\delta_{W2}\delta_{W1}\right)\ s_+\hat{W}_2\delta_{b1}+\delta_{b2}+s_+\delta_{W2}\left(\hat{b}_1+\delta_{b1}\right)\right].
$$

Therefore, the empirical risk $\hat{R}_S$ with respect to $\left\{\left[\hat{W}_i+\delta_{Wi}\right]_{i=1}^2,\left[\hat{b}_i+\delta_{bi}\right]_{i=1}^2\right\}$ is

$$
\begin{aligned}
&\hat{R}_S\left(\left[\hat{W}_i+\delta_{Wi}\right]_{i=1}^2,\left[\hat{b}_i+\delta_{bi}\right]_{i=1}^2\right) \\
&=\frac{1}{m}\sum_{i=1}^m l\left(Y_i,\left(\left(\tilde{W}+\delta\right)\tilde{X}\right)_{\cdot,i}\right) \\
&=\frac{1}{m}\sum_{i=1}^m l\left(Y_i,\left(\tilde{W}+\delta\right)\begin{bmatrix}x_i\\1\end{bmatrix}\right) \\
&=f(\tilde{W}+\delta).
\end{aligned}
$$

δ approaches zero when the disturbances $\{\delta_{W1},\delta_{W2},\delta_{b1},\delta_{b2}\}$ approach zero (element-wise). Since $\hat{W}$ is the local minimizer of $f(W)$, we have

$$
\hat{R}_S\left(\left[\hat{W}_i\right]_{i=1}^2,\left[\hat{b}_i\right]_{i=1}^2\right)=f(\hat{W})\le f(\hat{W}+\delta)=\hat{R}_S\left(\left[\hat{W}_i+\delta_{Wi}\right]_{i=1}^2,\left[\hat{b}_i+\delta_{bi}\right]_{i=1}^2\right). \tag{7.14}
$$

Because the disturbances $\{\delta_{W1},\delta_{W2},\delta_{b1},\delta_{b2}\}$ are arbitrary, Eq. (7.14) demonstrates that $\left\{\left[\hat{W}_i\right]_{i=1}^2,\left[\hat{b}_i\right]_{i=1}^2\right\}$ is a local minimizer.

The proof is completed. □

Step (b). Prove the constructed local minima are spurious.

Theorem 7.9 *Under the same conditions of Lemma 7.4 and Assumptions 7.1, 7.2, and 7.4, the constructed spurious local minima in Lemma 7.4 are spurious.*

Proof The minimizer $\tilde{W}$ is the solution of the following equation

$$\nabla_W f(W) = 0.$$

Specifically, we have

$$\frac{\partial f\left(\tilde{W}\right)}{\partial W_{i,j}} = 0, i \in \{1, \ldots, d_Y\}, j \in \{1, \ldots, d_X\},$$

Applying the definition of $f(W)$ (Eq. (7.9)),

$$\frac{\partial f\left(\tilde{W}\right)}{\partial W_{k,j}} = \sum_{i=1}^{m} \nabla_{\hat{Y}_i} l\left(Y_i, \tilde{W}\begin{bmatrix} x_i \\ 1 \end{bmatrix}\right) E_{k,j} \begin{bmatrix} x_i \\ 1 \end{bmatrix} = \sum_{i=1}^{m} \left(\nabla_{\hat{Y}_i} l\left(Y_i, \tilde{W}\begin{bmatrix} x_i \\ 1 \end{bmatrix}\right)\right)_k \left(\begin{bmatrix} x_i \\ 1 \end{bmatrix}\right)_j,$$

where $\hat{Y}_i = \tilde{W}\begin{bmatrix} x_i \\ 1 \end{bmatrix}$, $\nabla_{\hat{Y}_i} l\left(Y_i, \tilde{W}\begin{bmatrix} x_i \\ 1 \end{bmatrix}\right) \in \mathbb{R}^{1 \times d_Y}$. Since k, j are arbitrary in $\{1, \ldots, d_Y\}$ and $\{1, \ldots, d_X\}$, respectively, we have

$$\boldsymbol{V}\left[X^T \; \mathbf{1}_m\right] = \mathbf{0}, \tag{7.15}$$

where

$$\mathbf{V} = \left[\left(\nabla_{\hat{Y}_1} l\left(Y_1, \tilde{W}\begin{bmatrix} x_1 \\ 1 \end{bmatrix}\right)\right)^T \cdots \left(\nabla_{\hat{Y}_m} l\left(Y_m, \tilde{W}\begin{bmatrix} x_n \\ 1 \end{bmatrix}\right)\right)^T\right].$$

We then define $\tilde{Y} = \tilde{W}\tilde{X}$. Applying Assumption 7.1, we have

$$\tilde{Y} - Y = \left(\tilde{W}\tilde{X} - Y\right) \neq \mathbf{0},$$

Thus, there exists some k-th row of $\tilde{Y} - Y$ that does not equal to 0.

We can rearrange the rows of $\tilde{W}$ and Y simultaneously, while $\tilde{W}$ is maintained as the local minimizer of $f(W)$ and $f(\tilde{W})$ invariant.[1] Without loss of generality, we assume $k = 1$ (k is the index of the row). Set $\boldsymbol{u} = \boldsymbol{V}_{1,\cdot}$ and $v_i = \tilde{Y}_{1,i}$ in Lemma 7.7. There exists a non-empty separation $I = [1 : l']$ and $J = [l' + 1 : m]$ of $S = \{1, 2, \ldots, m\}$ and a vector $\beta \in \mathbb{R}^{d_X}$, such that

(1.1) for any positive constant α small enough, and $i \in I$, $j \in J$, $\tilde{Y}_{1,i} - \alpha\beta^T x_i < \tilde{Y}_{1,j} - \alpha\beta^T x_j$;

(1.2) $\sum_{i \in I} \boldsymbol{V}_{1,i} \neq 0$.

[1] f is also the function in term of Y.

Define

$$\eta_1 = \tilde{Y}_{1,l'} - \alpha\beta^T x_{l'} + \frac{1}{2}\left(\min_{i\in\{l'+1,\ldots,m\}}\left(\tilde{Y}_{1,i} - \alpha\beta^T x_i\right) - \left(\tilde{Y}_{1,l'} - \alpha\beta^T x_{l'}\right)\right).$$

Applying (1.1), for any $i \in I$

$$\begin{aligned}&\tilde{Y}_{1,i} - \alpha\beta^T x_i - \eta_1\\ =&\left(\tilde{Y}_{1,i} - \alpha\beta^T x_i - \tilde{Y}_{1,l'} + \alpha\beta^T x_{l'}\right) - \frac{1}{2}\left(\min_{i\in\{l'+1,\ldots,m\}}\left(\tilde{Y}_{1,i} - \alpha\beta^T x_i\right) - \left(\tilde{Y}_{1,l'} - \alpha\beta^T x_{l'}\right)\right)\\ <&0,\end{aligned}$$

while for any $j \in J$,

$$\begin{aligned}&\tilde{Y}_{1,j} - \alpha\beta^T x_j - \eta_1 > 0\\ =&\left(\tilde{Y}_{1,j} - \alpha\beta^T x_i - \tilde{Y}_{1,l'} + \alpha\beta^T x_{l'}\right) - \frac{1}{2}\left(\min_{i\in\{l'+1,\ldots,m\}}\left(\tilde{Y}_{1,i} - \alpha\beta^T x_i\right) - \left(\tilde{Y}_{1,l'} - \alpha\beta^T x_{l'}\right)\right)\\ \geq&\frac{1}{2}\left(\min_{i\in\{l'+1,\ldots,m\}}\left(\tilde{Y}_{1,i} - \alpha\beta^T x_i\right) - \left(\tilde{Y}_{1,l'} - \alpha\beta^T x_{l'}\right)\right) > 0.\end{aligned}$$

Define $\gamma \in \mathbb{R}$ which satisfies

$$|\gamma| = \begin{cases}\frac{1}{2}\min_{i\in\{l'+1,\ldots,s_{t+1}\}} \alpha\beta^T(x_l - x_i), & l' < s_{t+1}\\ \alpha, & l' = s_{t+1}\end{cases},$$

where s_{t+1} is defined in Lemma 7.7.

We argue that

$$\left|\frac{1}{2}\left(\min_{i\in\{l'+1,\ldots,m\}}\left(\tilde{Y}_{1,i} - \alpha\beta^T x_i\right) - \left(\tilde{Y}_{1,l'} - \alpha\beta^T x_{l'}\right)\right)\right| - |\gamma| > 0. \tag{7.16}$$

When $l' = s_{t+1}$, Eq. (7.58) stands. Also,

$$\lim_{\alpha\to 0^+} \gamma = 0,$$

$$\lim_{\alpha\to 0^+}\left(\min_{i\in\{l'+1,\ldots,m\}}\left(\tilde{Y}_{1,i} - \alpha\beta^T x_i\right) - \left(\tilde{Y}_{1,l'} - \alpha\beta^T x_{l'}\right)\right) = \min_{i\in\{l'+1,\ldots,m\}} \tilde{Y}_{1,i} - \tilde{Y}_{1,l'} > 0.$$

Therefore, we get Eq. (7.16) when α is small enough.

When $l' < s_{t+1}$, Eq. (7.57) stands. Therefore,

$$|\gamma| = \frac{1}{2}\left|\frac{1}{2}\left(\min_{i\in\{l'+1,\ldots,m\}}\left(\tilde{Y}_{1,i} - \alpha\beta^T x_i\right) - \left(\tilde{Y}_{1,l'} - \alpha\beta^T x_{l'}\right)\right)\right|,$$

which apparently leads to Eq. (7.16).

Therefore, for any $i \in I$, we have that

$$
\begin{aligned}
&\tilde{Y}_{1,i} - \alpha\beta^T x_i - \eta_1 + |\gamma| \\
\leq &-\frac{1}{2}\left(\min_{i\in\{l'+1,\ldots,m\}}\left(\tilde{Y}_{1,i} - \alpha\beta^T x_i\right) - \left(\tilde{Y}_{1,l'} - \alpha\beta^T x_{l'}\right)\right) + |\gamma| < 0,
\end{aligned}
$$

while for any $j \in J$,

$$
\begin{aligned}
&\tilde{Y}_{1,j} - \alpha\beta^T x_j - \eta_1 - |\gamma| \\
\geq &\frac{1}{2}\left(\min_{i\in\{l'+1,\ldots,m\}}\left(\tilde{Y}_{1,i} - \alpha\beta^T x_i\right) - \left(\tilde{Y}_{1,l'} - \alpha\beta^T x_{l'}\right)\right) - |\gamma| > 0.
\end{aligned}
$$

Furthermore, define η_i $(2 \leq i \leq d_Y)$ as negative reals with absolute value sufficiently large, such that for any $i \in [2 : d_Y]$ and any $j \in [1 : n]$,

$$
\tilde{Y}_{i,j} - \eta_i > 0.
$$

Now we construct a point in the parameter space whose empirical risk is smaller than the proposed local minimum in Lemma 7.4 as follows

$$
\tilde{W}_1 = \left(\tilde{W}^T_{1,[1:d_X]} - \beta\alpha^T, -\tilde{W}^T_{1,[1:d_X]} + \beta\alpha^T, \tilde{W}^T_{2,[1:d_X]}, \cdots \tilde{W}^T_{d_Y,[1:d_X]}, 0_{d_X\times(d_1-d_Y-1)}\right)^T, \tag{7.17}
$$

$$
\tilde{b}_1 = \left(\tilde{W}_{1,[d_X+1]} - \eta_1 + \gamma, \ldots, \tilde{W}_{2,[d_X+1]} - \eta_2, \cdots \tilde{W}_{d_Y,[d_X+1]} - \eta_{d_Y}, 0_{1\times(d_1-d_Y-1)}\right)^T, \tag{7.18}
$$

$$
\tilde{W}_2 = \begin{bmatrix}
\frac{1}{s_++s_-} & -\frac{1}{s_++s_-} & 0 & 0 & \cdots & 0 & 0 & \cdots & 0 \\
0 & 0 & \frac{1}{s_+} & \cdots & 0 & \vdots & \vdots & & \vdots \\
\vdots & \vdots & \vdots & \frac{1}{s_+} & \cdots & 0 & & \ddots & \\
\vdots & \vdots & \vdots & \vdots & \ddots & \vdots & \vdots & & \vdots \\
0 & 0 & 0 & 0 & \cdots & \frac{1}{s_+} & 0 & \cdots & 0
\end{bmatrix}, \tag{7.19}
$$

and

$$
\tilde{b}_2 = \left(\eta_1, \eta_2, \cdots, \eta_{d_Y}\right)^T, \tag{7.20}
$$

where $\tilde{W}_i$ and $\tilde{b}_i$ are the weight matrix and the bias of the i-th layer, respectively.

After some calculations, the network output of the first layer before the activation in terms of $\left\{\left[\tilde{W}_i\right]_{i=1}^2, \left[\tilde{b}_i\right]_{i=1}^2\right\}$ is

$$\tilde{Y}^{(1)} = \tilde{W}_1 X + \tilde{b}_1 \mathbf{1}_m^T = \begin{bmatrix} \tilde{W}_{1,\cdot}\tilde{X} - \alpha\beta^T X - \eta_1 \mathbf{1}_m^T + \gamma \mathbf{1}_m^T \\ -\tilde{W}_{1,\cdot}\tilde{X} + \alpha\beta^T X + \eta_1 \mathbf{1}_m^T + \gamma \mathbf{1}_m^T \\ \tilde{W}_{2,\cdot}\tilde{X} - \eta_2 \mathbf{1}_m^T \\ \vdots \\ \tilde{W}_{d_Y,\cdot}\tilde{X} - \eta_{d_Y} \mathbf{1}_m^T \\ \mathbf{0}_{(d_1-d_Y-1)\times m} \end{bmatrix}.$$

Therefore, the output of the whole neural network is

$$\hat{Y} = \tilde{W}_2 h_{s_-,s_+}\left(\begin{bmatrix} \tilde{W}_{1,\cdot}\tilde{X} - \alpha\beta^T X - \eta_1 \mathbf{1}_m^T + \gamma \mathbf{1}_m^T \\ -\tilde{W}_{1,\cdot}\tilde{X} + \alpha\beta^T X + \eta_1 \mathbf{1}_m^T + \gamma \mathbf{1}_m^T \\ \tilde{W}_{2,\cdot}\tilde{X} - \eta_2 \mathbf{1}_m^T \\ \vdots \\ \tilde{W}_{d_Y,\cdot}\tilde{X} - \eta_{d_Y} \mathbf{1}_m^T \\ \mathbf{0}_{(d_1-d_Y-1)\times m} \end{bmatrix}\right) + \tilde{b}_2 \mathbf{1}_m^T.$$

Specifically, if $j \leq l'$,

$$\begin{aligned} \left(\tilde{Y}^{(1)}\left(\left[\tilde{W}_i\right]_{i=1}^2, \left[\tilde{b}_i\right]_{i=1}^2\right)\right)_{1,j} &= \tilde{W}_{1,\cdot}\begin{bmatrix} x_j \\ 1 \end{bmatrix} - \alpha\beta^T x_j - \eta_1 + \gamma \\ &= \tilde{Y}_{1,j} - \alpha\beta^T x_j - \eta_1 + \gamma < 0, \\ \left(\tilde{Y}^{(1)}\left(\left[\tilde{W}_i\right]_{i=1}^2, \left[\tilde{b}_i\right]_{i=1}^2\right)\right)_{2,j} &= -\tilde{W}_{1,\cdot}\begin{bmatrix} x_j \\ 1 \end{bmatrix} + \alpha\beta^T x_j + \eta_1 + \gamma \\ &= -\tilde{Y}_{1,j} + \alpha\beta^T x_j + \eta_1 + \gamma > 0. \end{aligned}$$

Therefore, $(1, j)$-th component of $\hat{Y}\left(\left[\tilde{W}_i\right]_{i=1}^2, \left[\tilde{b}_i\right]_{i=1}^2\right)$ is

$$\begin{aligned} &\left(\hat{Y}\left(\left[\tilde{W}_i\right]_{i=1}^2, \left[\tilde{b}_i\right]_{i=1}^2\right)\right)_{1,j} \\ &= \left(\tfrac{1}{s_+ + s_-}, -\tfrac{1}{s_+ + s_-}, 0, \ldots, 0\right) h_{s_-,s_+}\left(\begin{bmatrix} \tilde{W}_{1,\cdot}X - \alpha\beta^T X - \eta_1 \mathbf{1}_m^T + \gamma \mathbf{1}_m^T \\ -\tilde{W}_{1,\cdot}X + \alpha\beta^T X + \eta_1 \mathbf{1}_m^T + \gamma \mathbf{1}_m^T \\ \tilde{W}_{2,\cdot}X - \eta_2 \mathbf{1}_m^T \\ \vdots \\ \tilde{W}_{d_Y,\cdot}X - \eta_{d_Y} \mathbf{1}_m^T \\ \mathbf{0}_{d_1-d_Y-1} \mathbf{1}_m^T \end{bmatrix}\right)_{\cdot,j} + \eta_1 \end{aligned}$$

$$
\begin{aligned}
&= \frac{s_-}{s_+ + s_-}(\tilde{Y}_{1,j} - \alpha\beta^T x_j - \eta_1 + \gamma) - \frac{s_+}{s_+ + s_-}(-\tilde{Y}_{1,j} + \alpha\beta^T x_j + \eta_1 + \gamma) + \eta_1 \\
&= \tilde{Y}_{1,j} - \alpha\beta^T x_j + \frac{s_- - s_+}{s_+ + s_-}\gamma; \qquad (7.21)
\end{aligned}
$$

Similarly, when $j > l'$, the $(1, j)$-th component is

$$
\begin{aligned}
&\left(\hat{Y}\left(\left[\tilde{W}_i\right]_{i=1}^2, \left[\tilde{b}_i\right]_{i=1}^2\right)\right)_{1,j} \\
= &\frac{s_+}{s_+ + s_-}(\tilde{Y}_{1,j} - \alpha\beta^T x_j - \eta_1 + \gamma) - \frac{s_-}{s_+ + s_-}(-\tilde{Y}_{1,j} + \alpha\beta^T x_j + \eta_1 + \gamma) + \eta_1 \\
= &\tilde{Y}_{1,j} - \alpha\beta^T x_j + \frac{s_+ - s_-}{s_+ + s_-}\gamma, \qquad (7.22)
\end{aligned}
$$

and

$$
\left(\hat{Y}\left(\left[\tilde{W}_i\right]_{i=1}^2, \left[\tilde{b}_i\right]_{i=1}^2\right)\right)_{i,j} = \frac{s_+}{s_+}(\tilde{Y}_{i,j} - \eta_i) + \eta_i = \tilde{Y}_{i,j}, i \geq 2. \qquad (7.23)
$$

Thus, the empirical risk of the neural network with parameters $\left\{\left[\tilde{W}_i\right]_{i=1}^2, \left[\tilde{b}_i\right]_{i=1}^2\right\}$ is

$$
\begin{aligned}
&\hat{R}_S\left(\left[\tilde{W}_i\right]_{i=1}^2, \left[\tilde{b}_i\right]_{i=1}^2\right) \\
=&\frac{1}{m}\sum_{i=1}^m l\left(Y_i, \tilde{W}_2\left(\tilde{W}_1 x_i + \tilde{b}_1 \mathbf{1}_m^T\right) + \tilde{b}_2 \mathbf{1}_m^T\right) \\
=&\frac{1}{m}\sum_{i=1}^m \left(l\left(Y_i, \tilde{W}\begin{bmatrix} x_i \\ 1 \end{bmatrix}\right) + \nabla_{\hat{Y}_i} l\left(Y_i, \tilde{W}\begin{bmatrix} x_i \\ 1 \end{bmatrix}\right)\left(\tilde{W}_2\left(\tilde{W}_1 x_i + \tilde{b}_1 \mathbf{1}_m^T\right) + \tilde{b}_2 \mathbf{1}_m^T - \tilde{W}\begin{bmatrix} x_i \\ 1 \end{bmatrix}\right)\right) \\
&+ \sum_{i=1}^m o\left(\left\|\tilde{W}_2\left(\tilde{W}_1 x_i + \tilde{b}_1 \mathbf{1}_m^T\right) + \tilde{b}_2 \mathbf{1}_m^T - \tilde{W}\begin{bmatrix} x_i \\ 1 \end{bmatrix}\right\|\right). \qquad (7.24)
\end{aligned}
$$

Applying Eqs. (7.21), (7.22), and (7.23), we have

$$
\begin{aligned}
&\sum_{i=1}^m \left(\tilde{W}_2\left(\tilde{W}_1 x_i + \tilde{b}_1\right) + \tilde{b}_2 - \tilde{W}\begin{bmatrix} x_i \\ 1 \end{bmatrix}\right)^T \nabla_{\hat{Y}_i} l\left(Y_i, \tilde{W}\begin{bmatrix} x_i \\ 1 \end{bmatrix}\right)^T \\
\overset{(*)}{=} &\sum_{i=1}^{l'} \mathbf{V}_{1,i}(-\alpha\beta^T x_i + \frac{s_+ - s_-}{s_+ + s_-}\gamma) + \sum_{i=l'+1}^m \mathbf{V}_{1,i}(-\alpha\beta^T x_i - \frac{s_+ - s_-}{s_+ + s_-}\gamma) \\
= &2\gamma\sum_{i=1}^{l'} \frac{s_+ - s_-}{s_+ + s_-}\mathbf{V}_{1,i},
\end{aligned}
$$

where Eq. $(*)$ is because

$$\left(\tilde{W}_2\left(\tilde{W}_1 x_i + \tilde{b}_1\right) + \tilde{b}_2 - \tilde{W}\begin{bmatrix} x_i \\ 1 \end{bmatrix}\right)_j = \begin{cases} -\alpha\beta^T x_j + \dfrac{s_- - s_+}{s_+ + s_-}\gamma, & j = 1, i \le l' \\ -\alpha\beta^T x_j - \dfrac{s_- - s_+}{s_+ + s_-}\gamma, & j = 1, i > l' \\ 0, & j \ge 2 \end{cases}.$$

Furthermore, note that $\alpha = O(\gamma)$ (from the definition of γ). We have

$$\begin{aligned} &\sum_{i=1}^{m} o\left(\left\|\tilde{W}_2\left(\tilde{W}_1 x_i + \tilde{b}_1\right) + \tilde{b}_2 - \hat{W}\begin{bmatrix} x_i \\ 1 \end{bmatrix}\right\|\right) \\ =&\sum_{i=1}^{m} o\left(\sqrt{\sum_{j=1}^{n}\left(\tilde{W}_2\left(\tilde{W}_1 x_i + \tilde{b}_1\right) + \tilde{b}_2 - \tilde{W}\begin{bmatrix} x_i \\ 1 \end{bmatrix}\right)_j^2}\right) \\ =&\, o(\gamma). \end{aligned}$$

Let α be sufficiently small while $\operatorname{sgn}(\gamma) = -\operatorname{sgn}\left(\sum_{i=1}^{l'} \frac{s_+ - s_-}{s_+ + s_-} \boldsymbol{V}_{1,i}\right)$. We have

$$\begin{aligned} &\sum_{i=1}^{m} l\left(Y_i, \tilde{W}_2\left(\tilde{W}_1 x_i + \tilde{b}_1\right) + \tilde{b}_2\right) - \sum_{i=1}^{m} l\left(Y_i, \hat{W}\begin{bmatrix} x_i \\ 1 \end{bmatrix}\right) \\ =&\, 2\gamma \sum_{i=1}^{l'} \frac{s_+ - s_-}{s_+ + s_-} \boldsymbol{V}_{1,i} + o(\gamma) \overset{(**)}{<} 0, \end{aligned}$$

where inequality $(**)$ comes from (1.2) (see Sect. 7.2.3.3).

From Lemma 7.4, there exists a local minimizer $\left\{\left[\hat{W}_i\right]_{i=1}^2, \left[\hat{b}_i\right]_{i=1}^2\right\}$ with empirical risk that equals to $f(\tilde{W})$. Meanwhile, we just construct a point in the parameter space with empirical risk smaller than $f(\tilde{W})$.

Therefore, $\left\{\left[\hat{W}_i\right]_{i=1}^2, \left[\hat{b}_i\right]_{i=1}^2\right\}$ is a spurious local minimum.

The proof is completed. □

7.2.3.4 Stage (2)

Stage (2) proves that neural networks with arbitrary hidden layers and two-piece linear activation h_{s_-,s_+} have spurious local minima. Here, we still assume $s_+ \neq 0$. We have justified this assumption in Stage (1).

This stage is organized similarly with Stage (1): (a) Lemma 7.5 constructs a local minimum; and (b) Theorem 7.10 proves the minimum is spurious.

Step (a). Construct local minima of the loss surface.

Lemma 7.5 *Suppose that all the conditions of Lemma 7.4 hold, while the neural network has $L-1$ hidden layers. Then, this network has a local minimum at*

$$\hat{W}_1' = \begin{bmatrix} \left[\tilde{W}\right]_{\cdot,[1:d_X]} \\ \mathbf{0}_{(d_1-d_Y)\times d_X} \end{bmatrix}, \hat{b}_1' = \begin{bmatrix} \left[\tilde{W}\right]_{\cdot,d_X+1} - \eta\mathbf{1}_{d_Y} \\ -\eta\mathbf{1}_{d_1-d_Y} \end{bmatrix},$$

$$\hat{W}_i' = \frac{1}{s_+}\sum_{j=1}^{d_Y} E_{j,j} + \frac{1}{s_+}\sum_{j=d_Y+1}^{d_i} E_{j,(d_Y+1)}, \hat{b}_i' = 0 (i = 2, 3, ..., L-1),$$

and

$$\hat{W}_L' = \left[\frac{1}{s_+} I_{d_Y}\ \mathbf{0}_{d_Y\times(d_{L-1}-d_Y)}\right], \hat{b}_L' = \eta\mathbf{1}_{d_Y},$$

where $\hat{W}_i'$ and $\hat{b}_i'$ are the weight matrix and the bias of the i-th layer, respectively, and η is a negative constant with absolute value sufficiently large such that

$$\tilde{W}\tilde{X} - \eta\mathbf{1}_{d_Y}\mathbf{1}_m^T > \mathbf{0}, \tag{7.25}$$

where $>$ is element-wise.

Proof Recall the discussion in Lemma 7.4 that all components of $\hat{W}_1 X + \hat{b}_1\mathbf{1}_m^T$ are positive. Specifically,

$$\hat{W}_1 X + \hat{b}_1\mathbf{1}_m^T = \begin{bmatrix} \tilde{Y} - \eta\mathbf{1}_{d_Y}\mathbf{1}_m^T \\ -\eta\mathbf{1}_{d_1-d_Y}\mathbf{1}_m^T \end{bmatrix},$$

where $\tilde{Y}$ is defined in Lemma 7.4.

Similar to the discussions in Lemma 7.4, when the parameters equal to $\left\{\left[\hat{W}_i'\right]_{i=1}^L, \left[\hat{b}_i'\right]_{i=1}^L\right\}$, the output of the first layer before the activation function is

$$\tilde{Y}^{(1)} = \hat{W}_1' X + \hat{b}_1'\mathbf{1}_m^T = \begin{bmatrix} \tilde{Y} - \eta\mathbf{1}_{d_Y}\mathbf{1}_m^T \\ -\eta\mathbf{1}_{d_1-d_Y}\mathbf{1}_m^T \end{bmatrix},$$

and

$$\tilde{Y} - \eta\mathbf{1}_{d_Y}\mathbf{1}_m^T > \mathbf{0}, \tag{7.26}$$

$$-\eta\mathbf{1}_{d_1-d_Y}\mathbf{1}_m^T > \mathbf{0}. \tag{7.27}$$

Here > is defined element-wise.

After the activation function, the output of the first layer is

$$Y^{(1)} = h_{s_-,s_+}(\hat{W}_1'X + \hat{b}_1'\mathbf{1}_m^T) = s_+(\hat{W}_1'X + \hat{b}_1'\mathbf{1}_m^T) = s_+ \begin{bmatrix} \tilde{Y} - \eta \mathbf{1}_{d_Y}\mathbf{1}_m^T \\ -\eta \mathbf{1}_{d_1 - d_Y}\mathbf{1}_m^T \end{bmatrix}.$$

We prove by induction that for all $i \in [1 : L - 1]$ that

$$\tilde{Y}^{(i)} > \mathbf{0}\text{ , element-wise,} \tag{7.28}$$

$$Y^{(i)} = s_+ \begin{bmatrix} \tilde{Y} - \eta \mathbf{1}_{d_Y}\mathbf{1}_m^T \\ -\eta \mathbf{1}_{d_i - d_Y}\mathbf{1}_m^T \end{bmatrix}. \tag{7.29}$$

Suppose that for $1 \leq k \leq L - 2$, $\tilde{Y}^{(k)}$ is positive (element-wise) and

$$Y^{(k)} = s_+ \begin{bmatrix} \tilde{Y} - \eta \mathbf{1}_{d_Y}\mathbf{1}_m^T \\ -\eta \mathbf{1}_{d_k - d_Y}\mathbf{1}_m^T \end{bmatrix}.$$

Then the output of the $(k + 1)$-th layer before the activation is

$$\begin{aligned} \tilde{Y}^{(k+1)} &= \hat{W}_{k+1}'Y^{(k)} + \hat{b}_{k+1}'\mathbf{1}_m^T \\ &= \frac{1}{s_+}\left(\sum_{j=1}^{d_Y} E_{j,j} + \sum_{j=d_Y+1}^{d_{k+1}} E_{j,(d_Y+1)}\right) s_+ \begin{bmatrix} \tilde{Y} - \eta \mathbf{1}_{d_Y}\mathbf{1}_m^T \\ -\eta \mathbf{1}_{d_k - d_Y}\mathbf{1}_m^T \end{bmatrix} \\ &= \begin{bmatrix} \tilde{Y} - \eta \mathbf{1}_{d_Y}\mathbf{1}_m^T \\ -\eta \mathbf{1}_{d_{k+1} - d_Y}\mathbf{1}_m^T \end{bmatrix}. \end{aligned}$$

Applying Eqs. (7.26) and (7.27), we have

$$\tilde{Y}^{(k+1)} = \begin{bmatrix} \tilde{Y} - \eta \mathbf{1}_{d_Y}\mathbf{1}_m^T \\ -\eta \mathbf{1}_{d_{k+1} - d_Y}\mathbf{1}_m^T \end{bmatrix} > \mathbf{0},$$

where > is defined element-wise. Therefore,

$$Y^{(k+1)} = h_{s_-,s_+}\left(\tilde{Y}^{(k+1)}\right) = s_+\tilde{Y}^{(k+1)} = s_+ \begin{bmatrix} \tilde{Y} - \eta \mathbf{1}_{d_Y}\mathbf{1}_m^T \\ -\eta \mathbf{1}_{d_{k+1} - d_Y}\mathbf{1}_m^T \end{bmatrix}.$$

We thereby prove Eqs. (7.28) and (7.29). Therefore, $Y^{(L)}$ can be calculated as

$$\hat{Y} = Y^{(L)} = \hat{W}'_L Y^{(L-1)} + \hat{b}'_L \mathbf{1}_m^T$$
$$= \frac{1}{s_+}\left[\, I_{d_Y} \;\; \mathbf{0}_{d_Y \times (d_{L-1}-d_Y)} \,\right] s_+ \begin{bmatrix} \tilde{Y} - \eta \mathbf{1}_{d_Y} \mathbf{1}_m^T \\ -\eta \mathbf{1}_{d_i - d_Y} \mathbf{1}_m^T \end{bmatrix} + \eta \mathbf{1}_{d_Y} \mathbf{1}_m^T = \tilde{Y}.$$

Then, we show the empirical risk is higher around $\left\{\left[\hat{W}'_i\right]_{i=1}^L, \left[\hat{b}'_i\right]_{i=1}^L\right\}$ in order to prove that $\left\{\left[\hat{W}'_i\right]_{i=1}^L, \left[\hat{b}'_i\right]_{i=1}^L\right\}$ is a local minimizer.

Let $\left\{\left[\hat{W}'_i + \delta'_{Wi}\right]_{i=1}^L, \left[\hat{b}'_i + \delta'_{bi}\right]_{i=1}^L\right\}$ be point in the parameter space which is close enough to the point $\left\{\left[\hat{W}'_i\right]_{i=1}^L, \left[\hat{b}'_i\right]_{i=1}^L\right\}$. Since the disturbances δ'_{Wi} and δ'_{bi} are both close to 0 (element-wise), all components of $\tilde{Y}^{(i)}\left(\left[\hat{W}'_i + \delta'_{Wi}\right]_{i=1}^L, \left[\hat{b}'_i + \delta'_{bi}\right]_{i=1}^L\right)$ remains positive. Therefore, the output of the neural network in terms of parameters $\left\{\left[\hat{W}'_i + \delta'_{Wi}\right]_{i=1}^L, \left[\hat{b}'_i + \delta'_{bi}\right]_{i=1}^L\right\}$ is

$$\hat{Y}\left(\left[\hat{W}'_i + \delta'_{Wi}\right]_{i=1}^L, \left[\hat{b}'_i + \delta'_{bi}\right]_{i=1}^L\right)$$
$$= (\hat{W}'_L + \delta'_{WL}) s_+ \left(\ldots s_+ \left(\left(\hat{W}'_1 + \delta'_{W1}\right) X + \left(\hat{b}'_1 + \delta'_{b1}\right) 1_m^T\right) \ldots\right) + \left(\hat{b}'_L + \delta'_{bL}\right) \mathbf{1}_m^T$$
$$= M_1 X + M_2 1_m^T,$$

where M_1 and M_2 can be obtained from $\left\{\left[\hat{W}'_i\right]_{i=1}^L, \left[\hat{b}'_i\right]_{i=1}^L\right\}$ and $\left\{\left[\delta'_{Wi}\right]_{i=1}^L, \left[\delta'_{bi}\right]_{i=1}^L\right\}$ through several multiplication and summation operations[2].

Rewrite the output as

$$M_1 X + M_2 \mathbf{1}_m^T = \left[M_1 \;\; M_2\right] \begin{bmatrix} X \\ \mathbf{1}_n^T \end{bmatrix}.$$

Therefore, the empirical risk $\hat{\mathcal{R}}$ before and after the disturbance can be expressed as $f(\tilde{W})$ and $f\left(\left[M_1 \;\; M_2\right]\right)$, respectively.

When the disturbances $\left\{\left[\delta'_{Wi}\right]_{i=1}^L, \left[\delta'_{bi}\right]_{i=1}^L\right\}$ approach 0 (element-wise), $\left[M_1 \;\; M_2\right]$ approaches $\tilde{W}$. Therefore, when $\left\{\left[\delta'_{Wi}\right]_{i=1}^L, \left[\delta'_{bi}\right]_{i=1}^L\right\}$ are all small enough, we have

[2] Since the exact form of M_1 and M_2 are not needed, we omit the exact formulations here.

$$\hat{R}_S\left(\left[\hat{W}_i' + \delta_{Wi}'\right]_{i=1}^L, \left[\hat{b}_i' + \delta_{bi}'\right]_{i=1}^L\right) = f([M_1\ M_2]) \geq f(\tilde{W}) = \hat{R}_S\left(\left[\hat{W}_i'\right]_{i=1}^L, \left[\hat{b}_i'\right]_{i=1}^L\right). \tag{7.30}$$

Since $\left\{\left[\hat{W}_i' + \delta_{Wi}'\right]_{i=1}^L, \left[\hat{b}_i' + \delta_{bi}'\right]_{i=1}^L\right\}$ are arbitrary within a sufficiently small neighbour of $\left\{\left[\hat{W}_i'\right]_{i=1}^L, \left[\hat{b}_i'\right]_{i=1}^L\right\}$, Eq. (7.30) yields that $\left\{\left[\hat{W}_i'\right]_{i=1}^2, \left[\hat{b}_i'\right]_{i=1}^2\right\}$ is a local minimizer. □

Step (b). Prove the constructed local minima are spurious.

Theorem 7.10 *Under the same conditions of Lemma 7.5 and Assumptions 7.1, 7.2, and 7.4, the constructed spurious local minima in Lemma 7.5 are spurious.*

Proof We first construct the weight matrix and bias of the i-th layer as follows,

$$\tilde{W}_1' = \tilde{W}_1, \tilde{b}_1' = \tilde{b}_1,$$
$$\tilde{W}_2' = \begin{bmatrix} \tilde{W}_2 \\ \mathbf{0}_{(d_2-d_Y)\times d_1} \end{bmatrix}, \tilde{b}_2' = \lambda \mathbf{1}_{d_2} + \begin{bmatrix} \tilde{b}_2 \\ \mathbf{0}_{(d_2-d_Y)\times 1} \end{bmatrix},$$
$$\tilde{W}_i' = \frac{1}{s_+}\sum_{i=1}^{d_Y} E_{i,i}, \tilde{b}_i' = 0 (i = 3, 4, ..., L-1),$$

and

$$\tilde{W}_L' = \frac{1}{s_+}\sum_{i=1}^{d_Y} E_{i,i}, \tilde{b}_L' = -\lambda \mathbf{1}_{d_Y},$$

where $\tilde{W}_1$, $\tilde{W}_2$, $\tilde{b}_1$ and $\tilde{b}_2$ are defined by Eqs. (7.17), (7.18), (7.19), and (7.20), respectively, and λ is a sufficiently large positive real such that

$$\hat{Y}\left(\left[\tilde{W}_i\right]_{i=1}^2, \left[\tilde{b}_i\right]_{i=1}^2\right) + \lambda \mathbf{1}_{d_2}\mathbf{1}_m^T > \mathbf{0}, \tag{7.31}$$

where $>$ is defined element-wise.

We argue that $\left\{\left[\tilde{W}_i'\right]_{i=1}^L, \left[\tilde{b}_i'\right]_{i=1}^L\right\}$ corresponds to a smaller empirical risk than $f(\tilde{W})$ which is defined in Lemma 7.4.

First, Theorem 7.9 has proved that the point $\left\{\tilde{W}_1, \tilde{W}_2, \tilde{b}_1, \tilde{b}_2\right\}$ corresponds to a smaller empirical risk than $f(\tilde{W})$.

We prove by induction that for any $i \in \{3, 4, ..., L-1\}$,

$$\tilde{Y}^{(i)}\left(\left[\tilde{W}_i'\right]_{i=1}^L, \left[\tilde{b}_i'\right]_{i=1}^L\right) \geq \mathbf{0}\text{, element-wise,} \tag{7.32}$$

$$Y^{(i)}\left(\left[\tilde{W}_i'\right]_{i=1}^L, \left[\tilde{b}_i'\right]_{i=1}^L\right) = s_+ \begin{bmatrix} \hat{Y}\left(\left[\tilde{W}_i\right]_{i=1}^2, \left[\tilde{b}_i\right]_{i=1}^2\right) + \lambda \mathbf{1}_{d_Y} \mathbf{1}_m^T \\ 0_{(d_i - d_Y)\times m} \end{bmatrix}. \tag{7.33}$$

Apparently the output of the first layer before the activation is

$$\tilde{Y}^{(1)}\left(\left[\tilde{W}_i'\right]_{i=1}^L, \left[\tilde{b}_i'\right]_{i=1}^L\right) = \tilde{W}_1' X + \tilde{b}_1' \mathbf{1}_m^T = \tilde{W}_1 X + \tilde{b}_1 \mathbf{1}_m^T = \tilde{Y}^{(1)}\left(\left[\tilde{W}_i\right]_{i=1}^2, \left[\tilde{b}_i\right]_{i=1}^2\right).$$

Therefore, the output of the first layer after the activation is

$$\begin{aligned} Y^{(1)}\left(\left[\tilde{W}_i'\right]_{i=1}^L, \left[\tilde{b}_i'\right]_{i=1}^L\right) &= h_{s_-,s_+}\left(\tilde{Y}^{(1)}\left(\left[\tilde{W}_i'\right]_{i=1}^L, \left[\tilde{b}_i'\right]_{i=1}^L\right)\right) \\ &= h_{s_-,s_+}\left(\tilde{Y}^{(1)}\left(\left[\tilde{W}_i\right]_{i=1}^L, \left[\tilde{b}_i\right]_{i=1}^L\right)\right) \\ &= Y^{(1)}\left(\left[\tilde{W}_i\right]_{i=1}^2, \left[\tilde{b}_i\right]_{i=1}^2\right). \end{aligned}$$

Thus, the output of the second layer before the activation is

$$\begin{aligned} \tilde{Y}^{(2)}\left(\left[\tilde{W}_i'\right]_{i=1}^L, \left[\tilde{b}_i'\right]_{i=1}^L\right) &= \tilde{W}_2' Y^{(1)}\left(\left[\tilde{W}_i'\right]_{i=1}^L, \left[\tilde{b}_i'\right]_{i=1}^L\right) + \tilde{b}_2' \mathbf{1}_m^T \\ &= \begin{bmatrix} \tilde{W}_2 \\ \mathbf{0}_{(d_2-d_Y)\times d_1} \end{bmatrix} Y^{(1)}\left(\left[\tilde{W}_i'\right]_{i=1}^L, \left[\tilde{b}_i'\right]_{i=1}^L\right) + \begin{bmatrix} \tilde{b}_2 \\ \mathbf{0}_{(d_2-d_Y)\times 1} \end{bmatrix} \mathbf{1}_m^T \\ &\quad + \lambda \mathbf{1}_{d_2} \mathbf{1}_m^T \\ &= \begin{bmatrix} \hat{Y}\left(\left[\tilde{W}_i\right]_{i=1}^2, \left[\tilde{b}_i\right]_{i=1}^2\right) + \lambda \mathbf{1}_{d_Y} \mathbf{1}_m^T \\ \lambda \mathbf{1}_{d_2 - d_Y} \mathbf{1}_m^T \end{bmatrix}. \end{aligned}$$

Applying the definition of λ (Eq. (7.31)),

$$\tilde{Y}^{(2)}\left(\left[\tilde{W}_i'\right]_{i=1}^L, \left[\tilde{b}_i'\right]_{i=1}^L\right) > \mathbf{0}\text{, element-wise.} \tag{7.34}$$

Therefore, the output of the second layer after the activation is

$$\begin{aligned} Y^{(2)}\left(\left[\tilde{W}_i'\right]_{i=1}^L, \left[\tilde{b}_i'\right]_{i=1}^L\right) &= h_{s_-,s_+}\left(\tilde{Y}^{(2)}\left(\left[\tilde{W}_i'\right]_{i=1}^L, \left[\tilde{b}_i'\right]_{i=1}^L\right)\right) \\ &= s_+\begin{bmatrix} \hat{Y}\left(\left[\tilde{W}_i\right]_{i=1}^2, \left[\tilde{b}_i\right]_{i=1}^2\right) + \lambda \mathbf{1}_{d_Y}\mathbf{1}_m^T \\ \lambda \mathbf{1}_{d_2-d_Y}\mathbf{1}_m^T \end{bmatrix}. \end{aligned}$$

Meanwhile, the output of the third layer before the activation is $\tilde{Y}^{(3)}\left(\left[\tilde{W}_i'\right]_{i=1}^L, \left[\tilde{b}_i'\right]_{i=1}^L\right)$ can be calculated based on $Y^{(2)}\left(\left[\tilde{W}_i'\right]_{i=1}^L, \left[\tilde{b}_i'\right]_{i=1}^L\right)$:

$$\begin{aligned} \tilde{Y}^{(3)}\left(\left[\tilde{W}_i'\right]_{i=1}^L, \left[\tilde{b}_i'\right]_{i=1}^L\right) &= \tilde{W}_3' Y^{(2)}\left(\left[\tilde{W}_i'\right]_{i=1}^L, \left[\tilde{b}_i'\right]_{i=1}^L\right) + \tilde{b}_3'\mathbf{1}_m^T \\ &= \frac{1}{s_+}\left(\sum_{i=1}^{d_Y} E_{i,i}\right) s_+\begin{bmatrix} \hat{Y}\left(\left[\tilde{W}_i\right]_{i=1}^2, \left[\tilde{b}_i\right]_{i=1}^2\right) + \lambda \mathbf{1}_{d_Y}\mathbf{1}_m^T \\ \lambda \mathbf{1}_{d_2-d_Y}\mathbf{1}_m^T \end{bmatrix} \\ &= \begin{bmatrix} \hat{Y}\left(\left[\tilde{W}_i\right]_{i=1}^2, \left[\tilde{b}_i\right]_{i=1}^2\right) + \lambda \mathbf{1}_{d_Y}\mathbf{1}_m^T \\ \mathbf{0}_{(d_3-d_Y)\times m} \end{bmatrix}. \end{aligned}$$

Applying Eq. (7.35),

$$\tilde{Y}^{(3)}\left(\left[\tilde{W}_i'\right]_{i=1}^L, \left[\tilde{b}_i'\right]_{i=1}^L\right) \geq \mathbf{0}\text{, element-wise.} \tag{7.35}$$

Therefore, the output of the third layer after the activation is

$$\begin{aligned} Y^{(3)}\left(\left[\tilde{W}_i'\right]_{i=1}^L, \left[\tilde{b}_i'\right]_{i=1}^L\right) &= h_{s_-,s_+}\left(\tilde{Y}^{(3)}\left(\left[\tilde{W}_i'\right]_{i=1}^L, \left[\tilde{b}_i'\right]_{i=1}^L\right)\right) \\ &= s_+\left(\tilde{Y}^{(3)}\left(\left[\tilde{W}_i'\right]_{i=1}^L, \left[\tilde{b}_i'\right]_{i=1}^L\right)\right) \\ &= s_+\begin{bmatrix} \hat{Y}\left(\left[\tilde{W}_i\right]_{i=1}^2, \left[\tilde{b}_i\right]_{i=1}^2\right) + \lambda \mathbf{1}_{d_Y}\mathbf{1}_m^T \\ \mathbf{0}_{(d_3-d_Y)\times m} \end{bmatrix}. \end{aligned}$$

Suppose Eqs. (7.32) and (7.33) hold for k ($3 \le k \le L-2$), when $k+1$,

$$\begin{aligned}
\tilde{Y}^{(k+1)}\left(\left[\tilde{W}_i'\right]_{i=1}^{L}, \left[\tilde{b}_i'\right]_{i=1}^{L}\right) &= \tilde{W}_{k+1}' Y^{(k)}\left(\left[\tilde{W}_i'\right]_{i=1}^{2}, \left[\tilde{b}_i'\right]_{i=1}^{2}\right) + \tilde{b}_{k+1}' \mathbf{1}_m^T \\
&= \frac{1}{s_+}\left(\sum_{i=1}^{d_Y} E_{i,i}\right) s_+ \begin{bmatrix} \hat{Y}\left(\left[\tilde{W}_i\right]_{i=1}^{2}, \left[\tilde{b}_i\right]_{i=1}^{2}\right) + \lambda \mathbf{1}_{d_Y} \mathbf{1}_m^T \\ \mathbf{0}_{(d_k - d_Y)\times m} \end{bmatrix} \\
&= \begin{bmatrix} \hat{Y}\left(\left[\tilde{W}_i\right]_{i=1}^{2}, \left[\tilde{b}_i\right]_{i=1}^{2}\right) + \lambda \mathbf{1}_{d_Y} \mathbf{1}_m^T \\ \mathbf{0}_{(d_{k+1} - d_Y)\times m} \end{bmatrix}.
\end{aligned}$$

Applying Eq. (7.35),

$$\tilde{Y}^{(k+1)}\left(\left[\tilde{W}_i'\right]_{i=1}^{L}, \left[\tilde{b}_i'\right]_{i=1}^{L}\right) \ge \mathbf{0}\ \text{, element-wise.} \tag{7.36}$$

Therefore, the output of the $(k+1)$-th layer after the activation is

$$\begin{aligned}
Y^{(k+1)}\left(\left[\tilde{W}_i'\right]_{i=1}^{L}, \left[\tilde{b}_i'\right]_{i=1}^{L}\right) &= h_{s_-,s_+}\left(\tilde{Y}^{(k+1)}\left(\left[\tilde{W}_i'\right]_{i=1}^{L}, \left[\tilde{b}_i'\right]_{i=1}^{L}\right)\right) \\
&= s_+ \tilde{Y}^{(k+1)}\left(\left[\tilde{W}_i'\right]_{i=1}^{L}, \left[\tilde{b}_i'\right]_{i=1}^{L}\right) \\
&= s_+ \begin{bmatrix} \hat{Y}\left(\left[\tilde{W}_i\right]_{i=1}^{2}, \left[\tilde{b}_i\right]_{i=1}^{2}\right) + \lambda \mathbf{1}_{d_Y} \mathbf{1}_m^T \\ \mathbf{0}_{(d_{k+1} - d_Y)\times m} \end{bmatrix}.
\end{aligned}$$

Therefore, Eqs. (7.32) and (7.33) hold for any $i \in \{3, 4, \ldots, L-1\}$.

Finally, the output of the network is

$$\begin{aligned}
\hat{Y}\left(\left[\tilde{W}_i'\right]_{i=1}^{L}, \left[\tilde{b}_i'\right]_{i=1}^{L}\right) &= Y^{(L)}\left(\left[\tilde{W}_i'\right]_{i=1}^{L}, \left[\tilde{b}_i'\right]_{i=1}^{L}\right) \\
&= \tilde{W}_L' Y^{(L-1)}\left(\left[\tilde{W}_i'\right]_{i=1}^{L}, \left[\tilde{b}_i'\right]_{i=1}^{L}\right) + \tilde{b}_L' \mathbf{1}_m^T \\
&= \left(\frac{1}{s_+}\sum_{i=1}^{d_Y} E_{i,i}\right) s_+ \begin{bmatrix} \hat{Y}\left(\left[\tilde{W}_i\right]_{i=1}^{2}, \left[\tilde{b}_i\right]_{i=1}^{2}\right) + \lambda \mathbf{1}_{d_Y} \mathbf{1}_m^T \\ 0_{(d_{L-1} - d_Y)\times m} \end{bmatrix} - \lambda \mathbf{1}_{d_Y} \mathbf{1}_m^T \\
&= \hat{Y}\left(\left[\tilde{W}_i\right]_{i=1}^{2}, \left[\tilde{b}_i\right]_{i=1}^{2}\right).
\end{aligned}$$

Applying Theorem 7.9, we have

$$\hat{R}_S\left(\left[\tilde{W}_i'\right]_{i=1}^{L}, \left[\tilde{W}_i'\right]_{i=1}^{L}\right) = \hat{R}_S\left(\left[\tilde{W}_i\right]_{i=1}^{2}, \left[\tilde{b}_i\right]_{i=1}^{2}\right) < f(\tilde{W}).$$

The proof is completed. □

7.2.3.5 Stage (3)

Finally, we prove Theorem 7.5. This stage also follows the two-step strategy.

Step (a). Construct local minima of the loss surface.

Lemma 7.6 *Suppose t is a non-differentiable point for the piece-wise linear activation function h and σ is a constant such that the activation h is differentiable in the intervals $(t-\sigma, t)$ and $(t, t+\sigma)$. Assume that M is a sufficiently large positive real such that*

$$\frac{1}{M}\left\|\hat{W}_1'X + \hat{b}_1'\mathbf{1}_m^T\right\|_F < \sigma. \tag{7.37}$$

Let α_i be any positive real such that

$$\begin{aligned} \alpha_1 &= 1 \\ 0 < \alpha_i &< 1, i = 2, \ldots, L-1. \end{aligned} \tag{7.38}$$

Then, under Assumption 7.3, any neural network with piecewise linear activations and $L-1$ hidden layers has local minima at

$$\hat{W}_1'' = \frac{1}{M}\hat{W}_1', \hat{b}_1'' = \frac{1}{M}\hat{b}_1' + t\mathbf{1}_{d_1},$$

$$\hat{W}_i'' = \alpha_i \hat{W}_i', \hat{b}_i'' = -\alpha_i \hat{W}_i' h(t)\mathbf{1}_{d_{i-1}} + t\mathbf{1}_{d_i} + \frac{\Pi_{j=2}^{i}\alpha_j}{M}\hat{b}_i', (i = 2, 3, ..., L-1),$$

and

$$\hat{W}_L'' = \frac{1}{\Pi_{j=2}^{L}\alpha_j} M \hat{W}_L', \hat{b}_L'' = -\frac{M}{\prod_{j=2}^{L-1}\alpha_j}\hat{W}_L' h(t)\mathbf{1}_{d_{L-1}} + \hat{b}_L'$$

where $\left\{\left[\hat{W}_i'\right]_{i=1}^{L}, \left[\hat{b}_i'\right]_{i=1}^{L}\right\}$ is the local minimizer constructed in Lemma 7.5. Also, the loss is continuously differentiable, whose derivative with respect to the prediction $\hat{Y}_i$ may equal to 0 only when the prediction $\hat{Y}_i$ and label Y_i are the same.

Proof Define $s_- = \lim_{\theta \to 0^-} h'(\theta)$ and $s_+ = \lim_{\theta \to 0^+} h'(\theta)$.

We then prove by induction that for all $i \in [1 : L-1]$, all components of the i-th layer output before the activation $\tilde{Y}^{(i)}\left(\left[\hat{W}_i''\right]_{i=1}^{L}, \left[\hat{b}_i''\right]_{i=1}^{L}\right)$ are in interval $(t, t+\sigma)$, and

$$Y^{(i)}\left(\left[\hat{W}_i''\right]_{i=1}^L, \left[\hat{b}_i''\right]_{i=1}^L\right) = h(t)\mathbf{1}_{d_i}\mathbf{1}_m^T + \frac{\Pi_{j=1}^i \alpha_j}{M} Y^{(i)}\left(\left[\hat{W}_i'\right]_{i=1}^L, \left[\hat{b}_i'\right]_{i=1}^L\right).$$

The first layer output before the activation is,

$$\tilde{Y}^{(1)}\left(\left[\hat{W}_i''\right]_{i=1}^L, \left[\hat{b}_i''\right]_{i=1}^L\right) = \hat{W}_1''X + \hat{b}_1''\mathbf{1}_m^T = \frac{1}{M}\hat{W}_1'X + \frac{1}{M}\hat{b}_1'\mathbf{1}_m^T + t\mathbf{1}_{d_1}\mathbf{1}_m^T. \tag{7.39}$$

We proved in Lemma 7.5 that $\hat{W}_1'X + \hat{b}_1'\mathbf{1}_m^T$ is positive (element-wise). Since the Frobenius norm of a matrix is no smaller than any component's absolute value, applying Eq. (7.37), we have that for all $i \in [1, d_1]$ and $j \in [1 : n]$,

$$0 < \frac{1}{M}\left(\hat{W}_1'X + \hat{b}_1'\mathbf{1}_m^T\right)_{ij} < \sigma. \tag{7.40}$$

Therefore, $\left(\frac{1}{M}\left(\hat{W}_1'X + \hat{b}_1'\mathbf{1}_m^T\right)_{ij} + t\right) \in (t, t+\sigma)$. So,

$$\begin{aligned} Y^{(1)}\left(\left[\hat{W}_i''\right]_{i=1}^L, \left[\hat{b}_i''\right]_{i=1}^L\right) &= h\left(\tilde{Y}^{(1)}\left(\left[\hat{W}_i''\right]_{i=1}^L, \left[\hat{b}_i''\right]_{i=1}^L\right)\right) \\ &\overset{(*)}{=} h_{s_-,s_+}\left(\frac{1}{M}\hat{W}_1'X + \frac{1}{M}\hat{b}_1'\mathbf{1}_m^T\right) + h(t)\mathbf{1}_{d_1}\mathbf{1}_m^T \\ &= \frac{1}{M}Y^{(1)}\left(\left[\hat{W}_i'\right]_{i=1}^L, \left[\hat{b}_i'\right]_{i=1}^L\right) + h(t)\mathbf{1}_{d_1}\mathbf{1}_m^T, \end{aligned}$$

where eq.(∗) is because for any $x \in (t-\sigma, t+\sigma)$,

$$h(x) = h(t) + h_{s_-,s_+}(x-t). \tag{7.41}$$

Suppose the above argument holds for k ($1 \le k \le L-2$). Then

$$\begin{aligned} &\tilde{Y}^{(k+1)}\left(\left[\hat{W}_i''\right]_{i=1}^L, \left[\hat{b}_i''\right]_{i=1}^L\right) \\ &= \hat{W}_{k+1}''Y^{(k)}\left(\left[\hat{W}_i''\right]_{i=1}^L, \left[\hat{b}_i''\right]_{i=1}^L\right) + \hat{b}_{k+1}''\mathbf{1}_m^T \\ &= (-\alpha_{k+1}\hat{W}_{k+1}'h(t)\mathbf{1}_{d_{k+1}} \\ &\quad + t\mathbf{1}_{d_{k+1}} + \frac{\Pi_{i=1}^{k+1}\alpha_i}{M}\hat{b}_{k+1}')\mathbf{1}_m^T + \alpha_{k+1}\hat{W}_{k+1}'Y^{(k)}\left(\left[\hat{W}_i''\right]_{i=1}^L, \left[\hat{b}_i''\right]_{i=1}^L\right) \end{aligned}$$

$$
\begin{aligned}
&= \alpha_{k+1}\hat{W}'_{k+1}\left(h(t)\mathbf{1}_{d_k}\mathbf{1}_n^T + \frac{\Pi_{i=1}^{k}\alpha_i}{M}Y^{(k)}\left(\left[\hat{W}'_i\right]_{i=1}^{L}, \left[\hat{b}'_i\right]_{i=1}^{L}\right)\right) \\
&\quad + \left(-\alpha_{k+1}\hat{W}'_{k+1}h(t)\mathbf{1}_{d_k} + t\mathbf{1}_{d_{k+1}} + \frac{\Pi_{i=1}^{k+1}\alpha_i}{M}\hat{b}'_{k+1}\right)\mathbf{1}_m^T \\
&= \frac{\Pi_{i=1}^{k+1}\alpha_i}{M}\hat{W}'_{k+1}Y^{(k)}\left(\left[\hat{W}'_i\right]_{i=1}^{L}, \left[\hat{b}'_i\right]_{i=1}^{L}\right) + \frac{\Pi_{i=1}^{k+1}\alpha_i}{M}\hat{b}'_{k+1}\mathbf{1}_m^T + t\mathbf{1}_{d_{k+1}}\mathbf{1}_m^T \\
&= t\mathbf{1}_{d_{k+1}}\mathbf{1}_m^T + \frac{\Pi_{i=1}^{k+1}\alpha_i}{M}\tilde{Y}^{(k+1)}\left(\left[\hat{W}'_i\right]_{i=1}^{L}, \left[\hat{b}'_i\right]_{i=1}^{L}\right).
\end{aligned}
$$

Lemma 7.5 has proved that all components of $\tilde{Y}^{(k+1)}\left(\left[\hat{W}'_i\right]_{i=1}^{L}, \left[\hat{b}'_i\right]_{i=1}^{L}\right)$ are contained in $\tilde{Y}^{(1)}\left(\left[\hat{W}'_i\right]_{i=1}^{L}, \left[\hat{b}'_i\right]_{i=1}^{L}\right)$. Combining

$$
t\mathbf{1}_{d_1}\mathbf{1}_m^T < t\mathbf{1}_{d_1}\mathbf{1}_m^T + \frac{1}{M}\tilde{Y}^{(1)}\left(\left[\hat{W}'_i\right]_{i=1}^{L}, \left[\hat{b}'_i\right]_{i=1}^{L}\right) < (t+\sigma)\mathbf{1}_{d_1}\mathbf{1}_m^T,
$$

we have

$$
t\mathbf{1}_{d_{k+1}}\mathbf{1}_m^T < t\mathbf{1}_{d_{k+1}}\mathbf{1}_m^T + \frac{\Pi_{i=1}^{k+1}\alpha_i}{M}\tilde{Y}^{(k+1)}\left(\left[\hat{W}'_i\right]_{i=1}^{L}, \left[\hat{b}'_i\right]_{i=1}^{L}\right) \overset{(*)}{<} (t+\sigma)\mathbf{1}_{d_{k+1}}\mathbf{1}_m^T.
$$

Here $<$ are all element-wise, and inequality $(*)$ comes from the property of α_i (Eq. (7.38)).

Furthermore, the $(k+1)$-th layer output after the activation is

$$
\begin{aligned}
Y^{(k+1)}\left(\left[\hat{W}''_i\right]_{i=1}^{L}, \left[\hat{b}''_i\right]_{i=1}^{L}\right) &= h\left(\tilde{Y}^{(k+1)}\left(\left[\hat{W}''_i\right]_{i=1}^{L}, \left[\hat{b}''_i\right]_{i=1}^{L}\right)\right) \\
&= h\left(t\mathbf{1}_{d_{k+1}}\mathbf{1}_m^T + \frac{1}{M}\tilde{Y}^{(k+1)}\left(\left[\hat{W}'_i\right]_{i=1}^{L}, \left[\hat{b}'_i\right]_{i=1}^{L}\right)\right) \\
&\overset{(*)}{=} h(t)\mathbf{1}_{d_{k+1}}\mathbf{1}_m^T + h_{s_-,s_+}\left(\frac{\Pi_{i=1}^{k+1}\alpha_i}{M}\tilde{Y}^{(k+1)}\left(\left[\hat{W}'_i\right]_{i=1}^{L}, \left[\hat{b}'_i\right]_{i=1}^{L}\right)\right) \\
&= h(t)\mathbf{1}_{d_{k+1}}\mathbf{1}_m^T + \frac{\Pi_{i=1}^{k+1}\alpha_i}{M}Y^{(k+1)}\left(\left[\hat{W}'_i\right]_{i=1}^{L}, \left[\hat{b}'_i\right]_{i=1}^{L}\right),
\end{aligned}
$$

where Eq. $(*)$ is because of Eq. (7.41). The above argument is proved for any index $k \in \{1, \ldots, L-1\}$.

Therefore, the output of the network is

$$
\begin{aligned}
& Y^{(L)}\left(\left[\hat{W}_i''\right]_{i=1}^{L}, \left[\hat{b}_i''\right]_{i=1}^{L}\right) \\
&= \hat{W}_L'' Y^{(L-1)}\left(\left[\hat{W}_i''\right]_{i=1}^{L}, \left[\hat{b}_i''\right]_{i=1}^{L}\right) + \hat{b}_L'' \mathbf{1}_m^T \\
&= \frac{M}{\prod_{i=1}^{L-1}\alpha_i} \hat{W}_L' \left(h(t)\mathbf{1}_{d_{L-1}}\mathbf{1}_m^T + \frac{\prod_{i=1}^{L-1}\alpha_i}{M} Y^{(L-1)}\left(\left[\hat{W}_i'\right]_{i=1}^{L}, \left[\hat{b}_i'\right]_{i=1}^{L}\right)\right) \\
&\quad + \left(-\frac{M}{\prod_{i=1}^{L-1}\alpha_i}\hat{W}_L' h(t)\mathbf{1}_{d_{L-1}} + \hat{b}_L'\right)\mathbf{1}_m^T \\
&= \hat{W}_L' Y^{(L-1)}\left(\left[\hat{W}_i'\right]_{i=1}^{L}, \left[\hat{b}_i'\right]_{i=1}^{L}\right) + \hat{b}_L' \mathbf{1}_m^T \\
&= Y^{(L)}\left(\left[\hat{W}_i'\right]_{i=1}^{L}, \left[\hat{b}_i'\right]_{i=1}^{L}\right).
\end{aligned}
$$

Therefore,

$$
\hat{R}_S\left(\left[\hat{W}_i''\right]_{i=1}^{L}, \left[\hat{b}_i''\right]_{i=1}^{L}\right) = \hat{R}_S\left(\left[\hat{W}_i'\right]_{i=1}^{L}, \left[\hat{b}_i'\right]_{i=1}^{L}\right) = f(\tilde{W}).
$$

We then introduce some small disturbances $\left\{\left[\delta_{Wi}''\right]_{i=1}^{L}, \left[\delta_{bi}''\right]_{i=1}^{L}\right\}$ into $\left\{\left[\hat{W}_i''\right]_{i=1}^{L}, \left[\hat{b}_i''\right]_{i=1}^{L}\right\}$ in order to check the local optimality.

Since all comonents of $Y^{(i)}$ are in interval $(t, t+\sigma)$, the activations in every hidden layers is realized at linear parts. Therefore, the output of network is

$$
\begin{aligned}
& \hat{Y}\left(\left[\hat{W}_i'' + \delta_{Wi}''\right]_{i=1}^{L}, \left[\hat{b}_i'' + \delta_{bi}''\right]_{i=1}^{L}\right) \\
&= \left(\hat{W}_L'' + \delta_{WL}''\right) s_+\left(\cdots s_+\left(\left(\hat{W}_1'' + \delta_{W1}''\right) X + \left(\hat{b}_1'' + \delta_{b1}''\right)\mathbf{1}_m^T\right) + f(t)\mathbf{1}_{d_1}\mathbf{1}_m^T \cdots\right) \\
&\quad + f(t)\mathbf{1}_{d_L}\mathbf{1}_m^T + \left(\hat{b}_L'' + \delta_{bL}''\right)\mathbf{1}_m^T \\
&= \left[M_1 \ M_2\right]\begin{bmatrix} X \\ \mathbf{1}_m^T \end{bmatrix}.
\end{aligned}
$$

Similar to Lemma 7.5, $\left[M_1\ M_2\right]$ approaches $\tilde{W}$ as disturbances $\left\{[\delta_{Wi}]_{i=1}^L, [\delta_{bi}]_{i=1}^L\right\}$ approach $\mathbf{0}$ (element-wise). Combining that $\tilde{W}$ is a local minimizer of $f(W)$, we have

$$\hat{R}_S\left(\left[\hat{W}_i'' + \delta_{Wi}''\right]_{i=1}^L, \left[\hat{b}_i'' + \delta_{bi}''\right]_{i=1}^L\right) = f\left(\left[M_1\ M_2\right]\right) \geq f(\tilde{W}) = \hat{R}_S\left(\left[\hat{W}_i''\right]_{i=1}^L, \left[\hat{b}_i''\right]_{i=1}^L\right).$$

The proof is completed. □

Step (b). Prove the constructed local minima are spurious.

Proof of Theorem 7.5 Without loss of generality, we assume that all activations are the same.

Let t be a non-differentiable point of the piece-wise linear activation function h with

$$s_- = \lim_{\theta \to 0^-} h'(\theta),$$
$$s_+ = \lim_{\theta \to 0^+} h'(\theta).$$

Let σ be a constant such that h is linear in interval $(t-\sigma, t)$ and interval $(t, t+\sigma)$. Then construct that

$$\tilde{W}_1'' = \frac{1}{M}\tilde{W}_1', \tilde{b}_1'' = \frac{1}{M}\tilde{b}_1' + t\mathbf{1}_{d_1},$$
$$\tilde{W}_2'' = \frac{1}{\tilde{M}}\tilde{W}_2', \tilde{b}_2'' = t\mathbf{1}_{d_2} - \frac{1}{\tilde{M}}h(t)\tilde{W}_2'\mathbf{1}_{d_2} + \frac{1}{M\tilde{M}}\tilde{b}_2',$$
$$\tilde{W}_i'' = \tilde{W}_i', \tilde{b}_i'' = -\tilde{W}_i'h(t)\mathbf{1}_{d_{i-1}} + t\mathbf{1}_{d_i} + \frac{1}{M\tilde{M}}\tilde{b}_i', (i = 3, 4, \ldots, L-1)$$

and

$$\tilde{W}_L'' = M\tilde{M}\tilde{W}_L', \tilde{b}_L'' = \tilde{b}_L - M\tilde{M}\tilde{W}_L'h(t)\mathbf{1}_{L-1},$$

where $\left\{\left[\tilde{W}_i'\right]_{i=1}^L, \left[\tilde{b}_i'\right]_{i=1}^L\right\}$ are constructed in Theorem 7.10, M is a large enough positive real such that

$$\frac{1}{M}\left\|\tilde{W}_1'X + \tilde{b}_1'\mathbf{1}_m^T\right\|_F < \sigma, \tag{7.42}$$

and $\tilde{M}$ a large enough positive real such that

$$\frac{1}{\tilde{M}}\left\|\frac{1}{M}\tilde{Y}^{(2)}\left(\left[\tilde{W}_i'\right]_{i=1}^L, \left[\tilde{b}_i'\right]_{i=1}^L\right)\right\|_F < \sigma. \tag{7.43}$$

Then, we prove by induction that for any $i \in [2 : L-1]$, all components of $\tilde{Y}^{(i)}\left(\left[\tilde{W}_i''\right]_{i=1}^L, \left[\tilde{b}_i''\right]_{i=1}^L\right)$ are in interval $(t, t+\delta)$, and

$$Y^{(i)}\left(\left[\tilde{W}_i''\right]_{i=1}^L, \left[\tilde{b}_i''\right]_{i=1}^L\right) = h(t)\mathbf{1}_{d_i}\mathbf{1}_m^T + \frac{1}{\tilde{M}M} Y^{(i)}\left(\left[\tilde{W}_i'\right]_{i=1}^L, \left[\tilde{b}_i'\right]_{i=1}^L\right).$$

First,

$$\tilde{Y}^{(1)}\left(\left[\tilde{W}_i''\right]_{i=1}^L, \left[\tilde{b}_i''\right]_{i=1}^L\right) = \tilde{W}_1''X + \tilde{b}_1''\mathbf{1}_m^T = \frac{1}{M}(\tilde{W}_1'X + \tilde{b}_1'\mathbf{1}_m^T) + t\mathbf{1}_{d_1}^T\mathbf{1}_m^T. \tag{7.44}$$

For any $i \in [1 : d_1]$ and $j \in [1 : m]$, Eq. (7.42) implies

$$\left|\left(\frac{1}{M}(\tilde{W}_1'X + \tilde{b}_1'\mathbf{1}_m^T)\right)_{ij}\right| \le \frac{1}{M}\left\|\tilde{W}_1'X + \tilde{b}_1'\mathbf{1}_m^T\right\|_F < \sigma.$$

Thus,

$$\left(\frac{1}{M}(\tilde{W}_1'X + \tilde{b}_1'\mathbf{1}_m^T) + t\mathbf{1}_{d_1}^T\mathbf{1}_m^T\right)_{ij} \in (t-\sigma, t+\sigma). \tag{7.45}$$

Therefore, the output of the first layer after the activation is

$$\begin{aligned}
Y^{(1)}\left(\left[\tilde{W}_i''\right]_{i=1}^L, \left[\tilde{b}_i''\right]_{i=1}^L\right) &= h\left(\tilde{Y}^{(1)}\left(\left[\tilde{W}_i''\right]_{i=1}^L, \left[\tilde{b}_i''\right]_{i=1}^L\right)\right) \\
&= h\left(\frac{1}{M}(\tilde{W}_1'X + \tilde{b}_1'\mathbf{1}_m^T) + t\mathbf{1}_{d_1}\mathbf{1}_m^T\right) \\
&\overset{(*)}{=} h(t)\mathbf{1}_{d_1}\mathbf{1}_m^T + h_{s_-,s_+}\left(\frac{1}{M}\left(\tilde{W}_1'X + \tilde{b}_1'\right)\right) \\
&= h(t)\mathbf{1}_{d_1}\mathbf{1}_m^T + \frac{1}{M}h_{s_-,s_+}\left(\left(\tilde{W}_1'X + \tilde{b}_1'\right)\right) \\
&= h(t)\mathbf{1}_{d_1}\mathbf{1}_m^T + \frac{1}{M}Y^{(1)}\left(\left[\tilde{W}_i'\right]_{i=1}^L, \left[\tilde{b}_i'\right]_{i=1}^L\right),
\end{aligned}$$

where Eq. $(*)$ is from Eq. (7.41) for any $x \in (t-\delta, t+\delta)$. Also,

$$\begin{aligned}
\tilde{Y}^{(2)}\left(\left[\tilde{W}_i''\right]_{i=1}^L, \left[\tilde{b}_i''\right]_{i=1}^L\right) &= \tilde{W}_2''Y^{(1)}\left(\left[\tilde{W}_i''\right]_{i=1}^L, \left[\tilde{b}_i''\right]_{i=1}^L\right) + \tilde{b}_2''\mathbf{1}_m^T \\
&= \frac{1}{\tilde{M}}\left(\tilde{W}_2'\right)\left(h(t)\mathbf{1}_{d_1}\mathbf{1}_m^T + \frac{1}{M}Y^{(1)}\left(\left[\tilde{W}_i'\right]_{i=1}^L, \left[\tilde{b}_i'\right]_{i=1}^L\right)\right) \\
&\quad + t\mathbf{1}_{d_2}\mathbf{1}_m^T - \frac{1}{\tilde{M}}h(t)\tilde{W}_2'\mathbf{1}_{d_1}\mathbf{1}_m^T + \frac{1}{M\tilde{M}}\tilde{b}_2'\mathbf{1}_m^T
\end{aligned}$$

$$= \frac{1}{M\tilde{M}} \tilde{Y}^{(2)} \left(\left[\tilde{W}_i' \right]_{i=1}^L, \left[\tilde{b}_i' \right]_{i=1}^L \right) + t \mathbf{1}_{d_2} \mathbf{1}_m^T.$$

Recall in Theorem 7.10 we prove all components of $\tilde{Y}^{(2)} \left(\left[\tilde{W}_i' \right]_{i=1}^L, \left[\tilde{b}_i' \right]_{i=1}^L \right)$ are positive. Combining the definition of $\tilde{M}$ (Eq. (7.43)), we have

$$\begin{aligned} t \mathbf{1}_{d_2} \mathbf{1}_m^T &< \tilde{Y}^{(2)} \left(\left[\tilde{W}_i'' \right]_{i=1}^L, \left[\tilde{b}_i'' \right]_{i=1}^L \right) \\ &= \frac{1}{\tilde{M} M} \tilde{Y}^{(2)} \left(\left[\tilde{W}_i' \right]_{i=1}^L, \left[\tilde{b}_i' \right]_{i=1}^L \right) + t \mathbf{1}_{d_2} \mathbf{1}_m^T \\ &< (t + \sigma) \mathbf{1}_{d_2} \mathbf{1}_m^T. \end{aligned}$$

Therefore,

$$\begin{aligned} Y^{(2)} \left(\left[\tilde{W}_i'' \right]_{i=1}^L, \left[\tilde{b}_i'' \right]_{i=1}^L \right) &= h \left(\tilde{Y}^{(2)} \left(\left[\tilde{W}_i'' \right]_{i=1}^L, \left[\tilde{b}_i'' \right]_{i=1}^L \right) \right) \\ &= h \left(\frac{1}{\tilde{M} M} Y^{(2)} \left(\left[\tilde{W}_i' \right]_{i=1}^L, \left[\tilde{b}_i' \right]_{i=1}^L \right) + t \mathbf{1}_{d_2} \mathbf{1}_m^T \right) \\ &= h(t) \mathbf{1}_{d_2} \mathbf{1}_m^T + h_{s_-, s_+} \left(\frac{1}{\tilde{M} M} \tilde{Y}^{(2)} \left(\left[\tilde{W}_i' \right]_{i=1}^L, \left[\tilde{b}_i' \right]_{i=1}^L \right) \right) \\ &= h(t) \mathbf{1}_{d_2} \mathbf{1}_m^T + \frac{1}{\tilde{M} M} h_{s_-, s_+} \left(\tilde{Y}^{(2)} \left(\left[\tilde{W}_i' \right]_{i=1}^L, \left[\tilde{b}_i' \right]_{i=1}^L \right) \right) \\ &= h(t) \mathbf{1}_{d_2} \mathbf{1}_m^T + \frac{1}{\tilde{M} M} Y^{(2)} \left(\left[\tilde{W}_i' \right]_{i=1}^L, \left[\tilde{b}_i' \right]_{i=1}^L \right). \end{aligned}$$

Suppose the above argument holds for k-th layer.

The output of $(k+1)$-th layer before the activation is

$$\begin{aligned} &\tilde{Y}^{(k+1)} \left(\left[\tilde{W}_i'' \right]_{i=1}^L, \left[\tilde{b}_i'' \right]_{i=1}^L \right) \\ &= \tilde{W}_{k+1}'' Y^{(k)} \left(\left[\tilde{W}_i'' \right]_{i=1}^L, \left[\tilde{b}_i'' \right]_{i=1}^L \right) + \tilde{b}_{k+1}'' \mathbf{1}_m^T \\ &= \tilde{W}_{k+1}' \left(h(t) \mathbf{1}_{d_k} \mathbf{1}_m^T + \frac{1}{\tilde{M} M} Y^{(k)} \left(\left[\tilde{W}_i' \right]_{i=1}^L, \left[\tilde{b}_i' \right]_{i=1}^L \right) \right) \\ &\quad + \left(-\tilde{W}_{k+1}' h(t) \mathbf{1}_{d_k} + t \mathbf{1}_{d_{k+1}} + \frac{1}{M\tilde{M}} \tilde{b}_{k+1}' \right) \mathbf{1}_m^T \\ &= \frac{1}{M\tilde{M}} \left(\tilde{W}_{k+1}' Y^{(k)} \left(\left[\tilde{W}_i' \right]_{i=1}^L, \left[\tilde{b}_i' \right]_{i=1}^L \right) + \tilde{b}_{k+1}' \mathbf{1}_m^T \right) + t \mathbf{1}_{d_{k+1}} \mathbf{1}_m^T \\ &= \frac{1}{M\tilde{M}} \tilde{Y}^{(k+1)} \left(\left[\tilde{W}_i' \right]_{i=1}^L, \left[\tilde{b}_i' \right]_{i=1}^L \right) + t \mathbf{1}_{d_{k+1}} \mathbf{1}_m^T. \end{aligned}$$

Recall proved in Theorem 7.10 that all components of $\tilde{Y}^{(k+1)}\left(\left[\tilde{W}_i'\right]_{i=1}^L, \left[\tilde{b}_i'\right]_{i=1}^L\right)$ except those that are 0 are contained in $\tilde{Y}^{(k)}\left(\left[\tilde{W}_i'\right]_{i=1}^L, \left[\tilde{b}_i'\right]_{i=1}^L\right)$. We have

$$t\mathbf{1}_{d_{k+1}}\mathbf{1}_m^T < \frac{1}{M\tilde{M}}\tilde{Y}^{(k+1)}\left(\left[\tilde{W}_i'\right]_{i=1}^L, \left[\tilde{b}_i'\right]_{i=1}^L\right) + t\mathbf{1}_{d_{k+1}}\mathbf{1}_m^T < (t+\sigma)\mathbf{1}_{d_{k+1}}\mathbf{1}_m^T.$$

Therefore,

$$\begin{aligned} Y^{(k+1)}\left(\left[\tilde{W}_i''\right]_{i=1}^L, \left[\tilde{b}_i''\right]_{i=1}^L\right) &= h\left(\tilde{Y}^{(k)}\left(\left[\tilde{W}_i''\right]_{i=1}^L, \left[\tilde{b}_i''\right]_{i=1}^L\right)\right) \\ &= h\left(\frac{1}{M\tilde{M}}\tilde{Y}^{(k+1)}\left(\left[\tilde{W}_i'\right]_{i=1}^L, \left[\tilde{b}_i'\right]_{i=1}^L\right) + t\mathbf{1}_{d_{k+1}}\mathbf{1}_m^T\right) \\ &= h(t)\mathbf{1}_{d_{k+1}}\mathbf{1}_m^T + \frac{1}{M\tilde{M}}Y^{(k+1)}\left(\left[\tilde{W}_i'\right]_{i=1}^L, \left[\tilde{b}_i'\right]_{i=1}^L\right). \end{aligned}$$

Thus, the argument holds for any $k \in \{2, \ldots, L-1\}$.

So, we obtain

$$\begin{aligned} Y^{(L)}\left(\left[\tilde{W}_i''\right]_{i=1}^L, \left[\tilde{b}_i''\right]_{i=1}^L\right) &= \tilde{W}_L'' Y^{(L-1)}\left(\left[\tilde{W}_i''\right]_{i=1}^L, \left[\tilde{b}_i''\right]_{i=1}^L\right) + \tilde{b}_L'' \\ &= M\tilde{M}\tilde{W}_L'\left(h(t)\mathbf{1}_{d_{L-1}}\mathbf{1}_m^T + \frac{1}{M\tilde{M}}Y^{(L-1)}\left(\left[\tilde{W}_i'\right]_{i=1}^L, \left[\tilde{b}_i'\right]_{i=1}^L\right)\right) \\ &\quad + \tilde{b}_L'\mathbf{1}_m^T - M\tilde{M}\tilde{W}_L' h(t)\mathbf{1}_{d_{L-1}}\mathbf{1}_m^T \\ &= Y^{(L)}\left(\left[\tilde{W}_i'\right]_{i=1}^L, \left[\tilde{b}_i'\right]_{i=1}^L\right). \end{aligned}$$

Therefore,

$$\hat{R}_S\left(\left[\tilde{W}_i''\right]_{i=1}^L, \left[\tilde{b}_i''\right]_{i=1}^L\right) = \hat{R}_S\left(\left[\tilde{W}_i'\right]_{i=1}^L, \left[\tilde{b}_i'\right]_{i=1}^L\right). \tag{7.46}$$

From Eq. (7.46) and Theorem 7.10, we have

$$\hat{R}_S\left(\left[\tilde{W}_i''\right]_{i=1}^L, \left[\tilde{b}_i''\right]_{i=1}^L\right) < f\left(\tilde{W}\right),$$

which completes the proof of local minimizer.

Furthermore, the parameter M used in Lemma 7.6 (not those in this proof) is arbitrary in a continuous interval (cf. Eq. (7.37)), we have actually constructed infinite spurious local minima. This complete the proof of Theorem 7.5.

Theorem 7.5 relies on Assumption 7.4. We can further remove it by replacing Assumption 7.3 by a mildly more restrictive variant Assumption 7.6.

Corollary 7.3 *Suppose that Assumptions 7.1, 7.2, and 7.6 hold. Neural networks with arbitrary depth and arbitrary piecewise linear activations (excluding linear functions) have infinitely many spurious local minima under arbitrary continuously differentiable loss whose derivative can equal* 0 *only when the prediction and label are the same.*

Proof The proof is delivered by modifications of Theorem 7.9 in Stage 1 of Theorem 7.5's proof. We only need to prove the corollary under the assumption that $s_- + s_+ = 0$.

Let the local minimizer constructed in Lemma 7.4 be $\left\{\left[\hat{W}_i\right]_{i=1}^2, \left[\hat{b}_i\right]_{i=1}^2\right\}$. Denote by two zero matrix $0_{\tilde{W}_1} := 0_{d_X \times (d_1 - d_Y - 1)}$ and $0_{\tilde{b}_1} := 0_{1 \times (d_1 - d_Y - 1)}$, and $\nu = \tilde{W}_{1,[d_X+1]} - \eta_1$. Then, we construct a point in the parameter space whose empirical risk is smaller as follows:

$$\tilde{W}_1 = (\tilde{W}_{1,[1:d_X]}^T - \beta\alpha^T, \tilde{W}_{1,[1:d_X]}^T, -\tilde{W}_{1,[1:d_X]}^T + \beta\alpha^T, \tilde{W}_{2,[1:d_X]}^T, \dots, \tilde{W}_{d_Y,[1:d_X]}^T, 0_{\tilde{W}_1})^T,$$

$$\tilde{b}_1 = \left(\nu + \gamma, \tilde{W}_{1,[d_X+1]} - \eta, -\nu + \gamma, \tilde{W}_{2,[d_X+1]} - \eta_2, \cdots \tilde{W}_{d_Y,[d_X+1]} - \eta_{d_Y}, 0_{\tilde{b}_1}\right)^T, \tag{7.47}$$

$$\tilde{W}_2 = \begin{bmatrix} \frac{1}{2s_+} & \frac{1}{s_+} & -\frac{1}{2s_+} & 0 & 0 & \cdots & 0 & 0 & \cdots & 0 \\ 0 & 0 & 0 & \frac{1}{s_+} & 0 & \cdots & 0 & 0 & \cdots & 0 \\ 0 & 0 & 0 & 0 & \frac{1}{s_+} & \cdots & 0 & 0 & \cdots & 0 \\ \vdots & \vdots & \vdots & \vdots & \vdots & \ddots & \vdots & \vdots & \ddots & \vdots \\ 0 & 0 & 0 & 0 & 0 & \cdots & \frac{1}{s_+} & 0 & \cdots & 0 \end{bmatrix}, \tag{7.48}$$

and

$$\tilde{b}_2 = \left(\eta, \eta_2, \cdots, \eta_{d_Y}\right)^T, \tag{7.49}$$

where α, β, and η_i are defined the same as those in Theorem 7.9, and η is defined by Eq. (7.25).

Then, the output of the first layer is

$$Y^{(1)}\left(\left[\tilde{W}_i\right]_{i=1}^2, \left[\tilde{b}_i\right]_{i=1}^2\right) = h_{s_-,s_+}\left(\tilde{W}_1 X + \tilde{b}_1 \mathbf{1}_m^T\right)$$

$$= h_{s_-,s_+}\left(\begin{bmatrix} \tilde{W}_{1,\cdot}X - \alpha\beta^T X - \eta_1 \mathbf{1}_m^T + \gamma \mathbf{1}_m^T \\ \tilde{W}_{1,\cdot}X - \eta \mathbf{1}_m^T \\ -\tilde{W}_{1,\cdot}X + \alpha\beta^T X + \eta_1 \mathbf{1}_m^T + \gamma \mathbf{1}_m^T \\ \tilde{W}_{2,\cdot}X - \eta_2 \mathbf{1}_m^T \\ \vdots \\ \tilde{W}_{d_Y,\cdot}X - \eta_{d_Y} \mathbf{1}_m^T \\ \mathbf{0}_{d_1-d_Y-2}\mathbf{1}_m^T \end{bmatrix}\right).$$

Further, the output of the whole network is

$$\hat{Y}\left(\left[\tilde{W}_i\right]_{i=1}^2, \left[\tilde{b}_i\right]_{i=1}^2\right) = \tilde{W}_2 h_{s_-,s_+}\left(\begin{bmatrix} \tilde{W}_{1,\cdot}X - \alpha\beta^T X - \eta_1 \mathbf{1}_m^T + \gamma \mathbf{1}_m^T \\ \tilde{W}_{1,\cdot}X - \eta \mathbf{1}_m^T \\ -\tilde{W}_{1,\cdot}X + \alpha\beta^T X + \eta_1 \mathbf{1}_m^T + \gamma \mathbf{1}_m^T \\ \tilde{W}_{2,\cdot}X - \eta_2 \mathbf{1}_m^T \\ \vdots \\ \tilde{W}_{d_Y,\cdot}X - \eta_{d_Y} \mathbf{1}_m^T \\ \mathbf{0}_{d_1-d_Y-2}\mathbf{1}_m^T \end{bmatrix}\right) + \tilde{b}_2 \mathbf{1}_m^T$$

Therefore, if $j \leq l'$, the $(1, j)$-th component of $\hat{Y}\left(\left[\tilde{W}_i\right]_{i=1}^2, \left[\tilde{b}_i\right]_{i=1}^2\right)$ is

$$\begin{aligned}
&\left(\tilde{W}_2\right)_1 \tilde{Y}^{(1)}\left(\left[\tilde{W}_i\right]_{i=1}^2, \left[\tilde{b}_i\right]_{i=1}^2\right)_j \\
=&\frac{1}{2s_+}\left(-s_+\left(\tilde{Y}_{1,j} - \alpha\beta^T x_j - \eta_1 + \gamma\right) + 2s_+\left(\tilde{Y}_{1,j} - \eta\right) - s_+\left(-\tilde{Y}_{1j} + \alpha\beta^T x_j + \eta_1 + \gamma\right)\right) \\
&+\eta \\
=&\frac{1}{2s_+}\left(2s_+\tilde{Y}_{1,j} - 2s_+\eta - 2s_+\gamma\right) + \eta \\
=&\tilde{Y}_{1,j} - \gamma.
\end{aligned}$$

Otherwise $(j > l')$, the $(1, j)$-th component of $\hat{Y}\left(\left[\tilde{W}_i\right]_{i=1}^2, \left[\tilde{b}_i\right]_{i=1}^2\right)$ is

$$\begin{aligned}
&\left(\tilde{W}_2\right)_1 \tilde{Y}^{(1)}\left(\left[\tilde{W}_i\right]_{i=1}^2, \left[\tilde{b}_i\right]_{i=1}^2\right)_j \\
=&\frac{1}{2s_+}\left(s_+\left(\tilde{Y}_{1,j} - \alpha\beta^T x_j - \eta_1 + \gamma\right) + 2s_+\left(\tilde{Y}_{1,j} - \eta\right) - s_-\left(-\tilde{Y}_{1j} + \alpha\beta^T x_j + \eta_1 + \gamma\right)\right) \\
&+\eta
\end{aligned}$$

$$
\begin{aligned}
&= \frac{1}{2s_+}\left(s_+\left(\tilde{Y}_{1,j} - \alpha\beta^T x_j - \eta_1 + \gamma\right) + 2s_+\left(\tilde{Y}_{1,j} - \eta\right) + s_+\left(-\tilde{Y}_{1j} + \alpha\beta^T x_j + \eta_1 + \gamma\right)\right) \\
&\quad + \eta \\
&= \frac{1}{2s_+}\left(2s_+\tilde{Y}_{1,j} - 2s_+\eta + 2s_+\gamma\right) + \eta \\
&= \tilde{Y}_{1,j} + \gamma,
\end{aligned}
$$

and the (i, j)-th $(i > 1)$ component of $\hat{Y}\left(\left[\tilde{W}_i\right]_{i=1}^2, \left[\tilde{b}_i\right]_{i=1}^2\right)$ is $\tilde{Y}_{i,j}$.

Therefore, we have

$$
\left(\tilde{W}_2\left(\tilde{W}_1 x_i + \tilde{b}_1\right) + \tilde{b}_2 - \tilde{W}\begin{bmatrix} x_i \\ 1 \end{bmatrix}\right)_j = \begin{cases} -\gamma, & j = 1, i \le l; \\ \gamma, & j = 1, i > l; \\ 0, & j \ge 2. \end{cases}
$$

Then, similar to Theorem 7.9, we have

$$
\begin{aligned}
&\hat{R}_S\left(\left[\tilde{W}_i\right]_{i=1}^L, \left[\tilde{b}_i\right]_{i=1}^L\right) - \hat{R}_S\left(\left[\hat{W}_i\right]_{i=1}^L, \left[\hat{b}_i\right]_{i=1}^L\right) \\
&= \frac{1}{m}\sum_{i=1}^{n} l\left(Y_i, \tilde{W}_2\left(\tilde{W}_1 x_i + \tilde{b}_1\right) + \tilde{b}_2\right) - \frac{1}{m}\sum_{i=1}^{m} l\left(Y_i, \hat{W}\begin{bmatrix} x_i \\ 1 \end{bmatrix}\right) \\
&= \frac{1}{m}\sum_{i=1}^{n} \nabla_{\hat{Y}_i} l\left(Y_i, \tilde{W}\begin{bmatrix} x_i \\ 1 \end{bmatrix}\right)\left(\tilde{W}_2\left(\tilde{W}_1 x_i + \tilde{b}_1 \mathbf{1}_m^T\right) + \tilde{b}_2 \mathbf{1}_m^T - \tilde{W}\begin{bmatrix} x_i \\ 1 \end{bmatrix}\right) \\
&\quad + \sum_{i=1}^{n} o\left(\left\|\tilde{W}_2\left(\tilde{W}_1 x_i + \tilde{b}_1 \mathbf{1}_m^T\right) + \tilde{b}_2 \mathbf{1}_m^T - \tilde{W}\begin{bmatrix} x_i \\ 1 \end{bmatrix}\right\|\right) \\
&= -\frac{2}{m}\sum_{i=1}^{l'} \boldsymbol{V}_{1,i}\gamma + o(\gamma),
\end{aligned}
$$

where $\boldsymbol{V}$ and l' are also defined the same as those in Theorem 7.9.

When γ is sufficiently small and $\mathrm{sgn}(\gamma) = \mathrm{sgn}\left(\sum_{i=1}^{l'} \boldsymbol{V}_{1,i}\right)$, we have that

$$
\hat{R}_S\left(\left(\left[\tilde{W}_i\right]_{i=1}^2, \left[\tilde{b}_i\right]_{i=1}^2\right)\right) < f(\tilde{W}).
$$

This complete the proof of Corollary 7.3. □

7.2.3.6 A Preparation Lemma

We now prove the preparation lemma used above.

Lemma 7.7 *Suppose* $\boldsymbol{u} = (u_1 \cdots u_m) \in \mathbb{R}^{1\times m}$ *which satisfies* $\boldsymbol{u} \neq \mathbf{0}$ *and*

$$\sum_{i=1}^{n} u_i = 0, \tag{7.50}$$

while $\{x_1,\ldots,x_m\}$ *is a set of vector* $\subset \mathbb{R}^{m\times 1}$. *Suppose index set* $S = \{1, 2, \ldots, m\}$. *Then for any series of real number* $\{v_1, \cdots, v_m\}$, *there exists a non-empty separation* I, J *of* S, *which satisfies* $I \cup J = S$, $I \cap J = \emptyset$ *and both* I *and* J *are not empty, a vector* $\beta \in \mathbb{R}^{m\times 1}$,*such that,*

(1.1) for any sufficiently small positive real α, $i \in I$, *and* $j \in J$, *we have* $v_i - \alpha\beta^T x_i < v_j - \alpha\beta^T x_j$;

(1.2) $\sum_{i\in I} u_i \neq 0$.

Proof If there exists a non-empty separation I and J of the index set S, such that when $\beta = 0$, (1.1) and (1.2) hold, the lemma is apparently correct.

Otherwise, suppose that there is no non-empty separation I and J of the index set S such that (1.1) and (1.2) hold simultaneously when $\beta = 0$.

Some number v_i in the sequence $(v_1, v_2, \ldots, v_m)$ are probably equal to each other. We rerarrange the sequence by the increasing order as follows,

$$v_1 = v_2 = \cdots = v_{s_1} < v_{s_1+1} = \cdots = v_{s_2} < \cdots < v_{s_{k-1}+1} = \cdots = v_{s_k} = v_m, \tag{7.51}$$

where $s_k = m$.

Then, for any $j \in \{1, 2, \ldots, k-1\}$, we argue that

$$\sum_{i=1}^{s_j} u_i = 0.$$

Otherwise, suppose there exists a s_j, such that

$$\sum_{i=1}^{s_j} u_i \neq 0.$$

Let $I = \{1, 2, \ldots, s_j\}$ and $J = \{s_j + 1, \ldots, m\}$. Then, when $\beta = 0$, we have

$$v_i - \alpha\beta^T x_i = v_i < v_j = v_j - \alpha\beta^T x_j,$$

and

$$\sum_{i\in I} u_i = \sum_{i=1}^{s_j} u_i \neq 0,$$

which are exactly the arguments (1.1) and (1.2). Thereby we construct a contrary example. Therefore, for any $j \in \{1, 2, \ldots, k-1\}$, we have

$$\sum_{i=1}^{s_j} u_i = 0.$$

Since we assume that $\mathbf{u} \neq 0$, there exists an index $t \in \{1, \ldots, k-1\}$, such that there exists an index $i \in \{s_t + 1, ..., s_{t+1}\}$ that $u_i \neq 0$.

Let $l \in \{s_t + 1, ..., s_{t+1}\}$ is the index such that x_l has the largest norm while $u_l \neq 0$:

$$l = \underset{j \in \{s_t+1,\ldots,s_{t+1}\}, u_j \neq 0}{\arg\max} \left\| x_j \right\|. \tag{7.52}$$

We further rearrange the sequence $(v_{s_t+1}, ..., v_{s_{t+1}})$ such that there is an index $l' \in \{s_t + 1, \ldots, s_{t+1}\}$,

$$\|x_{l'}\| = \max_{j \in \{s_t+1,\ldots,s_{t+1}\}, u_j \neq 0} \left\| x_j \right\|,$$

and

$$\forall i \in \{s_t + 1, ..., l'\}, \langle x_{l'}, x_i \rangle \geq \|x_{l'}\|^2; \tag{7.53}$$

$$\forall i \in \{l' + 1, \ldots, s_{t+1}\}, \langle x_{l'}, x_i \rangle < \|x_{l'}\|^2. \tag{7.54}$$

It is worth noting that it is probably $l' = s_{t+1}$, but it is a trivial case that would not influence the result of this lemma.

Let $I = \{1, ..., l'\}$, $J = \{l' + 1, ..., n\}$, and $\beta = x_{l'}$. We prove (1.1) and (1.2) as follows.

Proof of argument (1.1) We argue that for any $i \in I$, $v_i - \alpha\beta^T x_i \leq v_{l'} - \alpha\beta^T x_{l'}$ and for any $j \in J$, $v_j - \alpha\beta^T x_j > v_{l'} - \alpha\beta^T x_{l'}$.

There are three situations:

(A) $i \in \{1, \ldots, s_t\}$ and $j \in \{s_{t+1} + 1, \ldots, n\}$. Applying Eq. (7.51), for any $i \in \{1, \ldots, s_t\}$ and $j \in \{s_{t+1} + 1, \ldots, m\}$, we have that $v_i < v_{l'}$ and $v_j > v_{l'}$. Therefore, when α is sufficiently small, we have the following inequalities,

$$v_i - \alpha\beta^T x_i < v_{l'} - \alpha\beta^T x_{l'},$$

$$v_j - \alpha\beta^T x_j > v_{l'} - \alpha\beta^T x_{l'}.$$

(B) $i \in \{s_t + 1, \ldots, l'\}$. Applying Eq. (7.53) and because of $\alpha > 0$, we have

$$-\alpha\langle\beta, x_i\rangle \leq -\alpha \|\beta\|^2 = -\alpha\langle\beta, x_{l'}\rangle.$$

Since $v_i = v_{l'}$, it further leads to

$$v_i - \alpha\beta^T x_i \le v_{l'} - \alpha\beta^T x_{l'}.$$

(C) $j \in \{l' + 1, \ldots, s_{t+1}\}$. Similarly, applying Eq. (7.54) and because of $\alpha > 0$, we have

$$-\alpha\langle\beta, x_j\rangle > -\alpha\,\|\beta\|^2 = -\alpha\langle\beta, x_{l'}\rangle.$$

Since $v_j = v_{l'}$, it further leads to

$$v_j - \alpha\beta^T x_j > v_{l'} - \alpha\beta^T x_{l'},$$

which is exactly the argument (1.1).

Proof of argument (1.2) We argue that for any $i \in \{s_t + 1, \ldots, l' - 1\}$, $u_i = 0$. Otherwise, suppose there exists an $i \in \{s_t + 1, \ldots, l' - 1\}$ such that $u_i \neq 0$. From Eq. (7.52), we have $\|x_i\| \le \|x_{l'}\|$. Therefore,

$$\langle x_{l'}, x_i\rangle \le \|x_{l'}\|\|x_i\| \le \|x_{l'}\|^2,$$

where the first inequality strictly holds if the vector $x_{l'}$ and x_i have the same direction, while the second inequlity strictly holds when x_i and $x_{l'}$ have the same norm. Because $x_{l'} \neq x_i$, we have the following inequality,

$$\langle x_{l'}, x_i\rangle < \|x_{l'}\|^2,$$

which contradicts to Eq. (7.53), i.e.,

$$\langle x_{l'}, x_i\rangle \ge \|x_{l'}\|^2, \forall i \in \{s_t + 1, \ldots, l'\}.$$

Therefore,

$$\sum_{i\in I} u_i = \sum_{i=1}^{s_t} u_i + \sum_{i=s_t+1}^{l'-1} u_i + u_{l'} = u_{l'} \neq 0.$$

The proof is completed. □

Remark 7.3 For any $i \in \{l' + 1, ..., s_{t+1}\}$, we have

$$\left(v_i - \alpha\beta^T x_i\right) - \left(v_{l'} - \alpha\beta^T x_{l'}\right) = \alpha\beta^T (x_{l'} - x_i), \tag{7.55}$$

while for any $j \in \{s_{t+1} + 1, ..., m\}$, we have

$$\left(v_j - \alpha\beta^T x_j\right) - \left(v_{l'} - \alpha\beta^T x_{l'}\right) = v_j - v_{l'} + \alpha\beta^T (x_{l'} - x_j). \tag{7.56}$$

Because $v_j > v_{l'}$, when the real number α is sufficiently small, we have

$$\alpha\beta^T (x_{l'} - x_i) < v_j - v_{l'} + \alpha\beta^T (x_{l'} - x_j).$$

Applying Eqs. (7.55) and (7.56), we have

$$\left(v_i - \alpha\beta^T x_i\right) - \left(v_{l'} - \alpha\beta^T x_{l'}\right) < \left(v_j - \alpha\beta^T x_j\right) - \left(v_{l'} - \alpha\beta^T x_{l'}\right).$$

Therefore, if $l' < s_{t+1}$, we have

$$\min_{i\in\{l'+1,\ldots,m\}} \left(v_i - \alpha\beta^T x_i\right) - \left(v_{l'} - \alpha\beta^T x_{l'}\right) = \min_{i\in\{l'+1,\ldots,s_{t+1}\}} \alpha\beta^T (x_{l'} - x_i); \quad (7.57)$$

while if $l' = s_{t+1}$,

$$\min_{i\in\{l'+1,\ldots,m\}} \left(v_i - \alpha\beta^T x_i\right) - \left(v_{l'} - \alpha\beta^T x_{l'}\right) = \min_{i\in\{l'+1,\ldots,m\}} v_i - v_l + \alpha\beta^T (x_{l'} - x_i). \quad (7.58)$$

Equations (7.57) and (7.58) make senses because $l' < m$. Otherwise, from Lemma 7.7 we have $\sum_{i=1}^{m} u_i \neq 0$, which contradicts to the assumption.

7.3 Proofs of Theorems 7.7, 7.8, Corollaries 7.1, and 7.2

This section gives the proofs of Theorems 7.7, 7.8, Corollaries 7.1, and 7.2 omitted from Sect. 7.2.2.

7.3.1 Squared Loss

We first check that the squared loss is strictly convex, which is even restrictive than "convex".

Lemma 7.8 *The empirical risk $\hat{R}_S$ under squared loss is strictly convex with respect to the prediction $\hat{Y}$.*

Proof The second derivative of the empirical risk $\hat{R}_S$ under squared loss with respect to the prediction $\hat{Y}$ is

$$\frac{\partial^2 l_{ce}(Y, \hat{Y})}{\partial \hat{Y}^2} = \frac{\partial^2 (y - \hat{Y})^2}{\partial \hat{Y}^2} = 2 > 0.$$

Therefore, the empirical risk $\hat{R}_S$ under squared loss is strictly convex with respect to prediction $\hat{Y}$. □

7.3.2 Smooth and Multilinear Partition

In the context where all activations are linear functions, the neural network simplifies to a multilinear model characterized by a smooth and multilinear loss surface. However, when nonlinear activations are introduced, the landscape of the loss surface becomes more complex due to the nonlinearity introduced by these activation functions. When input data passes through the linear portions of activation functions, the resulting output resides in a smooth and multilinear region of the loss surface. This smoothness and linearity allow for predictable behavior under parameter changes, ensuring that each region expands into an open cell with continuous and differentiable characteristics.

Conversely, nonlinear points within the activations are non-differentiable, leading to non-smooth empirical risk concerning the parameters. These nonlinear points correspond to the non-differentiable boundaries between cells on the loss surface, where the loss function exhibits abrupt changes due to the inherent nonlinearity of the activations.

7.3.3 Every Local Minimum Is Globally Minimal Within a Cell

Proof of Theorem 7.7 In every cell, the input sample points flows through the same linear parts of the activations no matter what values the parameters are.

(1) We first proves that the empirical risk $\hat{R}_S$ equals to a convex function with respect to a variable $\hat{W}$ that is calculated from the parameters W.

Suppose (W_1, W_2) is a local minimum within a cell. We argue that

$$\sum_{i=1}^{m} l\left(y_i, W_2 \text{diag}\left(A_{\cdot,i}\right) W_1 x_i\right) = \sum_{i=1}^{m} l\left(y_i, A_{\cdot,i}^T \text{diag}(W_2) W_1 x_i\right), \tag{7.59}$$

where $A_{\cdot,i}$ is the i-th column of the following matrix

$$A = \begin{bmatrix} h'_{s_-,s_+}((W_1)_{1,\cdot}x_1) & \cdots & h'_{s_-,s_+}((W_1)_{1,\cdot}x_m) \\ \vdots & \ddots & \vdots \\ h'_{s_-,s_+}((W_1)_{d_1,\cdot}x_1) & \cdots & h'_{s_-,s_+}((W_1)_{d_1,\cdot}x_m) \end{bmatrix}. \tag{7.60}$$

The left-hand side (LHS) is as follows,

$$\text{LHS} = \sum_{i=1}^{m} l\left(y_i, W_2 \text{diag}\left(A_{\cdot,i}\right) W_1 x_i\right)$$

$$= \sum_{i=1}^{m} l\left(y_i, \left[(W_2)_{1,1} A_{1,i} \cdots (W_2)_{1,d_1} A_{d_1,i}\right] W_1 x_i\right).$$

Meanwhile, the right-hand side (RHS) is as follows,

$$\begin{aligned} \text{RHS} &= \sum_{i=1}^{m} l\left(y_i, A_{\cdot,i}^T \text{diag}(W_2) W_1 x_i\right), \\ &= \sum_{i=1}^{m} l\left(y_i, \left[(W_2)_{1,1} A_{1,i} \cdots (W_2)_{1,d_1} A_{d_1,i}\right] W_1 x_i\right). \end{aligned}$$

Apparently, LHS = RHS. Thereby, we proved Eq. (7.59).

Afterwards, we define

$$\hat{W}_1 = \text{diag}(W_2) W_1, \tag{7.61}$$

and then straighten the matrix $\hat{W}_1$ to a vector $\hat{W}$,

$$\hat{W} = \left((\hat{W}_1)_{1,\cdot} \cdots (\hat{W}_1)_{d_1,\cdot}\right). \tag{7.62}$$

Also define

$$\hat{X} = \left(A_{\cdot,1} \otimes x_1 \cdots A_{\cdot,m} \otimes x_m\right). \tag{7.63}$$

Then, we can prove that the following equations,

$$\begin{aligned} \left(A_{\cdot,1}^T \hat{W}_1 x_1 \cdots A_{\cdot,n}^T \hat{W}_1 x_m\right) &= \left((\hat{W}_1)_{1,\cdot} \cdots (\hat{W}_1)_{d_1,\cdot}\right) \left(A_{\cdot,1} \otimes x_1 \cdots A_{\cdot,n} \otimes x_m\right) \\ &= \hat{W} \hat{X}. \end{aligned} \tag{7.64}$$

Applying Eq. (7.64), the empirical risk is transferred to convex function,

$$\hat{R}_S(W_1, W_2) = \frac{1}{m} \sum_{i=1}^{m} l\left(y_i, (A_{\cdot,i})^T \text{diag}(W_2) W_1 x_i\right) = \frac{1}{m} \sum_{i=1}^{m} l\left(y_i, \hat{W} \hat{X}_i\right). \tag{7.65}$$

The empirical risk is rearranged as a convex function in terms of $\hat{W}$ which unite the two weight matrices W_1 and W_2 and the activation h are together as $\hat{W}$.

Applying Eqs. (7.61) and (7.62), we have

$$\hat{W} = \left[(W_2)_1 (W_1)_{1,\cdot} \cdots (W_2)_{d_1} (W_1)_{d_1,\cdot}\right].$$

(2) We then prove that the local minima (including global minima) of the empirical risk $\hat{R}_S$ with respect to the parameter W is also local minima with respect to the corresponding variable $\hat{W}$.

We first prove that for any $i \in [1 : d_1 d_2]$, we have $e_i \hat{X} \nabla = 0$, where ∇ is

$$\nabla = \left[\nabla_{\left(\hat{W}\hat{X}\right)_1} l\left(Y_1, \left(\hat{W}\hat{X}\right)_1\right) \cdots \nabla_{\left(\hat{W}\hat{X}\right)_m} l\left(Y_m, \left(\hat{W}\hat{X}\right)_m\right) \right]^T.$$

To see this, we divide i into two cases: $(W_2)_i \neq 0$ and $(W_2)_i = 0$.

Case 1: $(W_2)_i \neq 0$. The local minimizer of the empirical risk $\hat{R}_S$ with respect to the parameter W satisfies the following equation $\frac{\partial \hat{R}_S}{\partial (W_1)_{i,j}} = 0$. Therefore,

$$\begin{aligned} 0 =& \frac{\partial \hat{R}_S}{\partial (W_1)_{i,j}} \\ =& \frac{\partial \left(\sum_{k=1}^{n} l\left(Y_{\cdot,k}, \left(\hat{W}\hat{X}\right)_{\cdot,k}\right)\right)}{\partial (W_1)_{i,j}} \\ =& \sum_{k=1}^{n} [\underbrace{0 \cdots 0}_{d_X(i-1)+j-1} \, (W_2)_i \, \underbrace{0 \cdots 0}_{d_X d_1 - d_X(i-1)-j}] \hat{X}_k \nabla_{\left(\hat{W}\hat{X}\right)_{\cdot,k}} l\left(Y_{\cdot,k}, \left(\hat{W}\hat{X}\right)_{\cdot,k}\right), \end{aligned} \tag{7.66}$$

where (W_2) is a vector and $(W_2)_i$ is its i-th component.

By dividing the both hand sides of Eq. (7.66) with $(W_2)_i$, we can get

$$(e_{d_X(i-1)+j} \hat{X}) \nabla = 0.$$

Case 2: $(W_2)_i = 0$. Suppose $\mathbf{u}_1 \in \mathbb{R}^{d_0}$ is a unitary vector, $u_2 \in \mathbb{R}$ is a real number, and ε is a small enough positive constant. Define a disturbance of W_1 and W_2 as

$$\Delta W_1 = [\underbrace{0 \cdots 0}_{d_X(i-1)}, \varepsilon \mathbf{u}_1, \underbrace{0 \cdots 0}_{d_1 d_X - d_X i}],$$

and

$$\Delta W_2 = [\underbrace{0 \cdots 0}_{i-1}, \varepsilon^2 u_2, \underbrace{0 \cdots 0}_{d_1 - i}].$$

When ε is sufficiently small, ΔW_1 and ΔW_2 are also sufficiently small. Since (W_1, W_2) is a local minimum, we have

$$\begin{aligned} \frac{1}{m} \sum_{k=1}^{m} l\left(Y_k, \left(\left(\hat{W} + \Delta\right) \hat{X}\right)_k\right) =& \hat{R}_S(W_1 + \Delta W_1, W_2 + \Delta W_2) \\ \geq& \hat{R}_S(W_1, W_2) = \frac{1}{m} \sum_{k=1}^{m} l\left(Y_k, \left(\hat{W}\hat{X}\right)_k\right), \end{aligned} \tag{7.67}$$

where Δ is defined as follows,

$$\begin{aligned}\Delta = &[(W_2 + \Delta W_2)_1(W_1 + \Delta W_1)_1, \ldots, (W_2 + \Delta W_2)_{d_1}(W_1 + \Delta W_1)_{d_1}] \\ &- \left[(W_2)_1(W_1)_1, \ldots, (W_2)_{d_1}(W_1)_{d_1}\right] \\ &\overset{(*)}{=}[\underbrace{0 \cdots 0}_{d_X(i-1)}, \varepsilon^2 u_2 \left(\varepsilon \mathbf{u}_1 + (W_1)_i\right), \underbrace{0 \cdots 0}_{d_1 d_X - d_X i}]. \end{aligned} \tag{7.68}$$

Here, Eq. $(*)$ comes from $(W_2)_i = 0$. Rearrange Eq. (7.67) and apply the Taylor's Theorem, we can get that

$$\Delta \cdot \hat{X}\nabla + \mathbf{O}\left(\|\Delta \cdot \hat{X}\|^2\right) \geq 0.$$

Applying Eq. (7.68), we have

$$\begin{aligned}&[0 \cdots 0, \varepsilon^2 u_2 \left(\varepsilon \mathbf{u}_1 + (W_1)_i\right), 0 \cdots 0]\hat{X}\nabla \\ &+ \varepsilon^4 O(\|[0 \cdots 0, u_2 \left(\varepsilon \mathbf{u}_1 + (W_1)_i\right), 0 \cdots 0]\hat{X}\|^2) \\ &\overset{(**)}{=}[0 \cdots 0, \varepsilon^3 u_2 \mathbf{u}_1, 0 \cdots 0]\hat{X}\nabla + \varepsilon^4 O(\|[0 \cdots 0, u_2 \left(\varepsilon \mathbf{u}_1 + (W_1)_i\right), 0 \cdots 0]\hat{X}\|^2) \\ &= \varepsilon^3 [0 \cdots 0, u_2 \mathbf{u}_1, 0 \cdots 0]\hat{X}\nabla + \mathbf{o}(\varepsilon^3) \geq 0. \end{aligned} \tag{7.69}$$

Here, Eq. $(**)$ can be obtained from follows. Because W_2 is a local minimizer, for any component $(W_2)_i$ of W_2,

$$\frac{\partial \left(\sum_{k=1}^m l\left(Y_k, \left(\hat{W}\hat{X}\right)_k\right)\right)}{\partial (W_2)_i} = 0,$$

which leads to

$$[\underbrace{0 \cdots 0}_{d_X(i-1)}, (W_1)_i, \underbrace{0 \cdots 0}_{d_X d_1 - i d_X}]\hat{X}\nabla = 0.$$

When ε approaches 0, Eq. (7.69) leads to the following inequality,

$$[\underbrace{0 \cdots 0}_{d_X(i-1)}, u_2 \mathbf{u}_1, \underbrace{0 \cdots 0}_{d_X d_1 - i d_X}]\hat{X}\nabla \geq 0.$$

Since $\mathbf{u}_1$ and u_2 are arbitrarily picked (while the norms equal 1), the inequality above further leads to

$$\left[0 \cdots 0 \; e_j \; 0 \cdots 0\right] \hat{X}\nabla = 0, \tag{7.70}$$

which completes the proof of the argument.

Therefore, for any i and j, we have proven that

$$e_{d_0(i-1)+j}\hat{X}\nabla = 0,$$

which demonstrates that

$$\hat{X}\nabla = 0,$$

which means $\hat{W}$ is also a local minimizer of the empirical risk $\hat{\mathcal{R}}$,

$$\hat{R}_S(W) = \sum_{i=1}^{m} l(Y_i, W\hat{X}_i). \tag{7.71}$$

(3) Applying the property of convex function, $\hat{W}$ is a global minimizer of the empirical risk $\hat{R}_S$, which leads to (W_1, W_2) is a global minimum inside this cell.

The proof is completed.

7.3.4 Equivalence Classes of Local Minimum Valleys in Cells

***Proof of Theorem* 7.8 *and Corollary* 7.1** In the proof of Theorem 7.7, we constructed a map Q: $(W_1, W_2) \to \hat{W}$. Further, in any fixed cell, the represented hypothesis of a neural network is uniquely determined by $\hat{W}$.

We first prove that all local minima in a cell are concentrated as a local minimum valley. Since the loss function l is strictly convex, the empirical risk has one unique local minimum (which is also a global minimum) with respect to $\hat{W}$ in every cell, if there exists some local minimum in the cell. Meanwhile, we have proved that all local minima with respect to (W_1, W_2) are also local minima with respect to the corresponding $\hat{W}$. Therefore, all local minima with respect to (W_1, W_2) correspond one unique $\hat{W}$. Within a cell, when W_1 expands by a positive real factor α to W_1' and W_2 shrinks by the same positive real factor α to W_2', we have $Q(W_1, W_2) = Q(W_1', W_2')$, i.e., the $\hat{W}$ remains invariant.

Further, we argue that all local minima in a cell are connected with each other by a continuous path, on which the empirical risk is invariant. For every local minima pair (W_1, W_2) and (W_1', W_2'), we have

$$\text{diag}(W_2)W_1 = \text{diag}(W_2')W_1'. \tag{7.72}$$

Since $h'_{s_-,s_+}(W_1X) = h'_{s_-,s_+}(W_1'X)$ (element-wise), for every $i \in [1, d_1]$,

$$\text{sgn}\left((W_2)_i\right) = \text{sgn}\left(\left(W_2'\right)_i\right).$$

Therefore, a continuous path from (W_1, W_2) to (W_1', W_2') can be constructed by finite moves, each of which expands a component of W_2 by a real constant α and then shrinks the corresponding line of W_1 by the same constant α.

We then prove that all local minima in a cell constitute an equivalence class. Define an operation $\sim_R$ as follows,

$$(W_1^1, W_2^1) \sim_R (W_1^2, W_2^2),$$

if

$$Q(W_1^1, W_2^1) = Q(W_1^2, W_2^2).$$

We then argue that $\sim_R$ is an equivalence relation. The three properties of equivalence relations are checked as follows.

(1) Reflexivity:

For any (W_1, W_2), we have

$$Q(W_1, W_2) = Q(W_1, W_2).$$

Therefore,

$$(W_1, W_2) \sim_R (W_1, W_2).$$

(2) Symmetry:

For any pair (W_1^1, W_2^1) and (W_1^2, W_2^2), Suppose that

$$(W_1^1, W_2^1) \sim_R (W_1^2, W_2^2).$$

Thus,

$$Q(W_1^1, W_2^1) = Q(W_1^2, W_2^2).$$

Apparently,

$$Q(W_1^2, W_2^2) = Q(W_1^1, W_2^1).$$

Therefore,

$$Q(W_1^2, W_2^2) \sim_R Q(W_1^1, W_2^1).$$

(3) Transitivity:

For any (W_1^1, W_2^1), (W_1^2, W_2^2), and (W_1^3, W_2^3), suppose that

$$(W_1^1, W_2^1) \sim_R (W_1^2, W_2^2),$$
$$(W_1^2, W_2^2) \sim_R (W_1^3, W_2^3).$$

Then,

$$Q(W_1^1, W_2^1) = Q(W_1^2, W_2^2),$$

$$Q(W_1^2, W_2^2) = Q(W_1^3, W_2^3).$$

Apparently,

$$Q(W_1^1, W_2^1) = Q(W_1^2, W_2^2) = Q(W_1^3, W_2^3).$$

Therefore,

$$(W_1^1, W_2^1) \sim_R (W_1^3, W_2^3).$$

We then prove the mapping Q is the quotient map.
Define a map as follows,

$$T : (W_1, W_2) \to (\text{diag}(W_2)W_1, \mathbf{1}_{1\times d_1}).$$

We then define an operator $\oplus$ as,

$$(W_1^1, W_2^1) \oplus (W_1^2, W_2^2) = T(W_1^1, W_2^1) + T(W_1^2, W_2^2),$$

the inverse of (W_1, W_2) is defined to be $(-W_1, W_2)$ and the zero element is defined to be $(\mathbf{0}, \mathbf{1}_{1\times d_1})$.

Obviously, the following is a linear mapping:

$$Q : ((\mathbb{R}^{d_1\times dx}, \mathbb{R}^{1\times d_1}), \oplus) \to (\mathbb{R}^{1\times dxd_1}, +).$$

For any pair (W_1^1, W_2^1) and (W_1^2, W_2^2), we have

$$(W_1^1, W_2^1) \sim_R (W_1^2, W_2^2),$$

if and only if

$$(W_1^1, W_2^1) \oplus (-W_1^2, W_2^2) \in \text{Ker}(Q).$$

Therefore, the quotient space $(\mathbb{R}^{d_1\times dx}, \mathbb{R}^{1\times d_1})/\text{Ker}(Q)$ is a definition of the equivalence relation $\sim_R$.

The proof is completed.

7.4 Geometric Structure of the Loss Surface

What does the loss surface actually look like? Understanding the geometric structure of the loss surface could lead to significant advancements in our comprehension of deep learning.

Linear partition of the loss surface. Soudry and Hoffer (2018) introduced the concept of a smooth and multilinear partition within the loss surface of neural networks, highlighting the impact of nonlinearities in piecewise linear activations. These nonlinearities effectively segment the loss surface into distinct regions characterized by smooth and multilinear properties. Specifically, every nonlinear point within the activation functions contributes to a set of non-differentiable boundaries between cells, while the linear segments of the activations correspond to the smooth and multilinear interiors of these cells. This decomposition provides insights into the geometric structure of the loss surface and its relationship with neural network behavior and training dynamics.

He et al. (2020) demonstrated several intricate properties of mode connectivity: (1) Within an open cell, if local minima exist, they are equally optimal in terms of empirical risk, making all local minima global within that cell. This highlights a uniformity of performance among local minima within the same region of the loss surface. (2) The local minima within any open cell form an equivalence class and are concentrated in a valley, suggesting a concentrating effect of optimal solutions in specific regions of the loss landscape. (3) When all activations are linear, the partition collapses into a single cell, including linear neural networks as a special case. This observation suggests the value of nonlinear activations in shaping the complexity and structure of the loss surface. These three findings provide important insights into the behavior and landscape of neural network loss surfaces under different activation regimes, shedding light on the nature of optimization and generalization in deep learning models.

The property (2) introduced by He et al. (2020) elucidates the concept of mode connectivity, which suggests that solutions discovered through SGD or its variants are connected by a path in the weight space, where all points along the path exhibit nearly identical empirical risk. This mode connectivity phenomenon has been empirically observed in studies by Garipov et al. (2018) and Draxler et al. (2018). Recently, Kuditipudi et al. (2019) provided theoretical support by demonstrating that mode connectivity can be assured through dropout stability and noise stability. These findings contribute to the understanding of how optimization algorithms explore the weight space and the resilience of solutions under perturbations and dropout conditions.

Correspondence between the landscapes of the empirical risk and expected risk: Two seminal works by Zhou and Feng (2018) and Mei et al. (2018) have theoretically revealed a critical link between the landscapes of the empirical risk surface and the expected counterpart. This correspondence suggests that studying the geometric properties of empirical risk surfaces can provide insights into model generalizability, which pertains to the gap between expected and empirical risks. Notably, these studies demonstrated that the gradient of the empirical risk, the stationary points of the empirical risk, and the empirical risk itself can all be asymptotically approximated by their population equivalents. Moreover, they delivered a generalization bound for the nonconvex scenario: with probability at least $1-\delta$,

$$O\left(\tau\sqrt{\frac{9}{8}[1+c_r(D-1)]}\sqrt{\frac{s\log(mU/D)+\log(4/\delta)}{m}}\right),$$

where all data x are assumed to be τ-sub-Gaussian, $c_r = \max\left(r^2/16, \left(r^2/16\right)^{l-1}\right)$, and s is the number of nonzero components of the weights. Later, Mei et al. (2018)

proved a similar correspondence between the Hessian matrices of the expected risk and the empirical risk under the assumption that the sample size is greater than the number of parameters.

Eigenvalues of the Hessian. Sagun et al. in 2016 (Sagun et al. 2016) and 2018 (Sagun et al. 2018) and Papyan in 2018 (Papyan 2018) conducted experiments to study the eigenvalues of the Hessian of the loss surface. Sagun et al. (2016), Sagun et al. (2018) discovered that (1) a large bulk of the eigenvalues are centred close to zero and (2) several outliers are located far from this bulk. Papyan (2018) presented the full spectrum of the Hessian matrix. In Papyan (2018, pp. 2, Figs. 1(a) and 1(c), and pp. 3, Figs. 2(a) and 2(c)), the authors compare the spectra of the Hessian matrices obtained when training and testing a VGG-11 network on the MNIST and CIFAR-10 datasets.

References

Baldi, Pierre, and Kurt Hornik. 1989. Neural networks and principal component analysis: learning from examples without local minima. *Neural Networks* 2 (1): 53–58.

Choromanska, Anna, Mikael Henaff, Michael Mathieu, Gérard Ben Arous, and Yann LeCun. 2015. The loss surfaces of multilayer networks. In *International Conference on Artificial Intelligence and Statistics*.

Draxler, Felix, Kambis Veschgini, Manfred Salmhofer, and Fred Hamprecht. 2018. Essentially no barriers in neural network energy landscape. In *International Conference on Machine Learning*.

Freeman, C Daniel, and Joan Bruna. 2017. Topology and geometry of half-rectified network optimization. In *International Conference on Learning Representations*.

Garipov, Timur, Pavel Izmailov, Dmitrii Podoprikhin, Dmitry P Vetrov, and Andrew G Wilson. 2018. Loss surfaces, mode connectivity, and fast ensembling of dnns. In *Advances in Neural Information Processing Systems*.

Goldblum, Micah, Jonas Geiping, Avi Schwarzschild, Michael Moeller, and Tom Goldstein. 2020. Truth or backpropaganda? An empirical investigation of deep learning theory. In *International Conference on Learning Representations*.

Hanin, Boris, and David Rolnick. 2019. Complexity of linear regions in deep networks. In *International Conference on Machine Learning*.

He, Fengxiang, Bohan Wang, and Dacheng Tao. 2020. Piecewise linear activations substantially shape the loss surfaces of neural networks. In *International Conference on Learning Representations*.

He, Kaiming, Xiangyu Zhang, Shaoqing Ren, and Jian Sun. 2016. Deep residual learning for image recognition. In *IEEE Conference on Computer Vision and Pattern Recognition*.

Kawaguchi, Kenji. 2016. Deep learning without poor local minima. In *Advances in Neural Information Processing Systems*.

Kuditipudi, Rohith, Xiang Wang, Holden Lee, Yi Zhang, Zhiyuan Li, Wei Hu, Sanjeev Arora, and Rong Ge. 2019. Explaining landscape connectivity of low-cost solutions for multilayer nets. In *Advances in Neural Information Processing Systems*.

Laurent, Thomas, and James von Brecht. 2018. The multilinear structure of ReLU networks. In *International Conference on Machine Learning*.

LeCun, Yann, Yoshua Bengio, and Geoffrey Hinton. 2015. Deep learning. *Nature* 521 (7553): 436.

Litjens, Geert, Thijs Kooi, Babak Ehteshami Bejnordi, Arnaud Arindra Adiyoso Setio, Francesco Ciompi, Mohsen Ghafoorian, Jeroen Awm Van Der Laak, Bram Van Ginneken, and Clara I

Sánchez. 2017. A survey on deep learning in medical image analysis. *Medical Image Analysis* 42: 60–88.
Lu, Haihao, and Kenji Kawaguchi. 2017. Depth creates no bad local minima. *arXiv preprint* arXiv:1702.08580.
Mei, Song, Yu Bai, and Andrea Montanari. 2018. The landscape of empirical risk for nonconvex losses. *The Annals of Statistics* 46 (6A): 2747–2774.
Papyan, Vardan. 2018. The full spectrum of deep net hessians at scale: dynamics with sample size. *arXiv preprint* arXiv:1811.07062.
Safran, Itay, and Ohad Shamir. 2018. Spurious local minima are common in two-layer ReLU neural networks. In *International Conference on Machine Learning*.
Sagun, Levent, Léon Bottou, and Yann LeCun. 2016. Singularity of the hessian in deep learning. *arXiv preprint* arXiv:1611.07476.
Sagun, Levent, Utku Evci, V Ugur Guney, Yann Dauphin, and Leon Bottou. 2018. Empirical analysis of the hessian of over-parametrized neural networks. In *International Conference on Learning Representations Workshop*.
Silver, David, Aja Huang, Chris J Maddison, Arthur Guez, Laurent Sifre, George Van Den Driessche, Julian Schrittwieser, Ioannis Antonoglou, Veda Panneershelvam, and Marc Lanctot. 2016. Mastering the game of go with deep neural networks and tree search. *Nature* 529 (7587): 484–489.
Soudry, Daniel, and Elad Hoffer. 2018. Exponentially vanishing sub-optimal local minima in multilayer neural networks. In *International Conference on Learning Representations Workshop*.
Swirszcz, Grzegorz, Wojciech Marian Czarnecki, and Razvan Pascanu. 2016. Local minima in training of deep networks. *arXiv preprint* arXiv:1611:06310.
Witten, Ian H, Eibe Frank, Mark A Hall, and Christopher J Pal. 2016. *Data Mining: Practical Machine Learning Tools and Techniques*. Morgan Kaufmann.
Yun, Chulhee, Suvrit Sra, and Ali Jadbabaie. 2018. Global optimality conditions for deep neural networks. In *International Conference on Learning Representations*.
Yun, Chulhee, Suvrit Sra, and Ali Jadbabaie. 2019. Small nonlinearities in activation functions create bad local minima in neural networks. In *International Conference on Learning Representations*.
Zhou, Pan, and Jiashi Feng. 2018. Empirical risk landscape analysis for understanding deep neural networks. In *International Conference on Learning Representations*.
Zhou, Yi, and Yingbin Liang. 2018. Critical points of neural networks: analytical forms and landscape properties. In *International Conference on Learning Representations*.

Chapter 8
Linear Partition in the Input Space

Recent research has demonstrated that the input space of a rectified linear unit (ReLU) network, which exclusively uses ReLU-like (two-piece linear) activation functions, is divided into linear regions by the nonlinear activations.

Specifically, within these linear regions, the mapping induced by a ReLU network behaves linearly for input data. Conversely, at the boundaries between linear regions, the mapping becomes nonlinear and nonsmooth. Intuitively, the linear regions correspond to the linear segments of the ReLU activations, representing specific activation patterns. Meanwhile, the boundaries are defined by transition points where the activation pattern changes. As a result, each input example is associated with a neural code—a 0-1 matrix representing its activation pattern within the linear region it occupies.

This chapter introduces the concept of a neural code, which serves as a parameterized representation of activation patterns induced by a neural network for input examples. Sufficient experiments demonstrate that the neural code exhibits the interesting encoding properties, which are shared by hash code, in most common scenarios of deep learning for classification tasks.

8.1 Preliminaries

Suppose that a ReLU network $\mathcal{N}$ is trained to fit a dataset $S = \{(x_i, y_i), i = 1, \dots, m\}$ for classification, where $x_i \in \mathcal{X} \subset \mathbb{R}^{d_X}$, d_X is the dimensionality of x, $y_i \in \mathcal{Y} = \{1, \dots, d\}$, d is the number of potential categories, and m is the training sample size. Additionally, we assume that all examples (x_i, y_i) are independent and identically distributed (i.i.d.) random variables drawn from a data distribution $\mathcal{D}$. We denote the resulting well-trained model by $\mathcal{M}$. Here, "well-trained" refers to the condition in which the training procedure has converged.

F. He and D. Tao, *Foundations of Deep Learning*, Machine Learning: Foundations, Methodologies, and Applications, https://doi.org/10.1007/978-981-16-8233-9_8

Recent works have shown that the input space of a ReLU network $\mathcal{N}$ is partitioned into multiple linear regions, each of which corresponds to a specific activation pattern of the ReLU activation functions. In this section, we represent the activation pattern as a matrix $\boldsymbol{P} \in \mathcal{P} \subset \{0, 1\}^{l \times w}$, where l and w are the depth and largest width, respectively, of this neural network $\mathcal{N}$. Specifically, the (i, j)-th component characterizes the activation status of the j-th ReLU neuron in the i-th layer. If the (i, j)-th component is equal to 1, this represents that this neuron is activated, while a value of 0 means that this neuron is deactivated or invalid.[1] The matrix $\boldsymbol{P}$ is termed the *neural code*. We can also reformulate the neural code as a vector if there is no possibility for confusion of the depth and width.

It's essential to note that the boundaries separating linear regions have no physical width. This occurs because these boundaries align precisely with transition points in the activations, effectively resembling infinitely thin lines. In more straightforward terms, imagine these boundaries as razor-thin lines that separate different regions within the model. Consequently, the likelihood of an example precisely landing on one of these boundaries is virtually non-existent. Therefore, for practical analysis purposes, we assume that no example resides within these boundaries. This simplification greatly aids in our understanding and interpretation of the model's behavior. Consequently, upon fixing the weights w of the model $\mathcal{M}$, every example $x \in \mathcal{X}$ can be indexed by the neural code $\boldsymbol{P}$ of the corresponding linear region. Note that an example x can be seen in either the training sample or the test sample.

8.2 Neural Networks Act as Hash Encoders

In this section, we investigate the results of an empirical study, shedding light on two important discoveries. First, we find that the neural code, which represents data within a neural network, functions much like a hash code for the corresponding input data. Second, we show that the mapping from raw data to their neural codes, known as the activation mapping $\mathcal{X} \to \mathcal{P}$, behaves similarly to a hash function. What's intriguing is that unlike specialized methods for learning to hash, well-trained neural networks, typically used for tasks like classification, inherently exhibit hash encoding capabilities without any deliberate tuning. Specifically, the neural code displays two fundamental encoding properties: determinism and categorization. To rigorously evaluate the effectiveness of the activation mapping as a hash mapping, we employ the following two standard quantitative metrics, similar to those used in learning-to-hash approaches.

Similar to works on learning-to-hash, we adopt the following two measures to quantitatively evaluate the activation mapping as a hash mapping.

[1] Different layers may have different numbers of neurons. Therefore, there might be some indices (i, j) that are invalid. We represent the activation patterns of these neurons by 0 since they are never activated.

Redundancy ratio. We define the following *redundancy ratio* to measure the determinism property. Generally, a smaller redundancy ratio is preferred when we are evaluating the activation mapping as a hashing function.

Definition 8.1 (*Redundancy ratio*) Suppose that there are m examples in a dataset S. If they are located in n activation regions, the redundancy ratio is defined as $\frac{m-n}{m}$.

Categorization accuracy. In typical applications, hash codes are usually utilized for nearest neighbor searches. In our study, we employ straightforward algorithms like K-means, K-NN, and logistic regression on the neural codes. We gauge the encoding properties through training and test accuracy, where higher accuracy indicates superior encoding. Notably, K-means is traditionally employed for unsupervised learning, but here, we adapt it to verify our encoding properties within the realm of supervised learning. The modified pipeline is detailed in Sect. 8.4.

We examine the activation mappings generated by MLP (Multi-layer Perception), VGG-18, ResNet-18, ResNet-34, ResNeXt-26, and DenseNet-28 models trained on the MNIST dataset, as well as VGG-19, ResNet-18, ResNet-20, and ResNet-32 models trained on the CIFAR-10 dataset. Detailed information regarding the implementations is provided in Sect. 8.4. Across all cases, the redundancy ratio is nearly negligible (i.e. almost zero), and the categorization accuracy is commendable, as illustrated in Tables 8.1 and 8.2.

These results serve to confirm the reliability of both the determinism and categorization properties of our neural codes. Furthermore, we employ t-distributed stochastic neighbor embedding (t-SNE) (Maaten and Hinton 2008) to visually depict the neural codes, a technique showcased in Fig. 8.1. Through this visualization, we observe that data points belonging to the same category tend to cluster closely

Table 8.1 Accuracy of K-means and K-NN on the neural codes of convolutional neural networks (CNNs) trained on MNIST

Architecture	K-means acc (%)	K-NN acc (%)
VGG-18	99.95	99.33
ResNet-18	98.96	99.32
ResNet-34	99.66	99.49
ResNeXt-26	98.31	99.24
DenseNet-28	69.87	98.59

Table 8.2 Accuracy of logistic regression (LR) on the neural codes of CNNs trained on CIFAR-10

Architecture	LR acc (%)	Test acc (%)
VGG-19	92.19	91.43
ResNet-18	89.55	90.42
ResNet-20	88.76	90.44
ResNet-32	89.05	90.45

Fig. 8.1 t-SNE visualization of the neural codes of MNIST generated by a one-hidden-layer MLP of width 100

together, indicating a high degree of similarity. Conversely, distinct boundaries are discernible between data points from different categories, highlighting clear separations. This visual representation vividly illustrates the categorization property, demonstrating that neural codes effectively capture and differentiate between various categories of data.

8.3 Factors that Influence the Encoding Properties

The results presented in Tables 8.1 and 8.2 also suggest that the encoding properties exhibit variability in different scenarios. Through comprehensive experiments, we investigate which factors influence the encoding properties and how. The investigated factors include the model size, training time, sample size, three different popular regularizers, random data, and noisy labels. Some details of the experimental implementations are given in Sect. 8.4.

8.3.1 Relationship Between Model Size and Encoding Properties

We first study how the model size influences the encoding properties. We trained 115 one-hidden-layer MLPs of different widths on the MNIST dataset and 200 five-hidden-layer MLPs of different widths on the CIFAR-10 dataset (see Sect. 8.4 for the full list of widths considered in our experiments), while all irrelevant variables were strictly controlled. The experiments were repeated for 5 trials on MNIST and 10 trials on CIFAR-10.

Measurement of model size by width. When considering MLPs, a natural metric for evaluating model complexity is the layer width.[2] Using layer width as our measure of model size, we computed both the redundancy ratio and categorization accuracy across all scenarios, as illustrated in Fig. 8.2a, b. These plots reveal notable correlations between encoding properties and layer width: (1) Redundancy Ratio: Initially,

[2] Depth is also a natural measure of model size. However, the optimal training protocol (especially training time) significantly differs for networks of different depths. Thus, it is difficult to conduct experiments on depth while controlling other factors.

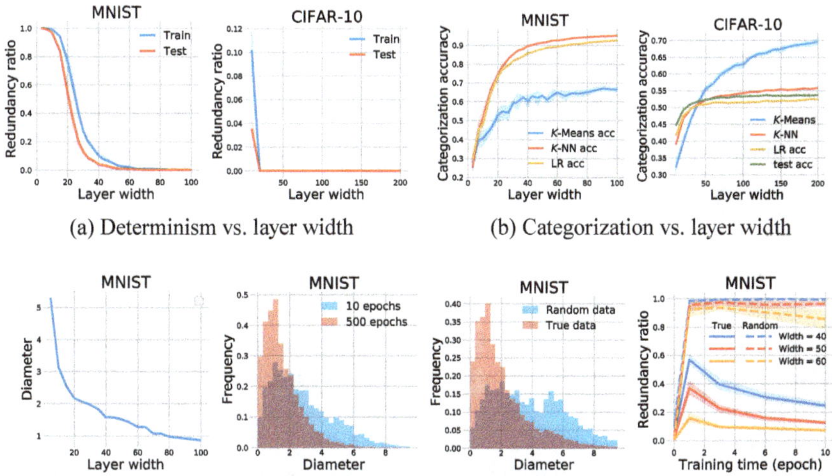

(a) Determinism vs. layer width (b) Categorization vs. layer width

(c) D vs. layer width (d) D at different times (e) D for random/true data (f) R vs. random/true data

Fig. 8.2 **a** Plots of the redundancy ratio as a function of the layer width of MLPs on both the training set (blue) and the test set (red). **b** Plots of the test accuracies of K-means (blue), K-NN (red), and logistic regression (LR, range) as functions of the MLP layer width. **c** Plots of the average stochastic activation diameter (D) as a function of the MLP layer width on MNIST. **d** Histograms of D calculated on MNIST for an MLP of width 50 trained on MNIST for 10 epochs (blue) and 500 epochs (red), respectively. **e** Histograms of D calculated on MNIST (blue) and randomly generated data of the same dimensions (red) for an MLP of width 50 trained on MNIST. The two red histograms are identical. **f** Plots of the redundancy ratio (R) calculated on MNIST ("True data", solid lines) and randomly generated data (dotted lines) as functions of the training time for MLPs of width 40 (blue), 50 (red), and 60 (orange). The dotted lines represent networks trained on the unaltered data and evaluated on random data. The models were trained 5 times on MNIST and 10 times on CIFAR-10 with different random seeds. The darker lines represent the averages over all seeds, and the shaded areas show the standard deviations

the redundancy ratio starts at relatively high values (almost 1 on both the training and test sets of MNIST, around 0.1 on the training set of CIFAR-10, and approximately 0.04 on the CIFAR-10 test set). However, it steadily decreases to nearly 0 across all cases as the layer width increases. (2) Categorization Accuracy: Initially, the categorization accuracy commences at relatively low values (about 25% on MNIST and 32-45% on CIFAR-10). However, as the width increases, the accuracy consistently improves across all scenarios, reaching relatively high values (approximately 70% for K-means on both datasets, exceeding 90% for K-NN and logistic regression on MNIST, and approximately 50% for K-NN and logistic regression on CIFAR-10, akin to the test accuracy on the raw data).

Measurement of model capacity in terms of the diameters of the linear regions. We devise a new measure to assess model capacity, termed the average stochastic activation diameter. This metric is computed through the following three steps: (1) we randomly sample a direction from a uniform distribution; (2) the stochastic diameter of a linear region is defined as the length of the longest line segment intersecting the linear region along the sampled direction; and (3) the average stochastic activation diameter is defined as the mean of these stochastic diameters across all linear regions containing data. In essence, a smaller average stochastic activation diameter indicates a finer division of the input space into smaller linear regions, enabling the representation of more intricate data structures. Therefore, this metric is an effective measure of the model capacity. Correspondingly, we observe a negative correlation between layer width and average stochastic activation diameter, as shown in Fig. 8.2c.

Similarly, Hanin and Rolnick (2019) introduced a concept termed the "typical distance from a random input to the boundary of its linear region". In contrast, our diameter conceptually represents the longest distance between any two points within a linear region. While (Hanin and Rolnick 2019) typical distance metric can be understood as the distance from a random input to the boundary of its linear region, our diameter measures the maximum possible separation within the linear region itself. When the linear region takes the form of an ideal ball, (Hanin and Rolnick 2019) typical distance will be equal to or smaller than the radius of the ball, which is half of our diameter. However, linear regions are often irregular in practice, as illustrated in Fig. 1 of Hanin and Rolnick (2019). Consequently, their distances typically turn out to be considerably smaller than our diameter. Consequently, the discrepancy between these two definitions can be significant, particularly depending on the level of irregularity. One may remain relatively constant while the other experiences significant variation. Consequently, Hanin and Rolnick (2019) distance metric may provide a lower bound on the linear region volume, whereas our diameter serves as an upper bound.

We further investigate the encoding properties beyond those of the data-generating distribution. To this end, we generate a set of examples following a uniform distribution across the unit ball centered at the original point. Additionally, we normalize the original data such that each pixel falls within the range [0, 1], ensuring comparable scales between the randomly generated and original data. Notably, we observe that the redundancy ratio exceeds 0.8 on the randomly generated data, as illustrated in Fig. 8.2f. This observation indicates that the determinism property is no longer upheld, implying that unique neural codes cannot effectively represent randomly generated data. Consequently, the categorization property becomes less discernible. To account for these findings, we propose the following hypothesis.

Hypothesis 8.1 The diameters of the linear regions in the support of a data distribution decrease as training progresses, while the diameters of regions far away show little change. □

We then collected the average stochastic activation diameters for each scenario, as illustrated in Fig. 8.2d, e. We observe that the stochastic diameters are more concentrated when the training time is longer; see Fig. 8.2d. Moreover, we observe the interesting result that the stochastic diameters for the true data are more concentrated at lower values than the stochastic diameters for the random data. Figure 8.2e shows the histograms of the stochastic diameters. The histograms for other scenarios are given in Sect. 8.5. These results fully support our hypothesis.

8.3.2 *Relationship Between Training Time and Encoding Properties*

We then explore how training time affects the encoding properties. Our experiments involved one-hidden-layer MLPs of varying widths trained on the MNIST dataset and five-hidden-layer MLPs of different widths trained on CIFAR-10. In total, we evaluated 810 models. To ensure the reliability of our findings, the experiments were repeated five times for MNIST and ten times for CIFAR-10, amounting to a comprehensive analysis of the influence of training time on encoding properties.

Throughout our experiments, we meticulously tracked the evolution of both the redundancy ratio and categorization accuracy at every epoch for all scenarios, documented in Fig. 8.3. These plots serve as a comprehensive visualization of the relationship between encoding properties and training time. Notably, our observations reveal a clear positive correlation: as the training time increases, we observe (1) a steady decline in the redundancy ratio and (2) a consistent improvement in categorization accuracy. This trend underscores the importance of training time in shaping the encoding capabilities of our models, shedding light on the dynamics of learning processes within neural networks.

We also make a noteworthy observation regarding the redundancy ratio of an untrained MLP on the MNIST dataset, which is nearly zero, as depicted in Fig. 8.3c. To provide a deeper understanding, let's investigate the rationale behind this finding: Upon random initialization, a neural network partitions the input space into numerous activation regions. If these regions are sufficiently small, each training data point effectively has its own dedicated activation region. However, it's essential to recognize that during this random initialization phase, the mapping from input data to output predictions lacks any meaningful structure. Consequently, the network may exhibit erratic behavior, producing vastly different predictions for data points originating from neighboring activation regions. As a consequence, categorization accuracy is notably poor, aligning with intuitive expectations. This phenomenon validates the notion that determinism alone is insufficient for adequately characterizing encoding properties. It also resonates with the concept of reservoir effects (Jaeger 2001; Maass et al. 2002), which emphasizes the dynamic and complex nature of neural network behavior during the initialization phase.

It's important to highlight a key distinction between our findings and those reported in Hanin and Rolnick (2019). While their research suggests that the count of linear regions increases with the training progress, our observations present a different perspective. We assert that the encoding properties we've examined do not extend beyond the training data distribution, regardless of any variations in

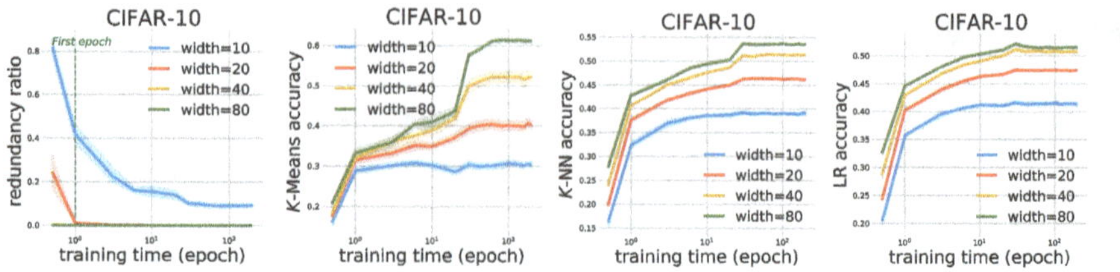

(a) R ratio vs. time (b) Categorization accuracy vs. time

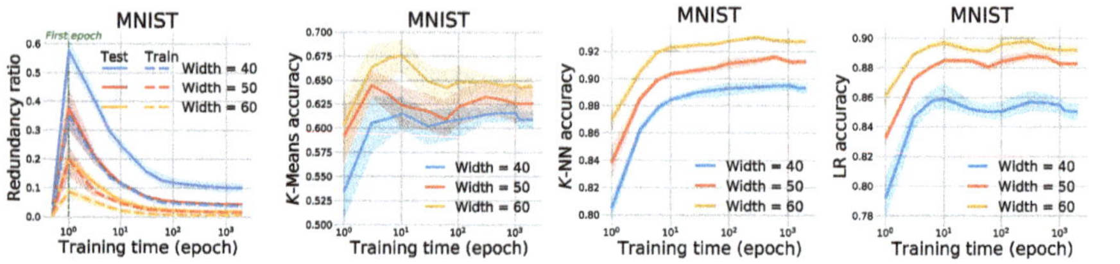

(c) R ratio vs. time (d) Test accuracies of $\boldsymbol{K}$-means, $\boldsymbol{K}$-NN, and logistic regression vs. time

Fig. 8.3 **a** The redundancy ratio (R ratio) as a function of the training time on CIFAR-10. **b** The test accuracies of K-means (left), K-NN (middle), and logistic regression (right) as functions of the training time. The models are MLPs of width 10 (blue), 20 (red), 40 (orange), and 80 (green). **c** The R ratio as a function of the training time on MNIST. **d** The test accuracies of K-means (left), K-NN (middle), and logistic regression (right) as functions of the training time. The models are MLPs of depth 40 (blue), 50 (red), and 60 (orange) on MNIST. The dotted lines represent networks trained on the unaltered data and evaluated on random data. All models obtained from MNIST were trained 5 times with different random seeds. The darker lines represent the averages over all seeds, and the shaded areas show the standard deviations

the count of linear regions. This assertion is supported by the data presented in Fig. 8.2f. What this implies is that while there may indeed be an increase in the count of linear regions, this phenomenon is localized to only a small fraction of the expansive input space. Consequently, even with this observed increase, there's no guarantee of a corresponding decrease in the redundancy ratio. This understanding features the intricate relationship between training dynamics and encoding properties within neural networks.

8.3.3 *Relationship Between Sample Size and Encoding Properties*

We now study the influence of sample size on the encoding properties of neural networks. To conduct this study, we trained a total of 210 one-hidden-layer MLPs, each with three different widths, and 480 five-hidden-layer MLPs, each with four different widths. These models were trained on training samples of varying sizes randomly sampled from the MNIST and CIFAR-10 datasets, respectively. It's essential to highlight our stringent experimental controls, ensuring that all irrelevant variables are meticulously managed to maintain experimental integrity (i.e. accurate and repeatable results). Additionally, instead of tracking epochs, we monitor the number of iterations. This decision is informed by the understanding that the number of iterations in one epoch scales proportionally with the sample size. To ensure the reliability and repeatability of our findings, each experiment is rigorously repeated five times for MNIST and ten times for CIFAR-10 datasets.

We conducted an extensive analysis by computing both the redundancy ratio and categorization accuracy across all scenarios, visually detailed in Fig. 8.4. Our findings unveil two significant trends: (1) The redundancy ratio, whether assessed on the training or test sample, exhibits a notable pattern. Upon initialization, it begins at a relatively high value, gradually diminishing to near-zero levels as the training sample size increases. This observation validates the diminishing redundancy as the model learns from a larger and more diverse dataset. (2) We observe a distinct positive correlation between the sample size and the test accuracies of all three algorithms. Specifically, as the sample size expands, the accuracy of K-means escalates from 20 to 40%, K-NN accuracy experiences a surge from 10 to 45%, and logistic regression accuracy shows a pronounced increase from 15 to 45%. These trends elucidate the profound impact of sample size on both redundancy ratio and classification accuracy and a negative correlation between redundancy ratio and categorization accuracy, providing useful insights into model performance under varying data conditions.

Surprisingly, we observe that the encoding properties on the test set also become stronger as the training sample size increases. Our hypothesis is as follows. Intuitively, a larger training sample size supports the neural network in attaining a higher expressive power, i.e., a finer linear partition in the input space. Meanwhile, a sample of larger size requires a finer linear partition to yield the same redundancy ratio. Our experiments show that the first effect is stronger than the second one. Thus, a larger sample size can help reduce the redundancy ratio.

8.3.4 Layerwise Ablation Study

We next study how different layers impact the encoding properties. We conducted a layerwise ablation study on the CIFAR-10 dataset based on five-hidden-layer MLPs, in which every layer was of width 40.

We conducted an exhaustive analysis, meticulously calculating both the redundancy ratio and categorization accuracy across every epoch, as illustrated in Fig. 8.5. This thorough examination yielded the following insights:

(1) The redundancy ratio of the neural code formed by the initial layer consistently remains close to zero, indicating a notable absence of redundancy. However, despite this, the corresponding categorization accuracy remains relatively poor. This observation suggests that while redundancy is minimized, the initial layer may not capture sufficient discriminative information for effective categorization.

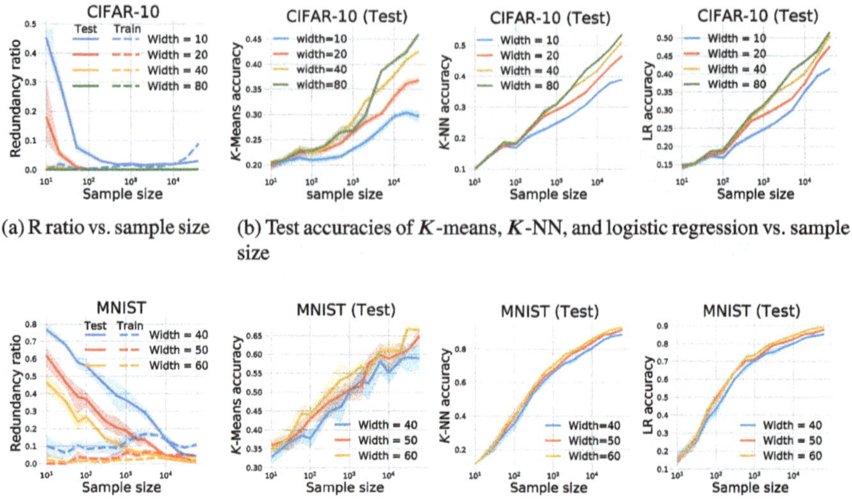

(a) R ratio vs. sample size (b) Test accuracies of K-means, K-NN, and logistic regression vs. sample size

(c) Determinism vs. sample size on both training and test sets (d) Test accuracies of K-means, K-NN, and logistic regression vs. sample size

Fig. 8.4 **a** The redundancy ratios (R ratios) on the training set (dotted lines) and test set (solid lines) of CIFAR-10 as functions of the sample size. **b** The test accuracies of K-means (left), K-NN (middle), and logistic regression (right) as functions of the sample size. The models are MLPs of width 10 (blue), 20 (red), 40 (orange), and 80 (green). **c** The R ratios on the training set (dotted lines) and test set (solid lines) of MNIST as functions of the sample size. **d** The test accuracies of K-means (left), K-NN (middle), and logistic regression (right) as functions of the sample size. The models are MLPs of width 40 (blue), 50 (red), and 60 (orange) on MNIST. The models were trained 10 times on CIFAR-10 or 5 times on MNIST, with a different random seed each time. The darker lines represent the averages over all seeds, and the shaded areas show the standard deviations

(2) As we progress through increasingly higher single layers, the redundancy ratio gradually increases, indicating a more diverse representation of data. Correspondingly, there is a noticeable improvement in categorization accuracy, suggesting that deeper layers encode more discriminative features.

(3) The impact of training time on the encoding properties of a single layer mirrors that of the entire network, emphasizing the importance of adequate training time in shaping encoding capabilities.

(4) As the neural code incorporates more layers, the redundancy ratio steadily decreases, indicating a more efficient encoding of data across multiple layers.

(5) Categorization accuracy exhibits a gradual enhancement as the neural code evolves from the first layer to encompass the entire network. This progression highlights the iterative refinement of features and representations throughout the network architecture.

(6) The observed pattern in (5) is disrupted when forming the neural code in reverse, starting from the last layer and progressing backward. This reversal suggests that the hierarchical organization of features may play a crucial role in information representation.

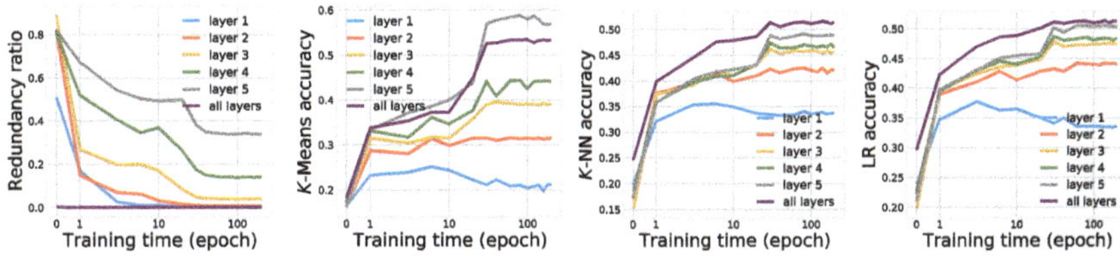

(a) Determinism vs. time (b) Categorization accuracy vs. training time for different single layers

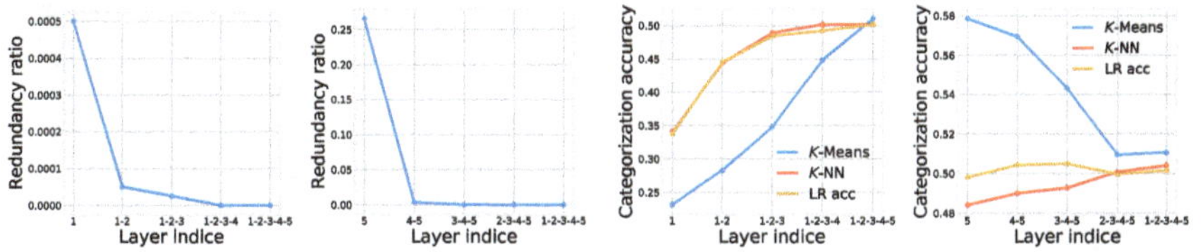

(c) Determinism for different layers (d) Categorization accuracy for different layers

Fig. 8.5 **a** Plots of the redundancy ratios of the neural codes formed by different single layers of MLPs trained on CIFAR-10 as functions of the training time. **b** The test accuracies of K-means (left), K-NN (middle), and logistic regression (right) as functions of the training time. **c** The redundancy ratios of the neural codes formed by multiple MLP layers as functions of the sample size. **d** The test accuracies of K-means (left), K-NN (middle), and logistic regression (right) as functions of the sample size. The models are MLPs of width 40 on CIFAR-10

(7) The categorization accuracy of the neural code formed solely by the last layer is comparable to that of the entire network, indicating that the final layer encapsulates critical discriminative features necessary for accurate classification.

The insight (2), particularly regarding the redundancy ratio, offers insight into the interplay between hashing properties and the generalizability of deep learning. The gradual concentration of data into fewer cells from the initial to the final layer validates the network's ability to extract increasingly informative and discriminative features, contributing to its overall effectiveness and generalization for categorization. This understanding sheds light on the intricate relationship between encoding properties and the broader principles governing deep learning architectures.

8.3.5 *Impacts of Regularization, Random Data, and Random Labels*

We also study the impact of regularization on the encoding properties. We trained 345 MLPs on the MNIST dataset with and without batch normalization, gradient clipping, and weight decay. The results suggest that regularization has an impact on the encoding properties but that this impact is smaller than those of the model size, training time, and sample size; see Fig. 8.6 (Figs. 8.7 and 8.8).

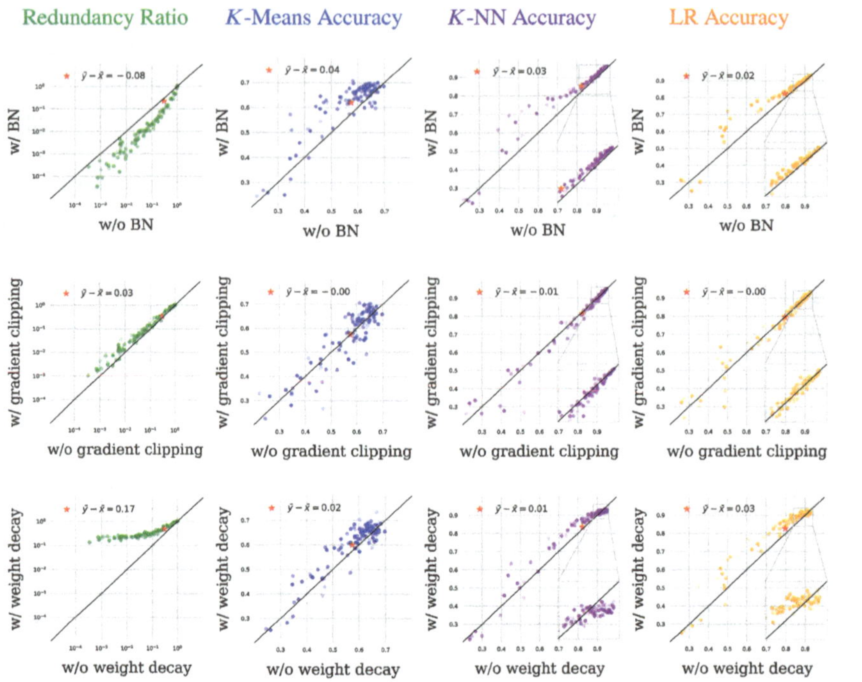

Fig. 8.6 **First row**: Scatter plots of the redundancy ratios and test accuracies of K-means (blue), K-NN (violet), and logistic regression (LR, orange) for MLPs with depths ranging from 3 to 100 with batch normalization (y-axis) and without gradient clipping (x-axis). A smaller $\bar{y} - \bar{x}$ is preferred for the redundancy ratio, and a larger one is preferred for the test accuracy. In total, 115 models are represented in one scatter plot. **Second row**: Scatter plots of the redundancy ratios and test accuracies of K-means (blue), K-NN (violet), and LR (orange) for MLPs with depths ranging from 3 to 100 with gradient clipping (y-axis) and without gradient clipping (x-axis). A smaller $\bar{y} - \bar{x}$ is preferred for the redundancy ratio, and a larger one is preferred for the test accuracy. In total, 115 models are represented in one scatter plot. **Third row**: Scatter plots of the redundancy ratios and test accuracies of K-means (blue), K-NN (violet), and LR (orange) for MLPs with depths ranging from 3 to 100 with weight decay (y-axis) and without weight decay (x-axis). A smaller $\bar{y} - \bar{x}$ is preferred for the redundancy ratio, and a larger one is preferred for the test accuracy. In total, 115 models are represented in one scatter plot

We also generated random data in which every pixel was generated from the uniform distribution $U(0, 1)$. We then trained MLPs and convolutional neural networks (CNNs) on the generated data. Unfortunately, the training process did not converge. We then added label noise to MNIST at different noise rates (0.1, 0.2, 0.3). The encoding properties still showed the same general trends, although they become somewhat worse. Our results suggest that the structure of the input data can influence the organization of the hash space (Table 8.3).

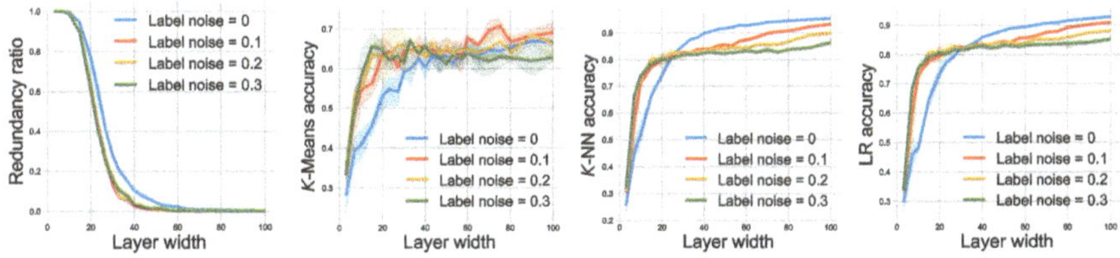

(a) Determinism with different levels of label noise

(b) Test accuracies of K-means, K-NN, and logistic regression with different levels of label noise

Fig. 8.7 **a** The redundancy ratios for MNIST with different levels of label noise as functions of the layer width. **b** The test accuracies of K-means (left), K-NN (middle), and logistic regression (right) with different levels of label noise as functions of the layer width. The models are MLPs trained on MNIST at different label noise rates of 0 (blue), 0.1 (red), 0.2 (orange) and 0.3 (green). All models were trained on MNIST with noisy labels for classification 5 times with different random seeds. The darker lines represent the averages over all seeds, and the shaded areas show the standard deviations

Fig. 8.8 An example of random data

Table 8.3 Training accuracy and training loss of a one-hidden-layer MLP of width 100 on random data

Epoch	0	100	300	500
Training acc (%)	10.92	11.24	11.24	11.24
Loss	230.56	230.13	230.13	230.13

8.3.6 Activation Hash Phase Chart

Finally, we can define an *activation hash phase chart* that characterizes the space formed by the redundancy ratio, categorization accuracy, model size, training time, and sample size. Summarizing the relationships discovered above, the activation hash phase chart is divided into three canonical regions, corresponding to the *under-*

expressive regime, the *critically expressive regime*, and the *sufficiently expressive regime*. This chart can provide guidance in hyperparameter tuning, the design of novel algorithms, and algorithm diagnosis. We note that the thresholds between the three regimes are currently unknown. Exploring them is a promising direction for future research.

8.4 Additional Experimental Implementation Details

Datasets. Our experiments are based on the MNIST dataset (LeCun et al. 1998) and the CIFAR-10 dataset (Krizhevsky and Hinton 2009): (1) MNIST contains $60,000$ training examples and $10,000$ test examples from 10 classes. This dataset can be downloaded at http://yann.lecun.com/exdb/mnist/. (2) CIFAR-10 consists of $50,000$ training images and $10,000$ test images that belong to 10 classes. CIFAR-10 can be downloaded at https://www.cs.toronto.edu/~kriz/cifar.html. The splits of the training and test sets follow the official versions. All images are normalized such that every pixel value is in the range [0, 1].

Training settings. (1) For MNIST, MLPs were trained with Adam for $2,000$ epochs with a batch size of 128 and a constant learning rate. VGG, ResNet, ResNet, ResNet, and DenseNet models were trained with Adam for 500 epochs with a batch size of 128. The learning rate was initialized as 0.01 and decayed to $1/10$ of the previous value every 100 epochs. For all models, the hyperparameter β_1 was set to 0.9, and the hyperparameter β_2 was set to 0.999. (2) For CIFAR-10, MLPs with 5 hidden layers were trained with Adam for 200 epochs with a batch size of 64. The learning rate was initialized as 0.01 and decayed to $1/10$ of the previous value every 20 epochs. VGG and ResNet models were trained with stochastic gradient descent (SGD) for 200 epochs with a batch size of 64. The learning rate was initialized as 0.01 and decayed to $1/10$ of the previous value every 50 epochs. On MNIST, the MLPs were trained five times with different random seeds. On CIFAR-10, the MLPs were trained ten times with different random seeds.

Average stochastic diameter. We first trained MLPs with widths of {5, 10, 15, 20, 25, 30, 35, 40, 45, 50, 55, 60, 65, 70, 75, 80, 90, 100} on MNIST. Then, we randomly selected 600 examples from the test set and calculated the mean of their corresponding stochastic diameters.

Network architectures. The details of the network architectures investigated in Sect. 8.2 are shown in Tables 8.4 and 8.5.

Experimental design for K-means. The pipeline for the experiments on K-means was designed as follows: (1) We set K equal to the number of classes. (2) We ran K-means on the neural codes to obtain K clusters. (3) Every cluster was assigned a label from $\{1, 2, \ldots, 10\}$. Thus, 90 (cluster, label) pairs were obtained. (4) For every (cluster, label) pair, we assigned the label to all data in the cluster and calculated the accuracy. (5) We selected the highest accuracy as the accuracy of the K-means algorithm.

Table 8.4 The detailed architectures of the neural networks trained on MNIST

VGG-18	ResNet-18	ResNet-34	ResNeXt-26	DenseNet-28
3×3, 32, stride 2	3×3, 32, stride 2	3×3, 32, stride 2	3×3, 32, stride 2	3×3, 6, stride 2
maxpool, 3×3	maxpool, 3×3	maxpool, 3×3	maxpool, 3×3	maxpool, 3×3
$(3 \times 3, 32) \times 4$	$\begin{bmatrix} 3 \times 3, 32 \\ 3 \times 3, 32 \end{bmatrix} \times 2$	$\begin{bmatrix} 3 \times 3, 32 \\ 3 \times 3, 32 \end{bmatrix} \times 3$	$\begin{bmatrix} 1 \times 1, 32 \\ 3 \times 3, 32, C = 8 \\ 1 \times 1, 64 \end{bmatrix} \times 2$	$\begin{bmatrix} 1 \times 1, 12 \\ 3 \times 3, 3 \end{bmatrix} \times 4$
$(3 \times 3, 64) \times 4$	$\begin{bmatrix} 3 \times 3, 64 \\ 3 \times 3, 64 \end{bmatrix} \times 2$	$\begin{bmatrix} 3 \times 3, 64 \\ 3 \times 3, 64 \end{bmatrix} \times 4$		$conv, 1 \times 1$ $avgpool, 2 \times 2$
$(3 \times 3, 128) \times 4$	$\begin{bmatrix} 3 \times 3, 128 \\ 3 \times 3, 128 \end{bmatrix} \times 2$	$\begin{bmatrix} 3 \times 3, 128 \\ 3 \times 3, 128 \end{bmatrix} \times 6$	$\begin{bmatrix} 1 \times 1, 64 \\ 3 \times 3, 64, C = 8 \\ 1 \times 1, 128 \end{bmatrix} \times 3$	$\begin{bmatrix} 1 \times 1, 12 \\ 3 \times 3, 3 \end{bmatrix} \times 4$
$(3 \times 3, 256) \times 4$	$\begin{bmatrix} 3 \times 3, 256 \\ 3 \times 3, 256 \end{bmatrix} \times 2$	$\begin{bmatrix} 3 \times 3, 256 \\ 3 \times 3, 256 \end{bmatrix} \times 3$		$conv, 1 \times 1$ $avgpool, 2 \times 2$
			$\begin{bmatrix} 1 \times 1, 128 \\ 3 \times 3, 128, C = 8 \\ 1 \times 1, 256 \end{bmatrix} \times 3$	$\begin{bmatrix} 1 \times 1, 12 \\ 3 \times 3, 3 \end{bmatrix} \times 4$
avgpool	avgpool	avgpool	avgpool	avgpool
fc-10, softmax	fc-10, softmax	fc-10, softmax	fc-10, softmax	fc-10, softmax

Table 8.5 The detailed architectures of the neural networks trained on CIFAR-10

VGG-19	ResNet-18	ResNet-20	ResNet-32
$(3 \times 3, 32) \times 2$ maxpool, 2×2	3×3, 64	3×3, 16	3×3, 16
$(3 \times 3, 128) \times 2$ maxpool, 2×2	$\begin{bmatrix} 3 \times 3, 64 \\ 3 \times 3, 64 \end{bmatrix} \times 2$	$\begin{bmatrix} 3 \times 3, 16 \\ 3 \times 3, 16 \end{bmatrix} \times 3$	$\begin{bmatrix} 3 \times 3, 16 \\ 3 \times 3, 16 \end{bmatrix} \times 5$
$(3 \times 3, 256) \times 4$ maxpool, 2×2	$\begin{bmatrix} 3 \times 3, 128 \\ 3 \times 3, 128 \end{bmatrix} \times 2$	$\begin{bmatrix} 3 \times 3, 32 \\ 3 \times 3, 32 \end{bmatrix} \times 3$	$\begin{bmatrix} 3 \times 3, 32 \\ 3 \times 3, 32 \end{bmatrix} \times 5$
$(3 \times 3, 512) \times 4$ maxpool, 2×2	$\begin{bmatrix} 3 \times 3, 256 \\ 3 \times 3, 256 \end{bmatrix} \times 2$	$\begin{bmatrix} 3 \times 3, 64 \\ 3 \times 3, 64 \end{bmatrix} \times 3$	$\begin{bmatrix} 3 \times 3, 64 \\ 3 \times 3, 64 \end{bmatrix} \times 5$
$(3 \times 3, 512) \times 4$ maxpool, 2×2	$\begin{bmatrix} 3 \times 3, 512 \\ 3 \times 3, 512 \end{bmatrix} \times 2$		
$fc - 4096$ $fc - 4096$	avgpool	avgpool	avgpool
fc-10, softmax	fc-10, softmax	fc-10, softmax	fc-10, softmax

Experiments on the relationship between the model size and the encoding properties. We trained MLPs with widths of $\{3, 7, 10, 15, 20, 23, 27, 30, 33, 37, 40, 43, 47, 50, 53, 57, 60, 65, 70, 75, 80, 90, 100\}$ on MNIST and $\{10, 20, 30, 40, 50, 60, 70, 80, 90, 100, 110, 120, 130, 140, 150, 160, 170, 180, 190, 200\}$ on CIFAR-10.

Experiments concerning the relationship between the training process and the model size. (1) For MNIST, we trained MLPs with widths of $\{40, 50, 60\}$. The redundancy ratios and test accuracies of K-means, K-NN, and logistic regression were calculated for the following training epochs: $\{1, 3, 6, 10, 30, 60, 100, 300, 600, 1000, 1200, 1500, 1800, 2000\}$. (2) For CIFAR-10, we trained MLPs with widths of $\{10, 20, 40, 80\}$. The redundancy ratios and test accuracies of K-means, K-NN, and logistic regression were calculated for the following training epochs: $\{1, 3, 6, 10, 20, 30, 40, 60, 80, 100, 120, 140, 160, 180, 200\}$.

Experiments concerning the relationship between the sample size and the model size. (1) For MNIST, we trained MLPs with widths of $\{40, 50, 60\}$ on training samples with sizes of $\{10, 30, 60, 100, 300, 600, 1000, 2000, 3000, 6000, 10000, 20000, 30000, 60000\}$ randomly drawn from the training set. (2) For CIFAR-10, we trained MLPs with widths of $\{10, 20, 40, 80\}$ on training samples with sizes of $\{10, 20, 50, 100, 200, 500, 1000, 2000, 5000, 10000, 20000, 40000\}$ randomly drawn from the training set.

Experiments concerning the relationship between regularization and the model size. Three regularizers were tested in our experiments:

- Batch normalization: Adding a batch normalization layer before every ReLU layer.
- Weight decay: Utilizing the L_2 weight regularizer with hyperparameter $\lambda = 0.01$.
- Gradient clipping: Setting the clip norm to 1.

Layerwise ablation study. We trained MLPs of width 40 on CIFAR-10. The training strategy was the same as that previously used for MLPs trained on CIFAR-10.

Experiments concerning random data. All pixels of the random data were individually generated from the uniform distribution $U(0, 1)$. Each random example had dimensions of 28×28, i.e., the same as the MNIST images.

Experiments concerning noisy labels. Specified numbers of training examples from MNIST were assigned random labels in accordance with label noise ratios of 0.1, 0.2, and 0.3. Then, we trained one-hidden-layer MLPs with widths of $\{3, 7, 10, 15, 20, 23, 27, 30, 33, 37, 40, 43, 47, 50, 53, 57, 60, 65, 70, 75, 80, 90, 100\}$ on each noisy training set.

8.5 Additional Experimental Results

This section collects experimental results omitted from the main analysis section for brevity. The following figure corresponds to the study of the diameters. Please refer to Sect. 8.3.1.

References

Hanin, Boris, and David Rolnick. 2019. Deep relu networks have surprisingly few activation patterns. In *Advances in Neural Information Processing Systems*, 361–370.

Jaeger, Herbert. 2001. The "cho state" approach to analysing and training recurrent neural networks-with an erratum note. *Bonn, Germany: German National Research Center for Information Technology GMD Technical Report* 148 (34): 13.

Krizhevsky, Alex, and Geoffrey Hinton. 2009. Learning multiple layers of features from tiny images. Technical report, Citeseer.

LeCun, Yann, Léon Bottou, Yoshua Bengio, and Patrick Haffner. 1998. Gradient-based learning applied to document recognition. *Proceedings of the IEEE* 86 (11): 2278–2324.

Maass, Wolfgang, Thomas Natschläger, and Henry Markram. 2002. Real-time computing without stable states: a new framework for neural computation based on perturbations. *Neural Computation* 14 (11): 2531–2560.

Nesterov, Yurii E. 1983. A method for solving the convex programming problem with convergence rate o ($1/k^2$). In *Dokl. Akad. Nauk Sssr*, vol. 269, 543–547.

van der Maaten, Laurens, and Geoffrey Hinton. 2008. Visualizing data using t-sne. *Journal of Machine Learning Research* 9 (Nov): 2579–2605.

Chapter 9
Reflecting on the Role of Overparameterization: Is it Solely Harmful?

One might blame the overparameterized nature of deep learning for its lack of theoretical foundations. However, recent works have discovered that this overparameterization also contributes to the success of deep learning. Notably, there is currently no standard definition of "overparameterization". In many cases, the definition is quite restrictive, such as requiring an infinite network width. A promising direction of future research is to relax the conditions of overparameterization.

9.1 Double Descent and Benign Overfitting

Double descent. Classical machine learning usually exhibits a U-shaped bias-variance trade-off: the test performance initially increases as the hypothesis complexity increases and then decreases after an inflection point. However, Belkin et al. (2019) empirically found that this "single-descent" curve shape holds only when the model is not able to perfectly fit (interpolate) the training data; the bias-variance curve may also exhibit a second descent after an interpolation point. Nakkiran et al. (2020) empirically showed that a similar double-descent phenomenon also occurs as a function of the number of training epochs. These findings motivate a new complexity measure, the effective model complexity, which unifies the two aforementioned curves as a double-descent curve of the test error versus the effective model complexity. Specifically, the effective model complexity $\text{EMC}_{\mathcal{D},\,\epsilon}(\mathcal{T})$ of model $\mathcal{T}$ on sample set S drawn from data distribution $\mathcal{D}$ is defined as follows:

$$\text{EMC}_{\mathcal{D},\,\epsilon}(\mathcal{T}) = \max\{m|\mathbb{E}_{S\sim\mathcal{D}^m}[\hat{R}_S(\mathcal{T})] \leq \epsilon\}.$$

Neural networks as interpolators. The double-descent phenomenon introduces a conflict with conventional statistical learning theory. A key factor in this phenomenon is the interpolation point. A related line of research is focused on studying neural

F. He and D. Tao, *Foundations of Deep Learning*, Machine Learning: Foundations, Methodologies, and Applications, https://doi.org/10.1007/978-981-16-8233-9_9

networks as interpolators. Belkin et al. (2018b) empirically demonstrated that kernel-based models with zero classification error or near-zero regression error also have very low test error, even when a high level of label noise exists. Belkin et al. (2018a) further constructed several examples to illustrate the "blessing of dimensionality": the asymptotic risk of the constructed interpolator approaches the Bayesian risk as the model dimensions increase to infinity. Muthukumar et al. (2020) studied the influence of a 0-1 loss and a squared loss on overparameterized neural networks. Liang and Rakhlin (2020) proved that a nonlinear kernel "ridgeless" regression that interpolates the training data generalizes well on test data. Belkin et al. (2020) calculated the peak points of the double-descent curves for two families of linear models, while Hastie et al. (2019) further studied the case of overparameterized linear regression models. Specifically, as the training sample size $m \to \infty$, the feature dimension $p \to \infty$, and the ratio $m/p \to \lambda \in (0, \infty)$,

$$R(\hat{\beta}) \to \begin{cases} \sigma^2 \frac{\gamma}{1-\gamma} & \gamma < 1, \\ r^2 \left(1 - \frac{1}{\gamma}\right) + \sigma^2 \frac{1}{\gamma - 1} & \gamma > 1, \end{cases} \tag{9.1}$$

where $\hat{\beta}$ is the minimum-norm least-squares estimator, β is the target mapping, and

$$R(\hat{\beta}) = \mathbb{E}\left[(x^\top \hat{\beta} - x^\top \beta)^2\right].$$

This line of research has also led to research on benign overfitting in linear regression (Bartlett et al. 2020), ridge regression (Tsigler and Bartlett 2020), the large-deviation regime (Chinot and Lerasle 2020), constant-step-size stochastic gradient descent (SGD) for regression (Zou et al. 2021), and neural networks (Li et al. 2021). Bartlett et al. (2020) characterized linear regression problems in which the minimum-norm interpolating prediction rule has near-optimal prediction accuracy. The authors demonstrated that for benign overfitting, overparameterization is crucial when the number of directions in the parameter space must be substantially larger than the sample size, and they derived nearly matching lower- and upper-bounds for the risk of the minimum-norm interpolating estimator, as follows (Fig. 9.1).

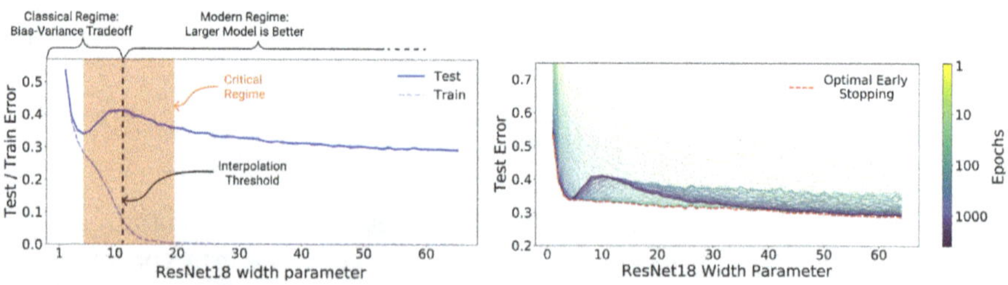

Fig. 9.1 Figure illustrating test loss and training loss versus parameter size for ResNet-18 on CIFAR-10. See Nakkiran et al. (2021)

Theorem 9.1 *Consider a linear regression problem defined in terms of a random covariate vector x, an outcome y and an optimal parameter vector θ^* satisfying $\mathbb{E}(y - x^T\theta^*) = \mathbb{E}_\theta(y - x^T\theta)^2$. For any positive constant σ_x, there exist quantities b, c, and c_1 for which the following holds. Let us define*

$$k^* = \min\{k \geq 0 : r_k(\sum) \geq bm\},$$

where $\sum = \mathbb{E}[xx^T]$ is the covariance operator and $r_k(\Sigma) = \Sigma_{i>k}\lambda_i/\lambda_{k+1}$. Suppose that $\delta < 1$ with $\log(1/\delta) < m/c$. If $k^* \geq m/c_1$, then $\mathbb{E}[R(\hat{\theta})] \geq \sigma^2/c$. Otherwise,

$$R(\hat{\theta}) \leq c\left(\|\theta^*\| \|\sum\| \max\left\{\sqrt{\frac{r_0(\sum)}{m}}, \frac{r_0(\sum)}{m}, \sqrt{\frac{\log(1/\delta)}{m}}\right\}\right) + c\log(1/\delta)\sigma_y^2\left(\frac{k^*}{m} + \frac{m}{R_{k^*}(\sum)}\right)$$

with probability at least $1 - \delta$ and

$$\mathbb{E}[R(\widehat{\theta})] \geq \frac{\sigma^2}{c}\left(\frac{k^*}{m} + \frac{m}{R_{k^*}(\sum)},\right).$$

where $R_k(\Sigma) = (\Sigma_{i>k}\lambda_i)^2/\Sigma_{i>k}\lambda_i^2$. Moreover, there exist universal constants a_1, a_2, and n_0 such that for all $m \geq m_0$, for all $\sum$, and for all $t \geq 0$, there exists a θ^* with $\|\theta^*\| = t$ such that for $x \sim N(0, \sum)$ and $y|x \sim N(x^T\theta^*, \|\theta^*\| \|\sum\|)$, with probability at least $1/4$,

$$R(\hat{\theta}) \geq \frac{1}{a_1}\|\theta^*\| \|\sum\| \mathbf{I}_{\frac{r_0(\sum)}{m\log(1+r_0(\sum))} \geq a_2}.$$

Cao et al. (2022) investigated benign overfitting in a two-layer convolutional neural network (CNN) and found that a two-layer CNN can achieve arbitrarily small training and test losses when the signal-to-noise ratio satisfies certain conditions.

Theorem 9.2 *Suppose that training sample size m, neural network width n and dimension d satisfy $d = \widetilde{\Omega}(n^8m^4)$, where $n, m = \Omega(polylog(d))$; the learning rate satisfies*

$$\eta \leq \widetilde{O}\left(n^{-2/q}\min\left\{\|\boldsymbol{\mu}\|_2^{-2}, 4\sigma_p^{-2}d^{-1}\right\}\right)$$

and $\widetilde{O}(md^{-1/2}) \cdot \min\left\{\left(\sigma_p\sqrt{d}\right)^{-1}, \|\boldsymbol{\mu}\|_2^{-1}\right\}\right) \leq$ the initialization level $\sigma_0 \leq \widetilde{O}(n^{-4}m^{-1}) \cdot \min\left\{(\sigma_p\sqrt{d})^{-1}, \|\boldsymbol{\mu}\|_2^{-1}\right\}$. For any $\epsilon > 0$, if $n \cdot \|\boldsymbol{\mu}\|_2^q = \widetilde{\Omega}(\sigma_p^q(\sqrt{d})^q)$, then there exists a $T = \widetilde{O}(\eta^{-1}\sigma_0^{-(q-2)}\|\boldsymbol{\mu}\|_2^{-q} + \eta^{-1}\epsilon^{-1}n^4\|\boldsymbol{\mu}\|_2^{-2})$ with probability at least $1 - d^{-1}$ such that 1) the training loss converges to ϵ, i.e., $\hat{R}_S(\mathbf{W}^{(T)}) \leq \epsilon$; 2) the CNN learns the signal: $\max_r \gamma_{j,r}^{(T)} \geq \Omega(n^{-1/q})$ for $j \in \{-1, +1\}$; and 3) the

CNN does not memorize the noise in the training data: $\max_{j,r,i} \zeta_{j,r,i}^{(T)} = \widetilde{\Omega}(\sigma_0\sigma_p\sqrt{d})$ *and* $\max_{j,r,i} |\omega_{j,r,i}^{(T)}| = \widetilde{\Omega}(\sigma_0\sigma_p\sqrt{d})$*. Here,* $\boldsymbol{\mu}$ *and* σ_0 *represent the signal strength and noise level respectively.*

This theorem illustrates that if the condition $n \cdot \|\boldsymbol{\mu}\|_2^q = \widetilde{\Omega}(\sigma_p^q(\sqrt{d})^q)$ holds, then a filter will exist that learns $\boldsymbol{\mu}$ by achieving $\gamma_{j,r_j^*} \geq \Omega(n^{-1/q})$.

In addition, these authors presented a theorem for noise memorization by a CNN.

Theorem 9.3 *Suppose that* $d = \widetilde{\Omega}(n^8 m^4)$*, where* $n, m = \Omega(polylog(d))$*; the learning rate satisfies* $\eta \leq \widetilde{O}\left(n^{-2/q} \min\left\{\|\boldsymbol{\mu}\|_2^{-2}, 4\sigma_p^{-2}d^{-1}\right\}\right)$*; and* $\widetilde{O}(md^{-1/2}) \cdot \min\left\{\left((\sigma_p\sqrt{d})^{-1}, \|\boldsymbol{\mu}\|_2^{-1}\right\}\right) \leq$ *the initialization level*

$$\sigma_0 \leq \widetilde{O}(n^{-4}m^{-1}) \min\left\{(\sigma_p\sqrt{d})^{-1}, \|\boldsymbol{\mu}\|_2^{-1}\right\}.$$

For any $\epsilon > 0$*, if* $\sigma_p^q(\sqrt{d})^q = \widetilde{\Omega}(m \cdot \|\boldsymbol{\mu}\|_2^q)$*, then there exists a* $T = \widetilde{O}\big(\eta^{-1} \cdot m(\sigma_p\sqrt{d})^{-q} \cdot \sigma_0^{-(q-2)} + \eta^{-1}\epsilon^{-1}mn^4 d^{-1}\sigma_p^{-2}\big)$ *with probability at least* $1 - d^{-1}$ *such that 1) the training loss converges to* ϵ*, i.e.,* $\hat{R}_S(\mathbf{W}^{(T)}) \leq \epsilon$*, and 2) the CNN memorizes the noise in the training data:* $\max_r \zeta_{y_i,r,i}^{(T)} \geq \Omega(n^{-1/q})$*.*

Lower and upper bounds on the population loss achieved by a CNN can be obtained based on the bounds on $\gamma_{j,r}^{(T)}$, $\zeta_{j,r,i}^{(T)}$ and $\omega_{j,r,i}^{(T)}$ in Theorems 9.2 and 9.3.

Theorem 9.4 *Suppose that* $d = \widetilde{\Omega}(n^8 m^4)$*, where* $n, m = \Omega(polylog(d))$*; the learning rate satisfies* $\eta \leq \widetilde{O}\left(n^{-2/q} \min\left\{\|\boldsymbol{\mu}\|_2^{-2}, 4\sigma_p^{-2}d^{-1}\right\}\right)$*; and* $\widetilde{O}(md^{-1/2}) \cdot \min\left\{\left((\sigma_p\sqrt{d})^{-1}, \|\boldsymbol{\mu}\|_2^{-1}\right\}\right) \leq$ *the initialization level*

$$\sigma_0 \leq \widetilde{O}(n^{-4}m^{-1}) \min\left\{(\sigma_p\sqrt{d})^{-1}, \|\boldsymbol{\mu}\|_2^{-1}\right\}.$$

For any $\epsilon > 0$*, 1) if* $n \cdot \|\boldsymbol{\mu}\|_2^q = \widetilde{\Omega}(\sigma_p^q(\sqrt{d})^q)$*, then gradient descent will yield a CNN with parameters* $\widehat{\mathbf{W}}$ *such that* $\hat{R}_S(\widehat{\mathbf{W}}) < \epsilon$ *and* $R(\widehat{\mathbf{W}}) < 6\epsilon + \exp(-n^2)$*, and 2) if* $\sigma_p^q(\sqrt{d})^q = \widetilde{\Omega}\left(n \cdot \|\boldsymbol{\mu}\|_2^q\right)$*, then gradient descent will yield a CNN with parameters* $\widehat{\mathbf{W}}$ *such that* $\hat{R}_S(\widehat{\mathbf{W}}) < \epsilon$ *and* $R(\widehat{\mathbf{W}}) = \Theta(1)$*, where* $R(\mathbf{W}) := \mathbb{E}_{(x,y)\sim\mathcal{D}}\ell\,[y \cdot f(\mathbf{W}, x)]$*.*

This theorem further illustrates the phenomenon of phase transition that is revealed in Theorems 9.2 and 9.3 with respect to the population loss.

9.2 Neural Tangent Kernels (NTKs)

Neal (1995, 1996), Williams (1996), and Hazan and Jaakkola (2015) laid the groundwork by proving the equivalence between an infinite-width shallow neural network and a Gaussian process. This fundamental insight provided a solid foundation for further exploration. Subsequently, researchers including Damianou and Lawrence (2013), Duvenaud et al. (2014), Hensman and Lawrence (2014), Lawrence and Moore (2007), Bui et al. (2016), and Lee et al. (2018) gradually expanded this equivalence to encompass infinite-width neural networks of arbitrary depths. These advancements broadened our understanding of the fundamental connections between neural networks and Gaussian processes, shedding light on their underlying mathematical properties. Furthermore, Lee et al. (2019) pushed the boundaries even further by demonstrating that infinite-width neural networks of any depth behave similarly to linear models when trained using gradient descent. This comprehensive body of work not only elucidates the intricate relationship between neural networks and Gaussian processes but also provides valuable insights into the behavior of deep learning models during training.

Suppose that f_θ is the network function induced by a neural network parameterized by θ. Further suppose that a realization function $F^{(L)}$ maps the parameters θ to the network function f_θ, where L is the network depth. The network function f_θ is optimized along the gradient with respect to the weights θ in the parameter space. The dynamics of f_θ in the function space (when the network width is infinitely large) have been proven by Jacot et al. (2018) to follow the gradient with respect to the following neural tangent kernel (NTK):

$$\Theta^{(L)}(\theta) = \sum_{p=1}^{P} \partial_{\theta_p} F^{(L)}(\theta) \otimes \partial_{\theta_p} F^{(L)}(\theta).$$

The finite-width NTK is defined as

$$\mathcal{H}\left(x, x'\right) = J\left(x\right) J\left(x'\right)^T, \tag{9.2}$$

where $J_{i\alpha}(x) = \partial_{\mathcal{H}_\alpha} z_i^L(x)$ is the Jacobian evaluated at a point x for parameter α and $z_i^L(x)$ is the i-th output of the last output layer. Jacot et al. (2018) has shown that the NTK converges to a deterministic kernel and remains constant over the course of training. Subsequently, the authors proved that the expected outputs of an infinitely wide network can be solved for by means of an ordinary differential equation (ODE) as follows:

$$\mu_t\left(X_{\text{train}}\right) = \left(\mathbf{Id} - e^{-\eta \mathcal{H}_{\text{train, train}}\, t}\right) Y_{\text{train}}\,. \tag{9.3}$$

Here, $\mathcal{H}_{\text{train, train}}$ denotes the NTK between the training inputs. As the number of steps t approaches infinity, Eq. 9.3 reduces to $\mu_t\left(X_{\text{train}}\right) = Y_{\text{train}}$. Equation 9.3 can be further written as

$$\tilde{\mu}_t\left(X_{\text{train}}\right)_i = \left(\mathbf{Id} - e^{-\eta \lambda_i t}\right) \tilde{Y}_{\text{train},\, i}, \tag{9.4}$$

where the λ_i are the eigenvalues of $\mathcal{H}_{\text{train, train}}$ and $\tilde{\mu}_t\left(X_{\text{train}}\right)$ and $\tilde{Y}_{\text{train}}$ are the mean predictions and labels, respectively. Given the ordered eigenvalues as $\lambda_0 \geq \cdots \lambda_m$, Lee et al. (2019) assumed that the maximum feasible learning rate scales as $\eta \sim 2/\lambda_0$, which has been empirically verified by Xiao et al. (2020). Plugging $\eta \sim 2/\lambda_0$ into Eq. 9.3, we find that λ_m will converge exponentially in a rate $\kappa = \lambda_0/\lambda_m$, i.e., the condition number. It shows that if the condition number of the NTK associated with a neural network diverges, then the network will become untrainable. Therefore, we can use κ as a metric for trainability. Investigation has shown that κ is inversely related to the performance of an architecture. Therefore, we generally wish to find a network with a smaller κ.

Jacot et al. (2018) later proved that as for the model learned for a least-squares regression problem, it is characterized by a linear differential equation in the infinite-width regime during the training process.

9.3 Loss Surface and Optimization

Influence of overparameterization on the loss surface. Overparameterization significantly influences the shapes of the loss surfaces of neural networks. Choromanska et al. (2015) empirically showed that (1) a large proportion of the local minima on the loss surface of an overparameterized network are "equivalent" to each other (they have the same empirical risk) and (2) neural networks of small or normal sizes have spurious local minima, but the proportion of spurious local minima decreases rapidly as the network size increases. Li et al. (2018) proved that overparameterized fully connected deep neural networks do not have any spurious local minima under the following assumptions: (1) the activation functions are continuous and (2) the loss functions are convex and differentiable. Nguyen et al. (2019) extended the "no bad local minima" property to networks trained under a cross-entropy loss. Nguyen (2019) further discovered that the global minima are connected to each other and concentrated in a unique valley if the neural network is sufficiently overparameterized.

Influence of overparameterization on optimization. It has been shown that overparameterization contributes to ensuring the optimization performance of gradient-based optimizers. Li and Liang (2018) proved that overparameterized neural networks trained via SGD on data drawn from a mixture of several distributions have a small generalization error. Du et al. (2019b) proved that, for a two-layer neural network as long as the network width is sufficiently large, gradient descent converges to a global minimum in a linear convergence rate, as follows:

$$\mathbb{P}[\|\hat{y}(t) - y\|_2^2 \leq \exp(-\lambda_0 t)\|\hat{y}(0) - y\|_2^2] \geq 1 - \delta,$$

where $y(t)$ is the prediction at iteration t, $\lambda_0 > 0$ is a constant real number, and $0 < \delta < 1$ is a real number. This result relies on the following overparameterization condition: the number of hidden units must satisfy $n = \Omega\left(m^6 \lambda_0^{-4} \delta^{-3}\right)$. The authors also show that overparameterization can restrict the weights to be close to the random initialization. In Chizat and Bach (2018), the optimization of a one-hidden-layer network was modelled to the minimization of a convex function of a measure discretized into a mixture of particles via continuous-time gradient descent. They proved that the gradient flow characterizing the optimization converges to a global minimizer.

Going beyond a single hidden layer, Allen-Zhu et al. (2019b), Du et al. (2019a), and Zou et al. (2020) concurrently proved that SGD converges to a globally optimal solution for an overparameterized deep neural network in polynomial time under slightly different assumptions on the network architecture and training data. Chen et al. (2019) proved that the convergence of optimization is guaranteed if the network width is polylogarithmic in the sample size m and $1/\epsilon$, where ϵ is the target error. A sample complexity of

$$n(\delta, R, L) = \tilde{O}(\text{poly}(R, L) \log^{4/3}(m/\delta))$$

is needed to ensure that with probability at least $1 - \delta$, SGD with a specific learning rate achieves an error no larger than

$$\frac{8L^2R^2}{m} + \frac{8\log(2/\delta)}{m} + 24\epsilon_{\text{NTRF}},$$

where ϵ_{NTRF} is the minimum achievable training loss, L is the network depth, R is the radius of the neural tangent random feature (NTRF) function class (Cao and Gu 2019) and $n(\delta, R, L)$ is network width.

9.4 Generalization and Learnability

Under the assumption that the nodes in the first layer all have linear functions while the hidden nodes in the last layer are nonlinear, Andoni et al. (2014) studied two-layer neural networks. These authors proved that if a sufficiently wide neural network is trained with a generic gradient descent algorithm and all weights are randomly initialized, it can learn any low-degree polynomial function. Brutzkus et al. (2018) also gave generalization guarantees for two-layer neural networks under specific assumptions. Arora et al. (2019) and Allen-Zhu et al. (2019a) proved that for neural networks with two or three layers, the sample complexity is almost independent of the parameter size. Arora et al. (2019) proved that any one-hidden-layer rectified linear unit (ReLU) network trained via gradient descent under a 1-Lipschitz loss has a generalization error of at most

$$\sqrt{\frac{2\boldsymbol{y}^\top(\boldsymbol{H}^\infty)^{-1}\boldsymbol{y}}{m}},$$

where $\boldsymbol{y} = (y_1, \ldots, y_m)^\top$, the y_i $(i = 1, \ldots, m)$ are the labels, and H^∞ is the Gram matrix, $\forall i, j = 1, \ldots, m$, defined as follows:

$$H_{i,j}^{\infty} = \mathbb{E}_{w \sim N(0,I)} \left[x_i^\top x_j^\top I_{w^\top x_i \geq 0, w^\top x_j \geq 0} \right] = \frac{x_i^\top x_j (\pi - \arccos(x_i^\top x_j))}{2\pi}.$$

Chen et al. (2019) showed that generalization is guaranteed if the network width is polylogarithmic in the sample size m and ϵ^{-1}, where ϵ is the target error. Cao and Gu (2019) also proved the generalization bounds for wide and deep neural networks. Wei et al. (2019) proved that regularizers can significantly influence the generalization and optimization properties.

Influence of the network depth on generalizability. Canziani et al. (2016) suggested that the deeper neural networks could have the better generalizability, summarizing various practical applications of neural networks. This understanding plays a major part in the reconciliation between the overparameterization of neural networks and their excellent generalizability. In addition to the previous results, there is another possible explanation that originates from information theory. Zhang et al. (2018) adopted the techniques developed by Xu and Raginsky (2017) and proved a generalization bound to characterize how the generalizability evolves as a network becomes deeper. The expectation of the generalization error, $\mathbb{E}[R - \hat{R}_S]$, has the following upper bound:

$$\exp\left(-\frac{L}{2}\log\frac{1}{\eta}\right)\sqrt{\frac{2\sigma^2}{m}I(S, W)},$$

where L is the network depth, $0 < \eta < 1$ is a real constant, S is the training sample, and W denotes the parameters of the learned hypothesis.

References

Allen-Zhu, Zeyuan, Yuanzhi Li, and Yingyu Liang. 2019a. Learning and generalization in overparameterized neural networks, going beyond two layers. In *Advances in Neural Information Processing Systems*, 6158–6169.

Allen-Zhu, Zeyuan, Yuanzhi Li, and Zhao Song. 2019b. A convergence theory for deep learning via over-parameterization. In *International Conference on Machine Learning*.

Andoni, Alexandr, Rina Panigrahy, Gregory Valiant, and Li Zhang. 2014. Learning polynomials with neural networks. In *International Conference on Machine Learning*, 1908–1916.

Arora, Sanjeev, Simon S Du, Wei Hu, Zhiyuan Li, and Ruosong Wang. 2019. Fine-grained analysis of optimization and generalization for overparameterized two-layer neural networks. *arXiv preprint* arXiv:1901.08584.

Bartlett, Peter L, Philip M Long, Gábor Lugosi, and Alexander Tsigler. 2020. Benign overfitting in linear regression. *Proceedings of the National Academy of Sciences* 117 (48): 30063–30070.

Belkin, Mikhail, Daniel J Hsu, and Partha Mitra. 2018a. Overfitting or perfect fitting? risk bounds for classification and regression rules that interpolate. *Advances in Neural Information Processing Systems* 31.

Belkin, Mikhail, Siyuan Ma, and Soumik Mandal. 2018b. To understand deep learning we need to understand kernel learning. In *International Conference on Machine Learning*, 541–549.

Belkin, Mikhail, Daniel Hsu, Siyuan Ma, and Soumik Mandal. 2019. Reconciling modern machine-learning practice and the classical bias-variance trade-off. *Proceedings of the National Academy of Sciences* 116 (32): 15849–15854.

Belkin, Mikhail, Daniel Hsu, and Xu. Ji. 2020. Two models of double descent for weak features. *SIAM Journal on Mathematics of Data Science* 2 (4): 1167–1180.

Brutzkus, Alon, Amir Globerson, Eran Malach, and Shai Shalev-Shwartz. 2018. SGD learns over-parameterized networks that provably generalize on linearly separable data. In *International Conference on Learning Representations*.

Bui, Thang, Daniel Hernández-Lobato, Jose Hernandez-Lobato, Yingzhen Li, and Richard Turner. 2016. Deep Gaussian processes for regression using approximate expectation propagation. In *International Conference on Machine Learning*, 1472–1481.

Canziani, Alfredo, Adam Paszke, and Eugenio Culurciello. 2016. An analysis of deep neural network models for practical applications. *arXiv preprint* arXiv:1605.07678.

Cao, Yuan, and Quanquan Gu. 2019. Generalization bounds of stochastic gradient descent for wide and deep neural networks. In *Advances in Neural Information Processing Systems*, 10836–10846.

Cao, Yuan, Zixiang Chen, Mikhail Belkin, and Quanquan Gu. 2022. Benign overfitting in two-layer convolutional neural networks. *arXiv preprint* arXiv:2202.06526.

Chen, Zixiang, Yuan Cao, Difan Zou, and Quanquan Gu. 2019. How much over-parameterization is sufficient to learn deep ReLU networks? *arXiv preprint* arXiv:1911.12360.

Chinot, Geoffrey, and Matthieu Lerasle. 2020. Benign overfitting in the large deviation regime. *arXiv preprint* arXiv:2003.05838, 1(5).

Chizat, Lenaic, and Francis Bach. 2018. On the global convergence of gradient descent for over-parameterized models using optimal transport. In *Advances in Neural Information Processing Systems*, 3036–3046.

Choromanska, Anna, Mikael Henaff, Michael Mathieu, Gérard Ben Arous, and Yann LeCun. 2015. The loss surfaces of multilayer networks. In *International Conference on Artificial Intelligence and Statistics*.

Damianou, Andreas, and Neil D Lawrence. 2013. Deep Gaussian processes. In *Artificial Intelligence and Statistics*, 207–215.

Du, Simon, Jason Lee, Haochuan Li, Liwei Wang, and Xiyu Zhai. 2019a. Gradient descent finds global minima of deep neural networks. In *International Conference on Machine Learning*, 1675–1685.

Du, Simon S., Xiyu Zhai, Barnabas Poczos, and Aarti Singh. 2019b. Gradient descent provably optimizes over-parameterized neural networks. In *International Conference on Learning Representations*.

Duvenaud, David, Oren Rippel, Ryan Adams, and Zoubin Ghahramani. 2014. Avoiding pathologies in very deep networks. In *Artificial Intelligence and Statistics*, 202–210.

Hastie, Trevor, Andrea Montanari, Saharon Rosset, and Ryan J Tibshirani. 2019. Surprises in high-dimensional ridgeless least squares interpolation. arxiv e-prints, page. *arXiv preprint* arXiv:1903.08560.

Hazan, Tamir, and Tommi Jaakkola. 2015. Steps toward deep kernel methods from infinite neural networks. *arXiv preprint* arXiv:1508.05133.

Hensman, James, and Neil D Lawrence. 2014. Nested variational compression in deep Gaussian processes. *arXiv preprint* arXiv:1412.1370.

Jacot, Arthur, Franck Gabriel, and Clément Hongler. 2018. Neural tangent kernel: convergence and generalization in neural networks. In *Advances in Neural Information Processing Systems*, 8571–8580.

Lawrence, Neil D, and Andrew J Moore. 2007. Hierarchical Gaussian process latent variable models. In *International Conference on Machine Learning*, 481–488.

Lee, Jaehoon, Lechao Xiao, Samuel Schoenholz, Yasaman Bahri, Roman Novak, Jascha Sohl-Dickstein, and Jeffrey Pennington. 2019. Wide neural networks of any depth evolve as linear models under gradient descent. In *Advances in Neural Information Processing Systems*, 8572–8583.

Lee, Jaehoon, Yasaman Bahri, Roman Novak, Samuel S Schoenholz, Jeffrey Pennington, and Jascha Sohl-Dickstein. 2018. Deep neural networks as Gaussian processes. In *International Conference on Learning Representations*.

Li, Dawei, Tian Ding, and Ruoyu Sun. 2018. Over-parameterized deep neural networks have no strict local minima for any continuous activations. *arXiv preprint* arXiv:1812.11039.

Li, Yaopeng, Ming Jia, Xu Han, and Xue-Song Bai. 2021. Towards a comprehensive optimization of engine efficiency and emissions by coupling artificial neural network (ann) with genetic algorithm (ga). *Energy* 225: 120331.

Li, Yuanzhi, and Yingyu Liang. 2018. Learning overparameterized neural networks via stochastic gradient descent on structured data. *Advances in Neural Information Processing Systems* 31: 8157–8166.

Liang, Tengyuan, and Alexander Rakhlin. 2020. Just interpolate: Kernel "ridgeless" regression can generalize. *The Annals of Statistics* 48 (3): 1329–1347.

Muthukumar, Vidya, Adhyyan Narang, Vignesh Subramanian, Mikhail Belkin, Daniel Hsu, and Anant Sahai. 2020. Classification vs regression in overparameterized regimes: does the loss function matter? *arXiv preprint* arXiv:2005.08054.

Nakkiran, Preetum, Behnam Neyshabur, and Hanie Sedghi. 2020. The deep bootstrap framework: Good online learners are good offline generalizers. *arXiv preprint* arXiv:2010.08127.

Nakkiran, Preetum, Gal Kaplun, Yamini Bansal, Tristan Yang, Boaz Barak, and Ilya Sutskever. 2021. Deep double descent: where bigger models and more data hurt. *Journal of Statistical Mechanics: Theory and Experiment* 2021 (12): 124003.

Neal, Radford M. 1995. *Bayesian Learning for Neural Networks*. PhD thesis, University of Toronto.

Neal, Radford M. 1996. Priors for infinite networks. In *Bayesian Learning for Neural Networks*, 29–53. Springer.

Nguyen, Quynh, Mahesh Chandra Mukkamala, and Matthias Hein. 2019. On the loss landscape of a class of deep neural networks with no bad local valleys. In *International Conference on Learning Representations*.

Nguyen, Quynh. 2019. On connected sublevel sets in deep learning. In *International Conference on Machine Learning*.

Tsigler, Alexander, and Peter L Bartlett. 2020. Benign overfitting in ridge regression. *arXiv preprint* arXiv:2009.14286.

Wei, Colin, Jason D Lee, Qiang Liu, and Tengyu Ma. 2019. Regularization matters: generalization and optimization of neural nets vs their induced kernel. In *Advances in Neural Information Processing Systems*, 9712–9724.

Williams, Christopher. 1996. Computing with infinite networks. *Advances in Neural Information Processing Systems* 9: 295–301.

Xiao, Lechao, Jeffrey Pennington, and Samuel Schoenholz. 2020. Disentangling trainability and generalization in deep neural networks. In *International Conference on Machine Learning*, 10462–10472.

Xu, Aolin, and Maxim Raginsky. 2017. Information-theoretic analysis of generalization capability of learning algorithms. In *Advances in Neural Information Processing Systems*, 2524–2533.

Zhang, Jingwei, Tongliang Liu, and Dacheng Tao. 2018. An information-theoretic view for deep learning. *arXiv preprint* arXiv:1804.09060.

Zou, Difan, Jingfeng Wu, Vladimir Braverman, Quanquan Gu, and Sham Kakade. 2021. Benign overfitting of constant-stepsize sgd for linear regression. In *Conference on Learning Theory*, 4633–4635.

Zou, Difan, Yuan Cao, Dongruo Zhou, and Quanquan Gu. 2020. Gradient descent optimizes over-parameterized deep ReLU networks. *Machine Learning* 109 (3): 467–492.

Chapter 10
Theoretical Foundations for Specific Architectures

In the previous chapters, we have presented an overview of the generalizability of neural networks. This chapter discusses the influence of some specific network structures. Convolutional neural networks (CNNs) introduce convolutional layers into deep learning, which have been widely applied in computer vision, natural language processing, and deep reinforcement learning. Recurrent neural networks (RNNs) possess a recurrent structure and shows its promising performance in the processing and analysis of sequential data. They have been applied to many real-world scenarios, including NLP and speech recognition. Recently, equivariant neural networks have shown its great successes in various areas, including 3D point cloud processing, chemistry, and astronomy. The intuition is that when the prior symmetry in the data can be maintained, networks can achieve better performance.

10.1 Convolutional Neural Networks (CNNs) and Recurrent Neural Networks (RNNs)

Convolutional neural networks. Convolutional neural networks (CNNs) introduce convolutional layers into deep learning, which have been widely applied in computer vision (Krizhevsky et al. 2012), natural language processing (Yu et al. 2018), and deep reinforcement learning (Silver et al. 2016).

Du et al. (2018) proved that to achieve a population prediction error of ϵ on d-dimensional input, a c-dimensional convolutional filter with a linear activation requires $\tilde{O}\left(c/\epsilon^2\right)$ training examples while a fully connected one requires $\Omega\left(d/\epsilon^2\right)$ training examples. Moreover, the authors further proved that for a one-hidden-layer CNN with output dimensionality r and the ratio of the stride size to the filter size being fixed, it requires $\tilde{O}\left((c+r)/\epsilon^2\right)$ training examples to achieve this. Zhou and Feng (2018) proved (1) given the magnitude of the parameters b_i in the i-th layer, an generalization bound for a CNN is $O\left(\log\left(\prod_{i=0}^{D} b_i\right)\right)$, and (2) a one-to-one correspondence between convergence guarantees for the stationary points and their

F. He and D. Tao, *Foundations of Deep Learning*, Machine Learning: Foundations, Methodologies, and Applications, https://doi.org/10.1007/978-981-16-8233-9_10

population counterparts. Lin and Zhang (2019) studied influence of the sparsity and permutation of convolutional layers on the spectral norm and generalizability. Other works have also studied the generalizability of CNNs with residual networks (He et al. 2016), including He et al. (2020) and Chen et al. (2019a).

Recurrent neural networks. Recurrent neural networks (RNNs) possess a recurrent structure and shows its promising performance in the processing and analysis of sequential data. They have been applied to many real-world scenarios, including NLP (Bahdanau et al. 2015; Sutskever et al. 2014) and speech recognition (Graves et al. 2006, 2013). Chen et al. (2019b) developed a generalization bound for RNNs based on works by Neyshabur et al. (2017) and Bartlett et al. (2017). Allen-Zhu and Li (2019) also analysed the generalizability of RNNs.

Based on the Fisher-Rao norm, Tu et al. (2020) developed a gradient measure and then derived a generalization bound as follows.

Theorem 10.1 *Let us fix a margin parameter α; then, for any $\delta > 0$, with probability at least $1-\delta$, the following holds for every RNN whose weight matrices $\theta = (U, W, V)$ satisfy $||V^T||_1 \leq \beta_V$, $||W^T||_1 \leq \beta_W$, $||U^T||_1 \leq \beta_U$ and $||\theta||_{fs} \leq r$:*

$$\mathbb{E}[\mathbf{I}_{\mathcal{M}_{y_L}(x,y)\leq 0}] \leq \frac{4k}{\alpha}\left(\frac{r||X||_F}{2m}\sqrt{\frac{1}{\lambda_{min}(\mathbb{E}(XX^T))}} + \frac{1}{m}\beta_V\beta_U||X^T||_1\Lambda\right)$$
$$+\frac{1}{m}\sum \mathbf{I}_{\mathcal{M}_{y_L}(x_i,y_i)\leq a} + 3\sqrt{\frac{\log\frac{2}{\delta}}{2m}}. \tag{10.1}$$

This gradient measure builds a relationship between generalizability and trainability and obtain a tighter bound. It shows that adding noise to the input data is a way to boost the generalizability. It also justifies the use of the gradient clipping technique (Merity et al. 2018; Gehring et al. 2017; Peters et al. 2018).

10.2 Equivariant Neural Networks

Recently, equivariant neural networks have shown its great successes in various areas, including 3D point cloud processing (Li et al. 2018), chemistry (Faber et al. 2016), and astronomy (Ntampaka et al. 2016; Ravanbakhsh et al. 2016). The intuition is that when the prior symmetry in the data can be maintained, networks can achieve better performance. As an example, objects in images of different classes can be oblique, while all rotated versions of the same image are expected to belong to the same class. Hence, there is some prior symmetry in such data, and if a model can preserve this symmetry, it will enjoy better performance.

This section is organized as follows. First, there are two main ways to design a new equivariant network. One way is to modify a traditional network; as an example, we consider group CNNs, which are impressive traditional models of this type. The other way is to limit the layers such that they always remain equivariant. Subsequently, we discuss the nonlinearity in equivariant neural networks. Finally, we present a theoretical analysis, including analyses of generalization and approximation, which show the benefits of equivariant neural networks.

10.2.1 Group CNNs

An equivariant neural network was generalized and developed by Cohen and Welling (2016a), who noted the following shift equivariance in CNNs:

$$\Phi(T_g x) = T'_g \Phi(x) \text{ for all } x, \tag{10.2}$$

where Φ is a convolutional layer, T_g is a shift transformation acting on the input x and T'_g is the corresponding transformation acting on the output. That is, transforming the input x by a transformation T_g is equivalent to transforming the output $\Phi(x)$ by a corresponding transformation T'_g, which is independent of the input x. In addition, the transformation T should be a linear representation of the group G, that is, for two elements g and h of the group G,

$$T_{gh} = T_g T_h. \tag{10.3}$$

However, convolution is shift equivariant but not rotation equivariant. To overcome this shortcoming, the authors defined the operation of group convolution, which is always equivariant with regular representations. Let f denote the input and ψ denote a filter; then, *group convolution* is defined as

$$f \star \psi(g) = \sum_{h \in G} \sum_{k} f_k(h) \psi_k(g^{-1}h). \tag{10.4}$$

For a regular representation L such that $L_h f(g) = f(h^{-1}g)$, it can be proven that

$$(L_u f) \star \psi = L_u(f \star \psi) \text{ for all } u \in G. \tag{10.5}$$

Despite the application of convolution, the equivariance of the nonlinearity and some operations should be maintained. For any pointwise nonlinearity C_v, it can be verified that

$$C_v L_h f = v \circ f \circ h^{-1} = (v \circ f) \circ h^{-1} = L_h(v \circ f) = L_h C_v f. \tag{10.6}$$

In addition, these authors defined subgroup pooling as follows:

$$P(f)(g) = \max_{h \in gU} f(h), \tag{10.7}$$

where U is a subgroup of G and $gU = \{gu : u \in U\}$ is a coset. The reader can prove that $PL_h f = L_h Pf$. As seen from the examples above, each operator in an equivariant neural network should itself be equivariant. Thus, when a new equivariant neural network is designed, all operators should be proven to be equivariant. It is sometimes necessary that the initial operator not be equivariant; in this case, it needs to be modified to an equivariant version.

10.2.2 Steerable Neural Networks

10.2.2.1 Steerable CNNs

In Cohen and Welling (2016b), the authors showed another way to achieve equivariance, that is, constraining the filter such that the corresponding layer is equivariant. A subgroup H of G is often considered first; for any $h \in H$, the filter satisfies

$$\psi_h \Psi = \Psi \rho_h \text{ for all } h \in H \tag{10.8}$$

for some linear representation ψ of H acting on the output of the layer and ρ acting on the input of the layer. Note that if the constraint is linear, then the solution space will be linear. The space of all solutions Ψ is denoted by $\mathrm{Hom}_H(\rho, \psi)$ because an equivariant map Ψ is a "homomorphism of group representations". Equivariant maps Ψ are also called *intertwiners* (Serre et al. 1977). Before training, a basis $\psi_1, \ldots, \psi_m$ of the solution space can be calculated, and all admitted weights can be represented as linear combinations of this basis of the form $\Psi = \sum_i \alpha_i \psi_i$. For any G, there are many choices of different linear representations, and thus, a steerable neural network is flexible and efficient. By training the coefficients of the linear combination, steerable neural networks can achieve excellent performance in various tasks; examples include graph neural networks (Maron et al. 2018), E_2 steerable CNNs (Weiler and Cesa 2019), and rotation-equivariant steerable CNNs (Weiler et al. 2018b). In the following, we will present some theoretical implementations of steerable neural networks.

A linear representation of G can be defined as an induced representation of the linear representation H. The following example comes from Cohen and Welling (2016b). The correlation $\Psi \star f$ can be computed as

$$(\Psi \star f)(x) = \Psi(\rho^{-1}(x) f), \tag{10.9}$$

where $x \in \mathbb{Z}^2$ is viewed as a translation. Let H represent all transformations acting on the input f, and let $t \in \mathbb{Z}^2$ and $r \in H$. It can be verified that

$$[\Psi \star \rho(tr) f](x) = \psi(r)[\Psi \star \rho]((tr)^{-1}x), \tag{10.10}$$

where t, r mean a translation and transformation, respectively. Thus, if we consider a representation ψ' such that

$$[\psi'(tr)f](x) = \psi(r)f((tr)^{-1}x), \tag{10.11}$$

then $\psi'(tr)\Psi \star f = \Psi \star \rho(tr)f$. The representation ψ' is the representation of G induced by ψ, which is denoted by $\mathrm{Ind}_H^G \psi$.

Any linear representation can be decomposed into a direct sum of irreducible representations, that is,

$$\psi = P^{-1} \bigoplus \psi_i P, \tag{10.12}$$

where P is an invertible matrix and ψ_i is an irreducible representation for all i. Similarly, if we write ρ as $Q^{-1} \bigoplus \rho_j Q$, then the equation $\psi\Psi = \Psi\rho$ can be written as

$$\bigoplus \psi_i \tilde{\Psi} = \tilde{\Psi} \bigoplus \rho_j, \tag{10.13}$$

where $\tilde{\Psi} = (P\Psi Q^{-1})$. Then, we can decompose $\tilde{\Psi}$ into $[\tilde{\Psi}_{ij}]_{ij}$ such that $\psi_i \tilde{\Psi}_{ij} = \tilde{\Psi}_{ij}\rho_j$. If ψ_i and ρ_j are not isomorphic, then $\tilde{\Psi}_{ij} = 0$.

10.2.2.2 Steerable Neural Networks

A general equivariant feedforward neural network (without bias) can be expressed as

$$W_L\sigma(W_{L-1}\sigma(\ldots\sigma(W_1x))), \tag{10.14}$$

where W_k satisfies

$$\psi^k \circ W_k = W_k \circ \psi^{k-1} \tag{10.15}$$

and σ satisfies that for any $k \in [L]$,

$$\psi^k \circ \sigma = \sigma \circ \psi^k. \tag{10.16}$$

One can verify that networks satisfying these two constraints are equivariant. We first assume that σ is a pointwise nonlinearity. Constraint (10.16) will be discussed further in the following section. We can prove that the image of the linear representation satisfying constraint (10.16) must be a set of generalized permutation matrices, where each ψ_g is a permutation matrix but the value of each nonzero entry need not be 1. In addition, the network can be expressed as

$$F(x) = \widetilde{W}_L\sigma(\widetilde{W}_{L-1}\sigma(\ldots\sigma(\widetilde{W}_1\tilde{x}))), \tag{10.17}$$

where $\widetilde{W}_L = [\psi^L_{g_1^{-1}} W_L \ldots \psi^L_{g_t^{-1}} W_L]$,

$$\tilde{x} = \begin{bmatrix} \psi^0_{g_1} \\ \psi^0_{g_2} \\ \vdots \\ \psi^0_{g_t} \end{bmatrix} x, \text{ and } \widetilde{W}_k = \begin{bmatrix} \psi^k_{g_1 g_1^{-1}} W_k & \dots & \psi^k_{g_1 g_t^{-1}} W_k \\ \psi^k_{g_2 g_1^{-1}} W_k & \dots & \psi^k_{g_2 g_t^{-1}} W_k \\ \vdots & \vdots & \vdots \\ \psi^k_{g_t g_1^{-1}} W_k & \dots & \psi^k_{g_t g_t^{-1}} W_k \end{bmatrix},$$

with any $k < L$ and any arbitrary W_k. Applications of steerable neural networks have also been reported. For example, Maron et al. (2018) considered equivariant graph neural networks and solved the solution space $\mathrm{Hom}_G(\rho, \psi)$.

10.2.2.3 The Dimensionality of $\mathrm{Hom}_G(\rho, \psi)$

Recall the definition of $\mathrm{Hom}_G(\rho, \psi) = \{W : W \circ \rho = \psi \circ W\}$. Here, we will introduce the result for the dimensionality of $\mathrm{Hom}_H(\rho, \psi)$. More details can be found in Reeder (2014). The dimensionality of $\mathrm{Hom}_H(\rho, \psi)$ is smaller than that of the weight matrix. The equation can be written as

$$\rho^T_g \otimes \psi^{-1}_g vec(W) = vec(W) \text{ for all } g \in G. \tag{10.18}$$

Let us define $\rho' : G \to \mathrm{GL}(\mathrm{Hom}(V, V'))$ as $\rho'_g W = \psi^{-1}_g W \rho_g$. If f is in $\mathrm{Hom}_G(\rho, \psi)$, then $\rho'_g f = f$ for any $g \in G$. Now, we consider that $P = \frac{1}{|G|} \sum_{g \in G} \rho'_g$ is a projection from $\mathrm{Hom}(V, V')$ to $\mathrm{Hom}_G(\rho, \psi)$. Then, we know that

$$\mathrm{Hom}(V, V') = \mathrm{Ker} P \oplus \mathrm{Hom}_G(\rho, \psi), \tag{10.19}$$

and in turn,

$$\mathrm{dimHom}_G(\rho, \psi) = \mathrm{tr}(P) = \frac{1}{|G|} \sum_{g \in G} \chi_{\rho'}(g), \tag{10.20}$$

where $\chi_{\rho'}(g) = \mathrm{tr}(\rho'_g)$ is the character. Equation (10.18) implies that $\chi_{\rho'}(g) = \chi_{\rho^T \oplus \psi^{-1}}(g) = \chi_\rho(g) \chi_\psi(g^{-1})$. Finally, we know that

$$\mathrm{dimHom}_G(\rho, \psi) = \frac{1}{|G|} \sum_{g \in G} \chi_\rho(g) \chi_\psi(g^{-1}) = \langle \chi_\rho, \chi_\psi \rangle. \tag{10.21}$$

As mentioned above, any representation can be decomposed into a direct sum of irreducible representations. For a finite group G, the number of different (non-isomorphic) irreducible representations is equal to the number of classes of G. Let all irreducible representations be denoted by $\psi_1, \dots, \psi_n$, and let $\rho = Q^{-1} \bigoplus m_i \psi_i Q$ and $\psi = P^{-1} \bigoplus m'_i \psi_i P$. Note that $\{\psi_1, \dots, \psi_n\}$ forms an orthonormal basis; then,

$$\dim \mathrm{Hom}_G(\rho, \psi) = \langle \chi_\rho, \chi_\psi \rangle = \sum_{i=1}^{n} m_i m_i'. \tag{10.22}$$

From this result, we know that the dimensionality of $\mathrm{Hom}_G(\rho, \psi)$ can be modified to any desired value. In addition, in practice, we need only to compute $\mathrm{Hom}_G \psi_i, \psi_j$ for all irreducible representations ψ_i and ψ_j. However, the linear representation should commute with the nonlinearity, and thus, the choice of the invertible matrix P is also important. We discuss this topic in the following section.

10.2.3 Nonlinearities in Equivariant Networks

As shown above, we can express linear representations as direct sums of irreducible representations. However, any linear representation should commute with the nonlinearity: $\sigma \circ \psi = \psi \circ \sigma$. More generally, this requirement can be extended to $\psi \circ \sigma = \sigma \circ \rho$. That means that the selection of group linear representations restricts the range of the admitted nonlinearities, or conversely, that a certain choice of nonlinearity allows for only particular linear representations. We consider the condition $\psi \circ \sigma = \sigma \circ \rho$.

For a pointwise nonlinearity, such as rectified linear unit (ReLU) activation, it can be proven that the image of the linear representation satisfying constraint (10.16) must consist of generalized permutation matrices, where each ψ_g is a permutation matrix but the value of each nonzero entry need not be 1. When the nonlinearity is ReLU, we can further prove that any such nonzero value is independent of the group element g. Two useful linear representations for pointwise nonlinearity are a regular representation and a quotient representation. In a regular representation, the basis is $\{e_g\}$, and $\rho_g e_{g'} = e_{gg'}$. Thus, it is a permutation representation with dimensions of $|G| \times |G|$. A quotient representation uses a subgroup H and the basis e_{gH}. Then, $\rho_g e_{g'H} = e_{gg'H}$, which is also a permutation matrix but with dimensions of $[G:H] \times [G:H]$. In the following section, we introduce the result that a network under this setting has a greater approximation ability.

With the exception of elementwise nonlinearities, there are also some special nonlinearities for which a given group representation achieves equivariance. For unitary representations, such that $\|\psi_g x\| = \|x\|$, nonlinearities that act on the norm of a feature but preserve its orientation are popular because they are always commutative. In general, the nonlinearity can be decomposed into $\eta(\|x\|)x/\|x\|$, where $\eta(\|x\|)$ can be $ReLU(\|x\| - b)$ for some positive b (Weiler et al. 2018a; Worrall et al. 2017), and squashing nonlinearities of the form $\|x\|^2/(\|x\|^2 + 1)$ (Sabour et al. 2017). These nonlinearities all offer excellent performance in different tasks. Meanwhile, the theory of steerable networks itself shows no preference for any specific nonlinearity.

10.2.4 Generalization of Equivariant Networks

In this section, we present two results to show the benefits of equivariant neural networks. First, however, let us begin with some generalization bounds. Intuitively, equivariant neural networks are relatively constrained in comparison to all possible neural networks. Specifically, the hypothesis set is smaller, implying better generalization. Some works have attempted to provide new generalization bounds for equivariant neural networks, including Kondor and Trivedi (2018), Sokolic et al. (2017), Sannai et al. (2021). However, generalization bounds cover only the worst case, and the results of the above works do not explicitly consider the benefits of equivariant models. To address this gap, we introduce the benefits of equivariant neural networks (Elesedy and Zaidi 2021; Lyle et al. 2020).

10.2.5 Generalization Bounds of Equivariant Networks

Some works have attempted to present generalization bounds for equivariant networks, such as Kondor and Trivedi (2018), Sannai et al. (2021). Here, we discuss the result in Sannai et al. (2021), which is the newest generalization bound. We first introduce the generalization bounds for invariant neural networks and equivariant neural networks, and we then share the method used to obtain them. Let $R(f) = \mathbb{E}_{(x,y)\sim\mathcal{D}}[l(f(x), y)]$ be the expected risk, and let $\hat{R}_S(f) = \frac{1}{m}\sum_{i=1}^{m} l(f(x_i), y_i)$ be the empirical risk. Then, for any invariant neural network f uniformly bounded by 1, the following inequality holds with probability at least $1 - 2\epsilon$:

$$R(f) - \hat{R}_S(f) \leq \sqrt{\frac{C}{|G|m^{2/n}}} + \sqrt{\frac{2\log(1/2\epsilon)}{m}}. \tag{10.23}$$

For any equivariant neural network f uniformly bounded by 1, the following inequality holds with probability at least $1 - 2\epsilon$:

$$R(f) - \hat{R}_S(f) \leq \sqrt{\frac{C}{|\mathrm{St}(G)|m^{2/n}}} + \sqrt{\frac{2\log(1/2\epsilon)}{m}}, \tag{10.24}$$

where C is independent of m, n and ϵ, and $\mathrm{St}(G) \subset G$ is a subgroup of elements whose first coordinates are fixed. These results follow from the Rademacher complexity bound, and they estimate the Rademacher complexity at the bound. Note that m is the number of examples and n is the dimensionality of the input, and it has been shown that the Rademacher complexity is $O(m^{-2/n})$.

Now, we introduce the method. For an orbit $\{gx : g \in G\}$, if $f(x)$ is determined, then $f(gx) = gf(x)$ is determined. This means that the whole space $[0, 1]^n$ can be divided into orbits. Then, a quotient map $\phi_G : [0, 1]^n \rightarrow [0, 1]^n/G$

can be defined as $\phi_G(x) = \{gx : g \in G\}$, and the hypothesis set $\{f : [0, 1]^n \rightarrow \mathbb{R}^M$ such that f is equivariant$\}$ can be covered by $\{f : [0, 1]^n/G \rightarrow \mathbb{R}^M\}$. The covering number of $[0, 1]^n/G$ can be bounded by $C/|G|\epsilon^n$. The results above can be proven in accordance with the Rademacher complexity bound.

In addition, note that $|G|$ is sometimes not independent of n. For example, when $G = S_n$, the generalization bound becomes

$$\sqrt{\frac{C}{n!m^{2/n}}} + \sqrt{\frac{2\log(1/2\epsilon)}{m}}. \tag{10.25}$$

10.2.5.1 Benefits of Equivariant Networks

Beyond traditional generalization bounds, some works have focused on the benefits of equivariant networks. In these works, the authors have shown the benefits of equivariant maps but not those of equivariant networks in practice. This means that these analyses offer no suggestions regarding how to design a new equivariant neural network. Thus, the theoretical analysis of equivariant neural networks is still immature.

Lyle et al. (2020) compared data augmentation and feature averaging for invariant targets. For a given dataset $\{(x_i, y_i)\}_{i=1}^m$ and a linear representation ρ transforming the data, suppose that the data have an invariant distribution, that is, $P_D(x, y) = P_D(\rho_g x, y)$ for all x, y, and $g \in G$. For data augmentation, the empirical risk is considered to be

$$\hat{R}_S(f) = \frac{1}{m}\sum_{i=1}^{m}\mathbb{E}_{g\sim\lambda} l(f(gx_i), y_i) \approx \frac{1}{mn}\sum_{i=1}^{m}\sum_{j=1}^{n} l(f(g_j x_i), y_i), \tag{10.26}$$

where g_j is sampled from a distribution λ. The empirical risk of data augmentation has been shown to have a relatively small variance because it relies on a better dataset with a better empirical distribution. However, this risk cannot guarantee that an invariant network will be learned. Another approach, feature averaging, is to take the average of the outputs of all transformations $\mathbb{E}_{g\sim\lambda} f_H(gx)$ as a new output of the first H layers. It is easy to find that in this approach, the empirical risk is the same as that in data augmentation. In the case of feature averaging, however, the network is always invariant. When the posterior Q is restricted to be invariant, the symmetrization gap $KL(P||Q)$ is smaller than that when Q is not invariant. The smaller symmetrization gap $KL(P||Q)$ indicates a better Probably Approximately Correct (PAC)-Bayes bound. In this sense, it reveals a benefit of invariant neural networks, *i.e.*, the invariant nature of the learned networks.

In addition, the orbit averaging technique was used in Elesedy and Zaidi (2021). The orbit averaging operation is defined as

$$Qf = \sum_{g \in G} \psi_g^{-1} \circ f \circ \rho_g \tag{10.27}$$

for finite groups and

$$Qf = \int_G \psi_g^{-1} \circ f \circ \rho_g \, d\lambda(g) \tag{10.28}$$

for compact groups and those with a Haar measure λ over G. Note that

$$Qf \circ \rho_h = \psi_h \int \psi_{gh}^{-1} \circ f \circ \rho_{gh} d\lambda(g) = \psi_h \circ Qf, \tag{10.29}$$

and thus, Qf is equivariant. Furthermore, Q is a projection map from the space of all functions to the space of all equivariant functions, and $Q^2 = Q$. When we consider the expected loss

$$R(f) = \mathbb{E}_{x \sim \mathcal{D}}[l(f(x), s(x))] = \mathbb{E}_{x \sim \mathcal{D}}[\|f(x) - s(x)\|_2^2], \tag{10.30}$$

if the target s is equivariant such that $\psi \circ s = s \circ \rho$, then $R(Qf) \leq R(f)$ for any f, and the gap is exactly

$$R(f) - R(Qf) = \int_X \langle f(x) - Qf(x), f(x) - Qf(x) \rangle d\mu(x) \geq 0.$$

In particular, when f is not equivariant, the gap satisfies $R(f) - R(Qf) > 0$. Finally, if f is the best predictor with the least expected risk, then f must be equivariant. If not, then Qf is a better predictor. However, a problem arises in this case: Qf may not be a neural network of the same width. If we use the predictor Qf, then the process will require more calculation, which is not desirable. In practice, an equivariant neural network is layerwise equivariant, and the linear representations in each of the layers are chosen prior to training. Thus, a real equivariant neural network has additional constraints.

10.2.6 Approximation of Equivariant Networks

Another aspect addressed by theoretical research is the approximation ability. Because equivariant neural networks can be viewed as constrained neural networks, a natural question is whether equivariant neural networks are able to approximate any equivariant function. To answer this question, some works have considered certain special cases and proved the approximation property under certain assumptions. Here, we introduce some related results from Ravanbakhsh (2020).

Two assumptions are adopted in the proof presented by the cited authors. One is that orbit averaging is a projection to all equivariant maps, and the other is that two-layer neural networks have uniform approximation properties. That is, for a given compact set C, any continuous function f supported on C can be uniformly approximated by a two-layer neural network F with ReLU nonlinearity. Any equivariant map f defined on C can be extended to be defined on $\tilde{C} = \{gx : x \in C\}$ as $\tilde{f}(gx) = gf(x)$ for any $x \in C$. Hence, without loss of generality, we can assume that $C = \tilde{C}$.

Now, there are two claims to be proven. On the one hand, when F is a two-layer universal approximator such that $\|f(x) - F(x)\| < \epsilon$ for any $x \in C$, then QF is also, and QF can be modified into a two-layer neural network. Note that

$$\|QF(x) - f(x)\| = \frac{1}{|G|}\|\sum_{g\in G} g^{-1}[F(gx) - f(gx)]\| \tag{10.31}$$

$$\leq \frac{1}{|G|}\sum_{g\in G}\|g^{-1}[F(x) - f(x)]\| \tag{10.32}$$

$$\leq \sup_{g\in G}\|g\|\epsilon, \tag{10.33}$$

which implies that QF is also a uniform approximator. On the other hand, for a two-layer neural network $F(x) = W_2\sigma W_1 \cdot x$, QF can be written as

$$QF(x) = \frac{1}{|G|}\left[{g_1}^{-1}W_2 \ldots {g_t}^{-1}W_2\right]\sigma\begin{bmatrix} W_1 g_1 \\ \vdots \\ W_1 g_t \end{bmatrix} x, \tag{10.34}$$

which is an equivariant neural network with a regular representation. By combining these two results, we prove the uniform approximation property for finite groups.

References

Allen-Zhu, Zeyuan, and Yuanzhi Li. 2019. Can SGD learn recurrent neural networks with provable generalization? In *Advances in Neural Information Processing Systems*, 10331–10341.

Bahdanau, Dzmitry, Kyunghyun Cho, and Yoshua Bengio. 2015. Neural machine translation by jointly learning to align and translate. In *International Conference on Learning Representations*.

Bartlett, Peter L, Dylan J Foster, and Matus J Telgarsky. 2017. Spectrally-normalized margin bounds for neural networks. In *Advances in Neural Information Processing Systems*, 6240–6249.

Chen, Hao, Zhanfeng Mo, Zhouwang Yang, and Xiao Wang. 2019a. Theoretical investigation of generalization bound for residual networks. In *International Joint Conference on Artificial Intelligence*, 2081–2087.

Chen, Minshuo, Xingguo Li, and Tuo Zhao. 2019b. On generalization bounds of a family of recurrent neural networks. *arXiv preprint* arXiv:1910.12947.

Cohen, Taco, and Max Welling. 2016a. Group equivariant convolutional networks. In *International Conference on Machine Learning*, 2990–2999.

Cohen, Taco S, and Max Welling. 2016b. Steerable cnns. *arXiv preprint* arXiv:1612.08498.

Du, Simon S, Yining Wang, Xiyu Zhai, Sivaraman Balakrishnan, Russ R Salakhutdinov, and Aarti Singh. 2018. How many samples are needed to estimate a convolutional neural network? In *Advances in Neural Information Processing Systems*, 373–383.

Elesedy, Bryn, and Sheheryar Zaidi. 2021. Provably strict generalisation benefit for equivariant models. In *International Conference on Machine Learning*, 2959–2969.

Faber, Felix A, Alexander Lindmaa, O Anatole Von Lilienfeld, and Rickard Armiento. 2016. Machine learning energies of 2 million elpasolite (a b c 2 d 6) crystals. *Physical Review Letters* 117 (13): 135502.

Gehring, Jonas, Michael Auli, David Grangier, Denis Yarats, and Yann N Dauphin. 2017. Convolutional sequence to sequence learning. In *International Conference on Machine Learning*, 1243–1252.

Graves, Alex, Abdel-rahman Mohamed, and Geoffrey Hinton. 2013. Speech recognition with deep recurrent neural networks. In *IEEE International Conference on Acoustics, Speech and Signal Processing*, 6645–6649.

Graves, Alex, Santiago Fernández, Faustino Gomez, and Jürgen Schmidhuber. 2006. Connectionist temporal classification: labelling unsegmented sequence data with recurrent neural networks. In *International Conference on Machine Learning*, 369–376.

He, Fengxiang, Tongliang Liu, and Dacheng Tao. 2020. Why resnet works? residuals generalize. *IEEE Transactions on Neural Networks and Learning Systems*.

He, Kaiming, Xiangyu Zhang, Shaoqing Ren, and Jian Sun. 2016. Deep residual learning for image recognition. In *IEEE Conference on Computer Vision and Pattern Recognition*.

Kondor, Risi, and Shubhendu Trivedi. 2018. On the generalization of equivariance and convolution in neural networks to the action of compact groups. In *International Conference on Machine Learning*, 2747–2755.

Krizhevsky, Alex, Ilya Sutskever, and Geoffrey E Hinton. 2012. Imagenet classification with deep convolutional neural networks. In *Advances in Neural Information Processing Systems*, 1097–1105.

Li, Yangyan, Rui Bu, Mingchao Sun, Wei Wu, Xinhan Di, and Baoquan Chen. 2018. Pointcnn: convolution on x-transformed points. In *Advances in Neural Information Processing Systems*, 820–830.

Lin, Shan, and Jingwei Zhang. 2019. Generalization bounds for convolutional neural networks. *arXiv preprint* arXiv:1910.01487.

Lyle, Clare, Mark van der Wilk, Marta Kwiatkowska, Yarin Gal, and Benjamin Bloem-Reddy. 2020. On the benefits of invariance in neural networks. *arXiv preprint* arXiv:2005.00178.

Maron, Haggai, Heli Ben-Hamu, Nadav Shamir, and Yaron Lipman. 2018. Invariant and equivariant graph networks. *arXiv preprint* arXiv:1812.09902.

Merity, Stephen, Nitish Shirish Keskar, and Richard Socher. 2018. Regularizing and optimizing lstm language models. In *International Conference on Learning Representations*.

Neyshabur, Behnam, Srinadh Bhojanapalli, and Nathan Srebro. 2017. A PAC-Bayesian approach to spectrally-normalized margin bounds for neural networks. *arXiv preprint* arXiv:1707.09564.

Ntampaka, Michelle, Hy Trac, Dougal J Sutherland, Sebastian Fromenteau, Barnabás Póczos, and Jeff Schneider. 2016. Dynamical mass measurements of contaminated galaxy clusters using machine learning. *The Astrophysical Journal* 831 (2): 135.

Peters, Matthew E, Mark Neumann, Mohit Iyyer, Matt Gardner, Christopher Clark, Kenton Lee, and Luke Zettlemoyer. 2018. Deep contextualized word representations. In *Annual Conference of the North American Chapter of the Association for Computational Linguistics*, 2227–2237.

Ravanbakhsh, Siamak, Junier Oliva, Sebastian Fromenteau, Layne Price, Shirley Ho, Jeff Schneider, and Barnabás Póczos. 2016. Estimating cosmological parameters from the dark matter distribution. In *International Conference on Machine Learning*, 2407–2416.

Ravanbakhsh, Siamak. 2020. Universal equivariant multilayer perceptrons. In *International Conference on Machine Learning*.

Reeder, Mark. 2014. Notes on group theory.

Sabour, Sara, Nicholas Frosst, and Geoffrey E Hinton. 2017. Dynamic routing between capsules. *Advances in Neural Information Processing Systems*, 30.

Sannai, Akiyoshi, Masaaki Imaizumi, and Makoto Kawano. 2021. Improved generalization bounds of group invariant/equivariant deep networks via quotient feature spaces. In *Uncertainty in Artificial Intelligence*, 771–780.

Serre, Jean-Pierre, et al. 1977. *Linear Representations of Finite Groups*, vol. 42. Springer.

Silver, David, Aja Huang, Chris J Maddison, Arthur Guez, Laurent Sifre, George Van Den Driessche, Julian Schrittwieser, Ioannis Antonoglou, Veda Panneershelvam, and Marc Lanctot. 2016. Mastering the game of go with deep neural networks and tree search. *Nature* 529 (7587): 484–489.

Sokolic, Jure, Raja Giryes, Guillermo Sapiro, and Miguel Rodrigues. 2017. Generalization error of invariant classifiers. In *Artificial Intelligence and Statistics*, 1094–1103.

Sutskever, Ilya, Oriol Vinyals, and Quoc V Le. 2014. Sequence to sequence learning with neural networks. In *Advances in Neural Information Processing Systems*, 3104–3112.

Tu, Zhuozhuo, Fengxiang He, and Dacheng Tao. 2020. Understanding generalization in recurrent neural networks. In *International Conference on Learning Representations*.

Weiler, Maurice, and Gabriele Cesa. 2019. General e (2)-equivariant steerable cnns. *Advances in Neural Information Processing Systems*, 32.

Weiler, Maurice, Mario Geiger, Max Welling, Wouter Boomsma, and Taco S Cohen. 2018a. 3d steerable cnns: learning rotationally equivariant features in volumetric data. *Advances in Neural Information Processing Systems*, 31.

Weiler, Maurice, Fred A Hamprecht, and Martin Storath. 2018b. Learning steerable filters for rotation equivariant cnns. In *IEEE Conference on Computer Vision and Pattern Recognition*, 849–858.

Worrall, Daniel E, Stephan J Garbin, Daniyar Turmukhambetov, and Gabriel J Brostow. 2017. Harmonic networks: deep translation and rotation equivariance. In *IEEE Conference on Computer Vision and Pattern Recognition*, 5028–5037.

Yu, Adams Wei, David Dohan, Minh-Thang Luong, Rui Zhao, Kai Chen, Mohammad Norouzi, and Quoc V Le. 2018. Qanet: combining local convolution with global self-attention for reading comprehension. In *International Conference on Learning Representations*.

Zhou, Pan, and Jiashi Feng. 2018. Understanding generalization and optimization performance of deep cnns. In *International Conference on Machine Learning*, 5960–5969.

Part III
Deep Learning Theory form the Trust-worthy Facet

Chapter 11
Privacy Preservation

In this chapter, we introduce an important line of research dedicated to measuring and providing protection for individual privacy in machine learning. We will start from a common standard for measuring privacy protection, differential privacy, and present related definitions, properties, and extended theories. We will examine various means of privacy protection that aim to hide personal information from hostile adversaries. To orient the topic in a context appropriate to this book, we also provide discussions concerning the relationship between privacy protection and other factors in trustworthy artificial intelligence.

11.1 Differential Privacy

Deep learning is deployed to process massive amounts of personal data, including financial data and medical information. The ability to extract highly valuable knowledge from the population while protecting each individual's privacy and sensitive data is of critical importance (Dwork and Mulligan 2013; Dwork and Roth 2014).

Differential privacy. Differential privacy is a major metric to measure the privacy-preserving ability of an algorithm (Dwork and Roth 2014). (ε, δ)-differential private means that, for a learning algorithm, upon exposure of the training data to a small disturbance, if its change in the output hypothesis is bounded as follows:

$$\log\left[\frac{\mathbb{P}_{\mathcal{A}(S)}(\mathcal{A}(S)\in B)-\delta}{\mathbb{P}_{\mathcal{A}(S')}(\mathcal{A}(S')\in B)}\right]\leq\varepsilon, \tag{11.1}$$

where B is an arbitrary subset of the hypothesis space and (S, S') is a pair of neighbouring sample sets in which S and S' differ by only one example. The left-hand side of Eq. (11.1) is also called the *privacy loss*.

In Eq. (11.1), a division operation is employed to measure changes in the output hypothesis. In recent years, numerous variants of the privacy-preserving ability metrics have been devised by tweaking this division operation or introducing

F. He and D. Tao, *Foundations of Deep Learning*, Machine Learning: Foundations, Methodologies, and Applications, https://doi.org/10.1007/978-981-16-8233-9_11

specific assumptions. The variants include: (1) Concentration Differential Privacy, which assumes privacy loss follows a sub-Gaussian distribution (Dwork and Rothblum 2016; Bun and Steinke 2016); (2) Mutual-Information Differential Privacy and Kullback-Leibler (KL) Differential Privacy, which replace the division operation with mutual information and the KL divergence, respectively (Cuff and Yu 2016; Wang et al. 2016; Liao et al. 2017; Chaudhuri et al. 2019); and (3) Rényi Differential Privacy, which substitutes the KL divergence with the Rényi divergence (Mironov 2017; Geumlek et al. 2017). These approaches offer nuanced perspectives on privacy preservation in machine learning models, prospecting privacy preservation in data-driven applications.

These techniques have also been applied to preserve privacy in deep learning (Abadi et al. 2016).

Generalization-privacy relationship. It is suggested that the generalizability is somehow equivalent to the privacy-preserving ability measured by differential privacy.

Dwork et al. (2015) proved a high-probability generalization bound for (ε, δ)-differentially private machine learning algorithms formulated as follows:

$$\mathbb{P}\left[R(\mathcal{A}(S)) - \hat{R}_S(\mathcal{A}(S)) < 4\varepsilon\right] > 1 - 8\delta^{\varepsilon}.$$

Oneto et al. (2017) improved this generalization bound as follows:

$$\mathbb{P}\left[\text{Diff } R < \sqrt{6\hat{R}_S(\mathcal{A}(S))\hat{\varepsilon}} + 6\left(\varepsilon^2 + 1/m\right)\right] > 1 - 3e^{-m\varepsilon^2}$$

and

$$\mathbb{P}\left[\text{Diff } R < \sqrt{4\hat{V}_S(\mathcal{A}(S))\hat{\varepsilon}} + \frac{5m}{m-1}\left(\varepsilon^2 + 1/m\right)\right] > 1 - 3e^{-m\varepsilon^2},$$

where

$$\text{Diff } R = R(\mathcal{A}(S)) - \hat{R}_S(\mathcal{A}(S)),$$
$$\hat{\varepsilon} = \varepsilon + \sqrt{1/m},$$

and $\hat{V}_S(\mathcal{A}(S))$ is the empirical variance of $l(\mathcal{A}(S), \cdot)$:

$$\hat{V}_S(\mathcal{A}(S)) = \frac{1}{2m(m-1)} \sum_{i \neq j} \left[\ell\left(\mathcal{A}(S), z_i\right) - \ell\left(\mathcal{A}(S), z_j\right)\right]^2.$$

Nissim and Stemmer (2015) further proved that

$$\mathbb{P}\left[R(\mathcal{A}(S)) - \hat{R}_S(\mathcal{A}(S)) < 13\varepsilon\right] > 1 - \frac{2\delta}{\varepsilon}\log\left(\frac{2}{\varepsilon}\right).$$

To date, the tightest generalization bound by far has been given by He et al. (2020) as follows:

$$\mathbb{P}\left[\left|\hat{R}_S(\mathcal{A}(S)) - R(\mathcal{A}(S))\right| < 9\varepsilon\right] > 1 - \frac{e^{-\varepsilon}\delta}{\varepsilon}\ln\left(\frac{2}{\varepsilon}\right). \tag{11.2}$$

This generalization bound was derived in three stages. The authors first proved an on-average generalization bound for any (ε, δ)-differentially private multidatabase learning algorithm

$$\tilde{\mathcal{A}} : \mathbf{S} \to \mathcal{H} \times \{1, \ldots, k\}$$

as follows:

$$\left|\underset{\mathbf{S}\sim\mathcal{D}^m}{\mathbb{E}}\left[\underset{\mathcal{A}(\mathbf{S})}{\mathbb{E}}\left[\hat{R}_{S_{i_{\mathcal{A}(\mathbf{S})}}}\left(h_{\mathcal{A}(\mathbf{S})}\right)\right] - \underset{\mathcal{A}(\mathbf{S})}{\mathbb{E}}\left[R\left(h_{\mathcal{A}(\mathbf{S})}\right)\right]\right]\right| \le e^{-\varepsilon}k\delta + 1 - e^{-\varepsilon}, \tag{11.3}$$

where the loss function satisfies $\|l\|_\infty \le 1$.

They then obtained a high-probability generalization bound for multidatabase algorithms as follows:

$$\mathbb{P}\left[\hat{R}_{S_{i_{\mathcal{A}(\mathbf{S})}}}\left(h_{\mathcal{A}(\mathbf{S})}\right) \le R\left(h_{\mathcal{A}(\mathbf{S})}\right) + ke^{-\varepsilon}\delta + 3\varepsilon\right] \ge \varepsilon. \tag{11.4}$$

Finally, they proved Eq. (11.2) through reduction to absurdity.

11.2 The Interplay of Generalizability and Privacy-Preserving Ability

The abilities of generalization to unseen data and privacy-preserving are becoming more and more important for machine learning. Specifically, a good generalization ability means that an algorithm learns the underlying patterns in the training data instead of just memorizing it (Vapnik 2013; Mohri et al. 2018). Therefore, good generalizability provides confidence for models learned on given data to readily apply to unseen data with similar distribution. Additionally, a massive amount of personal data has been collected for training machine learning models, such as financial and medical applications. Thus, extract the underlying highly valuable information in the data while preserving the highly sensitive information and individual privacy from leaking is profoundly important (Dwork and Mulligan 2013; Dwork and Roth 2014; Pittaluga and Koppal 2016).

This section focuses on the relationship between the privacy-preserving ability and generalization ability in iterative machine learning algorithms through the following steps: (1) exploring the relationship between privacy-preserving ability and generalization ability in any learning algorithm; and (2) analyzing how the iterative nature commonly existing in learning algorithms would influence the privacy-preserving ability and further the generalization ability.

We first drive two theorems to upper bound the generalization error of an learning algorithm via its differential privacy. Specifically, we prove a high-probability upper bound for the generalization error,

$$R(\mathcal{A}(S)) - \hat{R}_S(\mathcal{A}(S)),$$

where $\mathcal{A}(S)$ is the hypothesis learned by algorithm $\mathcal{A}$ on the training sample set S, $R(\mathcal{A}(S))$ is the expected risk, and $\hat{R}_S(\mathcal{A}(S))$ is the empirical risk.

To derive this bound, we employ a new on-average generalization bound for (ε, δ)-differentially private multi-database learning algorithms. Moreover, this bound can also be deduced from the proven high-probability generalization bound, indicating that differentially private machine learning algorithms are likely to be approximately correct (PAC)-learnable. These findings indicate a notable connection between an algorithm's privacy-preserving capabilities and its ability to generalize effectively. In essence, algorithms exhibiting strong privacy preservation also tend to demonstrate robust generalizability. Consequently, there exists an opportunity to enhance the generalizability of learning algorithms by bolstering their privacy-preserving mechanisms.

We then investigate the impact of the iterative nature inherent in learning algorithms on both their privacy-preserving capabilities and generalizability. Typically, the privacy-preserving effectiveness of an iterative algorithm diminishes over the course of training. This decline occurs due to the accumulation of leaked information as the algorithm iteratively processes data. To investigate this phenomenon, we further derived three composition theorems that measure the differential privacy of any iterative algorithm by assessing the privacy of each individual iteration. By intertwining these theorems with the established correlation between generalizability and privacy preservation, we gained insights into the generalizability of iterative learning algorithms. These composition theorems provide a framework for understanding how the iterative nature of algorithms influences both their privacy and their ability to generalize effectively.

These results provide an insight of the relationship between generalizability and privacy-preserving ability in iterative learning algorithms.

Existing works (Dwork et al. 2015; Nissim and Stemmer 2015; Oneto et al. 2017) have already proved some high-probability generalization bounds in the following form,

$$\mathbb{P}\left[R(\mathcal{A}(S)) - \hat{R}_S(\mathcal{A}(S)) > a\right] < b,$$

where a and b are two positive constant real numbers. The high-probability bound provided in this section is strictly tighter than the current best results proved in Nissim and Stemmer (2015) from two aspects: (1) our bound improves the term a from 13ε to 9ε; and (2) our bound improves the term b from $2\delta\varepsilon^{-1}\log(2/\varepsilon)$ to $2e^{-\varepsilon}\delta\varepsilon^{-1}\log(2\varepsilon)$. Besides, based on this bound we further derived a PAC-learnable guarantee for differentially private machine learning algorithms. Nissim and Stemmer have proved an on-average multi-database generalization bound, which is looser than ours by a factor of e^{ε}. Such improvements are significant in practice because ε can always be as large as 10 (Abadi et al. 2016). Besides, the bounds in Nissim and Stemmer (2015) are only applicable for binary classification, while ours are suitable for any differentially private learning algorithms.

Some works have also proved composition theorems (Dwork and Roth 2014; Kairouz et al. 2017). The approximation of factor δ in our composition theorems is tighter than the best existing result (Kairouz et al. 2017) by

$$\delta \frac{e^{\varepsilon} - 1}{e^{\varepsilon} + 1} \left(T - \left\lceil \frac{\varepsilon'}{\varepsilon} \right\rceil \right),$$

where T is the number of iterations, while the estimate of ε' remains the same. This improvement is significant in practice, where T is always considerably large and is helpful for further tightening our the generalization bounds for iterative learning algorithms considerably.

Our results are applicable to a wide range of machine learning algorithms. In this book, we consider the stochastic gradient MCMC scheme (Ma et al. 2015) and agnostic federated learning (Geyer et al. 2017) and take stochastic gradient Langevin dynamics (Welling and Teh 2011) as an example of application. Our results deliver generalization bounds for SGLD and agnostic federated learning. The obtained generalization bounds are not explicitly relied on the model size, which can be prohibitively large in modern methods, such as deep neural networks.

11.2.1 Preliminaries

Here, we slightly abuse the notations of distribution and its cumulative distribution function when no ambiguity is introduced because there is a one-one mapping between them if we ignore zero-probability events.

Definition 11.1 (*Max Divergence; cf.* Dwork and Roth (2014), *Definition 3.6*) For any random variables X and Y, the max divergence between X and Y is defined as

$$D_{\infty}(X \| Y) = \max_{S \subseteq \text{Supp}(X)} \left[\log \frac{\mathbb{P}(X \in S)}{\mathbb{P}(Y \in S)} \right].$$

Definition 11.2 (*δ-Approximate Max Divergence; cf.* (Dwork and Roth 2014), *Definition 3.6*) For any random variables X and Y, the δ-approximate max divergence between X to Y is defined as

$$D^{\delta}_{\infty}(X \| Y) = \max_{S \subseteq \text{Supp}(X) : \mathbb{P}(Y \in S) \geq \delta} \left[\log \frac{\mathbb{P}(X \in S) - \delta}{\mathbb{P}(Y \in S)} \right].$$

Definition 11.3 (*Statistical Distance; cf.* (Dwork and Roth 2014)) For any random variables X and Y, the statistical distance between X and Y is defined as

$$\Delta(X\|Y) = \max_S |\mathbb{P}(X \in S) - \mathbb{P}(Y \in S)|.$$

We then recall the following two lemmas.

Lemma 11.1 (cf. (Dwork and Rothblum 2016), Lemmas 3.9 and 3.10) *For any two distributions $\mathcal{D}$ and $\mathcal{D}'$, there exist distributions $\mathcal{M}$ and $\mathcal{M}'$ such that*

$$\max\{D_\infty(\mathcal{M}\|\mathcal{M}'), D_\infty(\mathcal{M}'\|\mathcal{M})\} = \max\{D_\infty(\mathcal{D}\|\mathcal{D}'), D_\infty(\mathcal{D}'\|\mathcal{D})\},$$

and

$$D_{KL}(\mathcal{D}\|\mathcal{D}') \le D_{KL}(\mathcal{M}\|\mathcal{M}') = D_{KL}(\mathcal{M}'\|\mathcal{M}).$$

Lemma 11.2 (cf. Dwork and Roth (2014), Theorem 3.17) *For any random variables Y and Z, we have*

$$D_\infty^\delta(Y\|Z) \le \varepsilon, \ D_\infty^\delta(Z\|Y) \le \varepsilon,$$

if and only if there exist random variables Y', Z' such that

$$\Delta(Y\|Y') \le \frac{\delta}{e^\varepsilon + 1}, \ \Delta(Z\|Z') \le \frac{\delta}{1 + e^\varepsilon},$$
$$D_\infty(Y'\|Z') \le \varepsilon, \ D_\infty(Z'\|Y') \le \varepsilon.$$

We finally recall Azuma Lemma (Boucheron et al. 2013) which gives a concentration inequality to the martingales.

Lemma 11.3 (Azuma Lemma; cf. (Mohri et al. 2018), p. 371) *Suppose $\{Y_i\}_{i=1}^T$ is a sequence of random variables, where $Y_i \in [-a_i, a_i]$. Let $\{X_i\}_{i=1}^T$ be a sequence of random variables such that,*

$$\mathbb{E}(Y_i|X_{i-1}, \ldots, X_1) \le C_i,$$

where $\{C_i\}_{i=1}^T$ is a sequence of constant real numbers. Then, we have the following inequality,

$$\mathbb{P}\left(\sum_{i=1}^T Y_i \ge \sum_{i=1}^T C_i + t\sqrt{\sum_{i=1}^T a_i^2}\right) \le e^{-\frac{t^2}{2}}.$$

11.2.2 Generalization Bounds for Iterative Differentially Private Algorithms

In this section, we ascertain the generalizability of iterative differentially private algorithms through the following two primary steps: (1) we establish generalization bounds applicable to any differentially private learning algorithm; and (2) we explore how the prevalent iterative nature inherent in learning algorithms affects both their privacy-preserving efficacy and their ability to generalize. This exploration is facilitated by three composition theorems. Additionally, we provide a brief outline of the theoretical proofs and illustrate their superiority over existing findings.

11.2.2.1 Bridging Generalization and Privacy Preservation

We first prove a high-probability generalization bound for any (ε, δ)-differentially private machine learning algorithm as follows.

Theorem 11.1 (High-Probability Generalization Bound via Differential Privacy) *Suppose algorithm $\mathcal{A}$ is (ε, δ)-differentially private, the training sample size $m \geq \frac{2}{\varepsilon^2} \ln\left(\frac{16}{e^{-\varepsilon}\delta}\right)$, and the loss function $\|l\|_\infty \leq 1$. Then, for any data distribution $\mathcal{D}$ over data space $\mathcal{Z}$, we have the following inequality,*

$$\mathbb{P}\left[\left|\hat{R}_S(\mathcal{A}(S)) - R(\mathcal{A}(S))\right| < 9\varepsilon\right] > 1 - \frac{e^{-\varepsilon}\delta}{\varepsilon} \ln\left(\frac{2}{\varepsilon}\right).$$

Theorem 11.1 presents a valuable finding: it establishes a direct link between a learning algorithm's privacy-preserving prowess and its generalizability. This implies that efforts aimed at improving privacy preservation can also lead to improved generalization performance. Moreover, the theorem suggests that (ε, δ) differentially private algorithms may offer a guarantee of probable approximate correctness (PAC) learnability. PAC-learnability ensures that the algorithm is likely to provide accurate predictions with high probability. This insight validates the potential synergy between privacy and performance in algorithm design. PAC-learnability is defined as follows.

Definition 11.4 (*PAC-Learnability; cf.* (Mohri et al. 2018), *Definition 2.4*) A concept class C is said to be PAC-learnable if there exists an algorithm $\mathcal{A}$ and a polynomial function $poly(\cdot, \cdot, \cdot, \cdot)$ such that for any $s > 0$ and $t > 0$, for all distributions $\mathcal{D}$ on the training example Z, any target concept $c \in C$, and any sample size

$$m \geq poly(1/s, 1/t, n, \text{size}(C)),$$

the following inequality holds,

$$\mathbb{P}_{S \sim \mathcal{D}^m}(R(\mathcal{A}(S)) < s) > 1 - t.$$

In Sect. 11.2.3, we show how our result leads to PAC-learnable guarantees by using SGLD and agnostic federated learning as examples.

Proof Skeleton

We now give the proof skeleton for Theorem 11.1. The proofs have three stages: (1) we first prove an on-average generalization bound for multi-database learning algorithms; (2) we then obtain a high-probability generalization bound for multi-database algorithms; and (3) we eventually prove Theorem 11.1 by reduction to absurdity.

Stage 1: Prove an on-average generalization bound for multi-database learning algorithms.

We first prove the following on-average generalization bound for multi-database learning algorithms which are defined as follows.

Definition 11.5 (*Multi-Database Learning Algorithms; cf.* (Nissim and Stemmer 2015)) Suppose the training sample set S is separated into k sub-databases $S_1, \ldots, S_k$, each of which has the size of m. For brevity, we rewrite the training sample set as follows.

$$\mathbf{S} = (S_1, \ldots, S_k).$$

The hypothesis $\tilde{\mathcal{A}}(\mathbf{S})$ learned by *multi-database algorithm* $\tilde{\mathcal{A}}$ on dataset $\mathbf{S}$ is defined as follows,

$$\tilde{\mathcal{A}}(\mathbf{S}) : \mathcal{Z}^{km}\mathcal{H} \times \{1, \ldots, k\}, \ \mathbf{S} \mapsto \left(h_{\mathcal{A}(\mathbf{S})}, i_{\mathcal{A}(\mathbf{S})}\right).$$

Theorem 11.2 (On-Average Multi-Database Generalization Bound) *Let algorithm,*

$$\tilde{\mathcal{A}} : \mathbf{S} \to \mathcal{H} \times \{1, \ldots, k\},$$

be (ε, δ)-differentially private and the loss function $\|l\|_\infty \leq 1$. Then, for any data distribution $\mathcal{D}$ over data space $\mathcal{Z}$, we have the following inequality,

$$\left| \mathop{\mathbb{E}}_{\mathbf{S}\sim\mathcal{D}^m} \left[\mathop{\mathbb{E}}_{\mathcal{A}(\mathbf{S})} \left[\hat{R}_{S_{i_{\mathcal{A}(\mathbf{S})}}} \left(h_{\mathcal{A}(\mathbf{S})}\right)\right] - \mathop{\mathbb{E}}_{\mathcal{A}(\mathbf{S})} \left[R\left(h_{\mathcal{A}(\mathbf{S})}\right)\right]\right]\right| \leq e^{-\varepsilon} k\delta + 1 - e^{-\varepsilon}. \tag{11.5}$$

Proof The left side of Eq. (11.5) can be rewritten as

$$\begin{aligned}
&\mathop{\mathbb{E}}_{\mathbf{S}\sim\mathcal{D}^{km}} \left[\mathop{\mathbb{E}}_{\mathcal{A}(\mathbf{S})} \left[\hat{R}_{S_{i_{\mathcal{A}(\mathbf{S})}}} \left(h_{\mathcal{A}(\mathbf{S})}\right)\right]\right] = \mathop{\mathbb{E}}_{\mathbf{S}\sim\mathcal{D}^{km}} \left[\mathop{\mathbb{E}}_{\mathcal{A}(\mathbf{S})} \left[\mathbb{E}_{z\sim S_{i_{\mathcal{A}(\mathbf{S})}}} \left[l\left(h_{\mathcal{A}(\mathbf{S})}, z\right)\right]\right]\right] \\
\overset{(*)}{=} &\mathop{\mathbb{E}}_{\mathbf{S}\sim\mathcal{D}^{km}} \left[\mathop{\mathbb{E}}_{\mathcal{A}(\mathbf{S})} \left[\mathop{\mathbb{E}}_{\mathbf{z}\sim\mathbf{S}} \left[l\left(h_{\mathcal{A}(\mathbf{S})}, z_{i_{\mathcal{A}(\mathbf{S})}}\right)\right]\right]\right] = \mathop{\mathbb{E}}_{\mathbf{S}\sim\mathcal{D}^{km}} \left[\mathop{\mathbb{E}}_{\mathbf{z}\sim\mathbf{S}} \left[\mathop{\mathbb{E}}_{\mathcal{A}(\mathbf{S})} \left[l\left(h_{\mathcal{A}(\mathbf{S})}, z_{i_{\mathcal{A}(\mathbf{S})}}\right)\right]\right]\right] \\
= &\mathop{\mathbb{E}}_{\mathbf{S}\sim\mathcal{D}^{km}} \left[\mathop{\mathbb{E}}_{\mathbf{z}\sim\mathbf{S}} \left[\mathop{\mathbb{E}}_{\mathcal{A}(\mathbf{S})} \left[l\left(h_{\mathcal{A}(\mathbf{S})}, z_{i_{\mathcal{A}(\mathbf{S})}}\right)\right]\right]\right]
\end{aligned}$$

$$
\begin{aligned}
&= \mathop{\mathbb{E}}_{\mathbf{S}\sim\mathcal{D}^{km}} \left[\mathop{\mathbb{E}}_{\mathbf{z}\sim\mathbf{S}} \left[\int_0^1 \mathbb{P}\left(l\left(h_{\mathcal{A}(\mathbf{S})}, z_{i_{\mathcal{A}(\mathbf{S})}}\right) \le t\right) dt \right]\right] \\
&= \mathop{\mathbb{E}}_{\mathbf{S}\sim\mathcal{D}^{km}} \left[\mathop{\mathbb{E}}_{\mathbf{z}\sim\mathbf{S}} \left[\sum_{i=1}^{k} \int_0^1 \mathbb{P}\left(l\left(h_{\mathcal{A}(\mathbf{S})}, z_i\right) \le t, i_{\mathcal{A}(\mathbf{S})} = i\right) \mathrm{d}t \right]\right],
\end{aligned}
$$

where $\mathbf{z}$ in the right side of $(*)$ is defined as $\{z_1, \dots, z_k\}$, z_i is uniformly selected from S_i. Since $\mathcal{A}$ is (ε, δ)-differentially private, we further have

$$
\begin{aligned}
&\mathop{\mathbb{E}}_{\mathbf{S}\sim\mathcal{D}^{km}} \left[\mathop{\mathbb{E}}_{\mathbf{z}\sim\mathbf{S}} \left[\sum_{i=1}^{k} \int_0^1 \mathbb{P}\left(l\left(h_{\mathcal{A}(\mathbf{S})}, z_i\right) \le t, i_{\mathcal{A}(\mathbf{S})} = i\right) \mathrm{d}t \right]\right] \\
\le &\mathop{\mathbb{E}}_{\mathbf{S}\sim\mathcal{D}^{km}} \left[\mathop{\mathbb{E}}_{\mathbf{z}\sim\mathbf{S}, z_0\sim D} \left[\sum_{i=1}^{k} \int_0^1 e^{\varepsilon}\mathbb{P}\left(l\left(h_{\mathcal{A}(\mathbf{S}^{z_i:z_0})}, z_i\right) \le t, i_{\mathcal{A}(\mathbf{S}^{z_i:z_0})} = i\right) + \delta \, \mathrm{d}t \right]\right] \\
= &e^{\varepsilon} \mathop{\mathbb{E}}_{\mathbf{S}\sim\mathcal{D}^{km}} \left[\mathop{\mathbb{E}}_{\mathbf{z}\sim\mathbf{S}, z_0\sim D} \left[\sum_{i=1}^{k} \int_0^1 \mathbb{P}\left(l\left(h_{\mathcal{A}(\mathbf{S}^{z_i:z_0})}, z_i\right) \le t, i_{\mathcal{A}(\mathbf{S}^{z_i:z_0})} = i\right) \mathrm{d}t \right]\right] + k\delta \\
= &\sum_{i=1}^{k} e^{\varepsilon} \mathop{\mathbb{E}}_{\mathbf{S}\sim\mathcal{D}^{km}} \left[\mathop{\mathbb{E}}_{\mathbf{z}\sim\mathbf{S}, z_0\sim D} \left[\int_0^1 \mathbb{P}\left(l\left(h_{\mathcal{A}(\mathbf{S}^{z_i:z_0})}, z_i\right) \le t, i_{\mathcal{A}(\mathbf{S}^{z_i:z_0})} = i\right) \mathrm{d}t \right]\right] + k\delta \\
= &\sum_{i=1}^{k} e^{\varepsilon} \mathop{\mathbb{E}}_{\mathbf{S}'\sim\mathcal{D}^{km-1}} \left[\mathop{\mathbb{E}}_{z_i\sim D, z_0\sim D} \left[\int_0^1 \mathbb{P}\left(l\left(h_{\mathcal{A}(\mathbf{S}'\cup\{z_0\})}, z_i\right) \le t, i_{\mathcal{A}(\mathbf{S}'\cup\{z_0\})} = i\right) \mathrm{d}t \right]\right] + k\delta.
\end{aligned}
$$

Let $\mathbf{S} = \mathbf{S}' \cup \{z_0\}$ and $z = z_i$ (it is without less of generality since all z_i is i.i.d. drawn from $\mathcal{D}$). Since $\mathbf{S}' \cup \{z_0\} \sim \mathcal{D}^{km}$, we have

$$
\begin{aligned}
&\sum_{i=1}^{k} e^{\varepsilon} \mathop{\mathbb{E}}_{\mathbf{S}'\sim\mathcal{D}^{km-1}} \left[\mathop{\mathbb{E}}_{z_i\sim D, z_0\sim D} \left[\int_0^1 \mathbb{P}\left(l\left(h_{\mathcal{A}(\mathbf{S}'\cup\{z_0\})}, z_i\right) \le t, i_{\mathcal{A}(\mathbf{S}'\cup\{z_0\})} = i\right) \mathrm{d}t \right]\right] + k\delta \\
= &\sum_{i=1}^{k} e^{\varepsilon} \mathop{\mathbb{E}}_{\mathbf{S}\sim\mathcal{D}^{km}} \left[\mathop{\mathbb{E}}_{z\sim\mathcal{D}} \left[\int_0^1 \mathbb{P}\left(l\left(h_{\mathcal{A}(\mathbf{S})}, z\right) \le t, i_{\mathcal{A}(\mathbf{S})} = i\right) \mathrm{d}t \right]\right] + k\delta \\
= &e^{\varepsilon} \mathop{\mathbb{E}}_{\mathbf{S}\sim\mathcal{D}^{km}} \left[\mathop{\mathbb{E}}_{z\sim\mathcal{D}} \left[\int_0^1 \mathbb{P}\left(l\left(h_{\mathcal{A}(\mathbf{S})}, z\right) \le t\right) \mathrm{d}t \right]\right] + k\delta \\
= &e^{\varepsilon} \mathop{\mathbb{E}}_{\mathbf{S}\sim\mathcal{D}^{km}} \left[\mathop{\mathbb{E}}_{z\sim\mathcal{D}} \left[\mathbb{E}_{\mathcal{A}(S)} \left[l\left(h_{\mathcal{A}(\mathbf{S})}, z\right)\right]\right]\right] + k\delta.
\end{aligned}
$$

Therefore, we have

$$
\mathop{\mathbb{E}}_{\mathbf{S}\sim\mathcal{D}^{km}} \left[\mathop{\mathbb{E}}_{\mathcal{A}(\mathbf{S})} \left[\hat{R}_{S_{i_{\mathcal{A}(\mathbf{S})}}}\left(h_{\mathcal{A}(\mathbf{S})}\right)\right]\right] \le k\delta + e^{\varepsilon} \mathop{\mathbb{E}}_{\mathbf{S}\sim\mathcal{D}^{km}} [\mathop{\mathbb{E}}_{\mathcal{A}(\mathbf{S})} [R\left(h_{\mathcal{A}(\mathbf{S})}\right)]]. \tag{11.6}
$$

Rearranging Eq. (11.6), we have

$$e^{-\varepsilon} \mathop{\mathbb{E}}_{\mathbf{S}\sim\mathcal{D}^{km}} \left[\mathop{\mathbb{E}}_{\mathcal{A}(\mathbf{S})} \left[\hat{R}_{S_{i_{\mathcal{A}(\mathbf{S})}}} (h_{\mathcal{A}(\mathbf{S})}) \right] \right] \leq e^{-\varepsilon} k\delta + \mathop{\mathbb{E}}_{\mathbf{S}\sim\mathcal{D}^{km}} \left[\mathop{\mathbb{E}}_{\mathcal{A}(\mathbf{S})} \left[R\left(h_{\mathcal{A}(\mathbf{S})}\right) \right] \right]$$

$$- \mathop{\mathbb{E}}_{\mathbf{S}\sim\mathcal{D}^{km}} [\mathop{\mathbb{E}}_{\mathcal{A}(\mathbf{S})} [R\left(h_{\mathcal{A}(\mathbf{S})}\right)]] \leq e^{-\varepsilon} k\delta - e^{-\varepsilon} \mathop{\mathbb{E}}_{\mathbf{S}\sim\mathcal{D}^{km}} \left[\mathop{\mathbb{E}}_{\mathcal{A}(\mathbf{S})} \left[\hat{R}_{S_{i_{\mathcal{A}(\mathbf{S})}}} (h_{\mathcal{A}(\mathbf{S})}) \right] \right],$$

which leads to

$$\begin{aligned}
&\mathop{\mathbb{E}}_{\mathbf{S}\sim\mathcal{D}^{km}} \left[\mathop{\mathbb{E}}_{\mathcal{A}(\mathbf{S})} \left[\hat{R}_{S_{i_{\mathcal{A}(\mathbf{S})}}} (h_{\mathcal{A}(\mathbf{S})}) \right] \right] - \mathop{\mathbb{E}}_{\mathbf{S}\sim\mathcal{D}^{km}} [\mathop{\mathbb{E}}_{\mathcal{A}(\mathbf{S})} [R\left(h_{\mathcal{A}(\mathbf{S})}\right)]] \\
\leq &e^{-\varepsilon} k\delta - e^{-\varepsilon} \mathop{\mathbb{E}}_{\mathbf{S}\sim\mathcal{D}^{km}} \left[\mathop{\mathbb{E}}_{\mathcal{A}(\mathbf{S})} \left[\hat{R}_{S_{i_{\mathcal{A}(\mathbf{S})}}} (h_{\mathcal{A}(\mathbf{S})}) \right] \right] + \mathop{\mathbb{E}}_{\mathbf{S}\sim\mathcal{D}^{km}} \left[\mathop{\mathbb{E}}_{\mathcal{A}(\mathbf{S})} \left[\hat{R}_{S_{i_{\mathcal{A}(\mathbf{S})}}} (h_{\mathcal{A}(\mathbf{S})}) \right] \right] \\
\leq &1 - e^{-\varepsilon} + e^{-\varepsilon} k\delta.
\end{aligned}$$

The other side of the inequality can be similarly obtained.

The proof is completed. □

Since $1 - e^{-\varepsilon} \leq \varepsilon$, we have the following corollary.

Corollary 11.1 *Suppose all the conditions in Theorem 11.2 hold, then we have the following inequality,*

$$\mathop{\mathbb{E}}_{\mathbf{S}\sim\mathcal{D}^{km}} \left[\mathop{\mathbb{E}}_{\mathcal{A}(\mathbf{S})} \left[\hat{R}_{S_{i_{\mathcal{A}(\mathbf{S})}}} (h_{\mathcal{A}(\mathbf{S})}) \right] \right] \leq e^{-\varepsilon} k\delta + \varepsilon + \mathop{\mathbb{E}}_{\mathbf{S}\sim\mathcal{D}^{km}} \left[\mathop{\mathbb{E}}_{\mathcal{A}(\mathbf{S})} \left[R\left(h_{\mathcal{A}(\mathbf{S})}\right) \right] \right].$$

Stage 2: Prove a high-probability generalization bound for multi-database algorithms.

Markov bound (cf. Mohri et al. 2018, Theorem C.1) is an important concentration inequality in learning theory. Here, we slightly modify the original version as follows,

$$\mathbb{E}_x \left[h(x)\right] \geq \mathbb{E}_x \left[h(x)\mathbf{I}_{h(x)\geq g(x)}\right] \geq \mathbb{E}_x \left[g(x)\mathbf{I}_{h(x)\geq g(x)}\right].$$

Then, combining it with Theorem 11.2, we derive the following high-probability generalization bound for multi-database algorithms.

Theorem 11.3 (High-Probability Multi-Database Generalization Bound) *Let the following algorithm,*

$$\mathcal{A} : \mathcal{Z}^{km} \to \mathcal{Y}^{\mathcal{X}} \times \{1, \ldots, k\},\ \mathbf{S} \mapsto \left(h_{\mathcal{A}(\mathbf{S})}, i_{\mathcal{A}(\mathbf{S})}\right),$$

be (ε, δ)-differential private, where km is the size of the whole dataset $\mathbf{S}$ and $\mathcal{Y}^{\mathcal{X}} = \{f : \mathcal{X} \to \mathcal{Y}\}$. Then, for any data distribution $\mathcal{D}$ over data space $\mathcal{Z}$, any database set $\mathbf{S} = \{S_i\}_{i=1}^k$, where S_i is a database contains m i.i.d. examples drawn from $\mathcal{D}$, we have the following generalization bound,

$$\mathbb{P}\left[\hat{R}_{S_{i_{\mathcal{A}(\mathbf{S})}}} \left(h_{\mathcal{A}(\mathbf{S})}\right) \leq R\left(h_{\mathcal{A}(\mathbf{S})}\right) + ke^{-\varepsilon}\delta + 3\varepsilon \right] \geq \varepsilon. \tag{11.7}$$

Proof By Corollary 11.1, we have that

$$\mathbb{E}_{\mathbf{S}\sim\mathcal{D}^{km}}\left[\mathbb{E}_{\mathcal{A}(\mathbf{S})}\left[\hat{R}_{S_{i_{\mathcal{A}(\mathbf{S})}}}(h_{\mathcal{A}(\mathbf{S})})\right]\right] \leq e^{-\varepsilon}k\delta + \varepsilon + \mathbb{E}_{\mathbf{S}\sim\mathcal{D}^{km}}\left[\mathbb{E}_{\mathcal{A}(\mathbf{S})}\left[R\left(h_{\mathcal{A}(\mathbf{S})}\right)\right]\right].$$

Since $\hat{R}_{S_{i_{\mathcal{A}(\mathbf{S})}}}\left(h_{\mathcal{A}(\mathbf{S})}\right) \geq 0$, we have that for any $\alpha > 0$,

$$\begin{aligned}
&\mathbb{E}_{\mathbf{S}\sim\mathcal{D}^{km}}\left[\mathbb{E}_{\mathcal{A}(\mathbf{S})}\left[\hat{R}_{S_{i_{\mathcal{A}(\mathbf{S})}}}(h_{\mathcal{A}(\mathbf{S})})\right]\right]\\
\geq &\mathbb{E}_{\mathbf{S}\sim\mathcal{D}^{km}}\left[\mathbb{E}_{\mathcal{A}(\mathbf{S})}\left[\hat{R}_{S_{i_{\mathcal{A}(\mathbf{S})}}}(h_{\mathcal{A}(\mathbf{S})})\mathbf{I}_{\hat{R}_{S_{i_{\mathcal{A}(\mathbf{S})}}}(h_{\mathcal{A}(\mathbf{S})})\geq R\left(h_{\mathcal{A}(\mathbf{S})}\right)+\alpha}\right]\right]\\
\geq &\mathbb{E}_{\mathbf{S}\sim\mathcal{D}^{km}}\left[\mathbb{E}_{\mathcal{A}(\mathbf{S})}\left[(\alpha + R\left(h_{\mathcal{A}(\mathbf{S})}\right))\mathbf{I}_{\hat{R}_{S_{i_{\mathcal{A}(\mathbf{S})}}}(h_{\mathcal{A}(\mathbf{S})})\geq R\left(h_{\mathcal{A}(\mathbf{S})}\right)+\alpha}\right]\right].
\end{aligned}$$

Furthermore, by splitting $\mathbb{E}_{\mathbf{S}\sim\mathcal{D}^{km}}[\mathbb{E}_{\mathcal{A}(\mathbf{S})}[R\left(h_{\mathcal{A}(\mathbf{S})}\right)]]$ into two parts, we have

$$\begin{aligned}
&\mathbb{E}_{\mathbf{S}\sim\mathcal{D}^{km}}\left[\mathbb{E}_{\mathcal{A}(\mathbf{S})}\left[\hat{R}_{S_{i_{\mathcal{A}(\mathbf{S})}}}(h_{\mathcal{A}(\mathbf{S})})\right]\right] - \mathbb{E}_{\mathbf{S}\sim\mathcal{D}^{km}}\left[\mathbb{E}_{\mathcal{A}(\mathbf{S})}\left[R\left(h_{\mathcal{A}(\mathbf{S})}\right)\right]\right]\\
\geq &\mathbb{E}_{\mathbf{S}\sim\mathcal{D}^{km}}\left[\mathbb{E}_{\mathcal{A}(\mathbf{S})}\left[(\alpha + R\left(h_{\mathcal{A}(\mathbf{S})}\right))\mathbf{I}_{\hat{R}_{S_{i_{\mathcal{A}(\mathbf{S})}}}(h_{\mathcal{A}(\mathbf{S})})\geq R\left(h_{\mathcal{A}(\mathbf{S})}\right)+\alpha}\right]\right]\\
&-\left(\mathbb{E}_{\mathbf{S}\sim\mathcal{D}^{km}}\left[\mathbb{E}_{\mathcal{A}(\mathbf{S})}\left[R\left(h_{\mathcal{A}(\mathbf{S})}\right)\mathbf{I}_{\hat{R}_{S_{i_{\mathcal{A}(\mathbf{S})}}}(h_{\mathcal{A}(\mathbf{S})})\geq R\left(h_{\mathcal{A}(\mathbf{S})}\right)+\alpha}\right]\right]\right.\\
&\left.+\mathbb{E}_{\mathbf{S}\sim\mathcal{D}^{km}}\left[\mathbb{E}_{\mathcal{A}(\mathbf{S})}\left[R\left(h_{\mathcal{A}(\mathbf{S})}\right)\mathbf{I}_{\hat{R}_{S_{i_{\mathcal{A}(\mathbf{S})}}}(h_{\mathcal{A}(\mathbf{S})})< R\left(h_{\mathcal{A}(\mathbf{S})}\right)+\alpha}\right]\right]\right)\\
\geq &\mathbb{E}_{\mathbf{S}\sim\mathcal{D}^{km}}\left[\mathbb{E}_{\mathcal{A}(\mathbf{S})}\left[\alpha\mathbf{I}_{\hat{R}_{S_{i_{\mathcal{A}(\mathbf{S})}}}(h_{\mathcal{A}(\mathbf{S})})\geq R\left(h_{\mathcal{A}(\mathbf{S})}\right)+\alpha}\right]\right]\\
&-\mathbb{E}_{\mathbf{S}\sim\mathcal{D}^{km}}\left[\mathbb{E}_{\mathcal{A}(\mathbf{S})}\left[R\left(h_{\mathcal{A}(\mathbf{S})}\right)\mathbf{I}_{\hat{R}_{S_{i_{\mathcal{A}(\mathbf{S})}}}(h_{\mathcal{A}(\mathbf{S})})< R\left(h_{\mathcal{A}(\mathbf{S})}\right)+\alpha}\right]\right]\Bigg)\\
\geq &\alpha\mathbb{P}\left(\hat{R}_{S_{i_{\mathcal{A}(\mathbf{S})}}}(h_{\mathcal{A}(\mathbf{S})}) > R\left(h_{\mathcal{A}(\mathbf{S})}\right)+\alpha\right) - \mathbb{P}\left(\hat{R}_{S_{i_{\mathcal{A}(\mathbf{S})}}}(h_{\mathcal{A}(\mathbf{S})}) \leq R\left(h_{\mathcal{A}(\mathbf{S})}\right)+\alpha\right).
\end{aligned}$$

Let $\alpha = e^{-\varepsilon}k\delta + 3\varepsilon$. We have

$$\mathbb{P}\left(\hat{R}_{S_{i_{\mathcal{A}(\mathbf{S})}}}(h_{\mathcal{A}(\mathbf{S})}) \leq R\left(h_{\mathcal{A}(\mathbf{S})}\right)+\alpha\right) \leq \frac{\alpha - (e^{-\varepsilon}k\delta + \varepsilon)}{1+\alpha} \leq \varepsilon.$$

The proof is completed. □

Stage 3: Prove Theorem 11.1 **by Reduction to Absurdity.**

We eventually prove Theorem 11.1 by *reduction to absurdity*. We assume that there exists an algorithm $\mathcal{A}$ which conflicts with Theorem 11.1. We can then construct an algorithm $\mathcal{B}$ based on the exponential mechanism which is defined as follows.

Definition 11.6 (*Exponential Mechanism; cf.* (Nissim and Stemmer 2015), *p. 3, and* (McSherry and Talwar 2007)) Suppose that S is a sample set, $u : (S, r) \mapsto \mathbb{R}^+$ is the utility function, R is an index set, ε is the privacy parameter, and Δu is the sensitivity of u defined by

$$\Delta u \triangleq \max_{r \in R} \max_{S, S' \text{ adjacent}} |u(S, r) - u(S', r)|.$$

Then, the exponential mechanism $q(S, u, R, \varepsilon)$ is defined as $(S, u, R, \varepsilon) \mapsto r$, where $r \in R$.

Then, we can prove the following lemma.

Lemma 11.4 *We define an algorithm $\mathcal{A} : \mathcal{Z}^m \to \mathcal{Y}^{\mathcal{X}}$, where m is the training sample size, $\mathcal{Z}$ is the data space, $\mathcal{Z}^m$ is the space of the training sample set, and $\mathcal{Y}^{\mathcal{X}} = \{f : \mathcal{X} \to \mathcal{Y}\}$. Suppose $k = \lceil \frac{\varepsilon}{e^{-\varepsilon}\delta} \rceil$ and*

$$m \geq \frac{2}{\varepsilon^2} \ln \left(\frac{16}{e^{-\varepsilon}\delta} \right).$$

If we have

$$\mathbb{P}\left[\hat{R}_S(\mathcal{A}(S)) \leq e^{-\varepsilon} k\delta + 8\varepsilon + R(\mathcal{A}(S)) \right] < 1 - \frac{e^{-\varepsilon}\delta}{\varepsilon} \ln \left(\frac{2}{\epsilon} \right), \tag{11.8}$$

then there exists an algorithm

$$\mathcal{B} : \mathcal{Z}^{km} \to \mathcal{Y}^{\mathcal{X}} \times \{1, \ldots, k\},$$

is $(2\varepsilon, \delta)$-differentially private and

$$\mathbb{P}\left[\hat{R}_{S_{i_{\mathcal{B}(\mathbf{S})}}} \left(h_{\mathcal{B}(\mathbf{S})} \right) \leq R \left(h_{\mathcal{B}(\mathbf{S})} \right) + k e^{-\varepsilon}\delta + 3\varepsilon \right] < \varepsilon, \tag{11.9}$$

where $\mathbf{S} = \{S_i\}_{i=1}^k$ and S_i is a database containing m i.i.d. sampled from $\mathcal{D}$.

Proof of Lemma 11.4 Construct algorithm $\mathcal{B}$ with input $\mathbf{S} = \{S_i\}_{i=1}^k$ and T (where $S_i, T \in \mathcal{Z}^m$) as follows:

Step 1. Run $\mathcal{A}$ on S_i, $i = 1, \ldots, k$. We denote the output as $h_i = \mathcal{A}(S_i)$.

Step 2. Let the utility function as $q(\mathbf{S}, T, i) = m \left(\hat{R}_{S_i}(h_i) - \hat{R}_T(h_i) \right)$. We apply the utility q to an ε-differential private exponential mechanism $\mathcal{M}(h_i, \mathbf{S}, T)$ and return the output.

We then prove that $\mathcal{B}$ satisfies

$$\mathbb{P}\left[l \left(h_{\mathcal{B}(\mathbf{S})}, S_{i_{\mathcal{B}(\mathbf{S})}} \right) \leq R \left(h_{\mathcal{B}(\mathbf{S})} \right) + k e^{-\varepsilon}\delta + 3\varepsilon \right] < \varepsilon.$$

By using Eq. (11.8), we have

$$\mathbb{P}\left(\forall i,\, \hat{R}_{S_i}(\mathcal{A}(S_i)) \leq e^{-\varepsilon}k\delta + 8\epsilon + R(\mathcal{A}(S_i))\right) \leq \left(1 - \frac{e^{-\varepsilon}\delta}{\varepsilon}\ln\left(\frac{2}{\epsilon}\right)\right)^k,$$

which leads to

$$\mathbb{P}\left(\exists i,\, \hat{R}_{S_i}(\mathcal{A}(S_i)) > e^{-\varepsilon}k\delta + 8\epsilon + R(\mathcal{A}(S_i))\right) > 1 - \left(1 - \frac{1}{k}\ln\left(\frac{2}{\epsilon}\right)\right)^k \geq 1 - \frac{\varepsilon}{2}. \tag{11.10}$$

Furthermore, since T is independent with $\mathbf{S}$, by using the Hoeffding inequality, we have

$$\mathbb{P}\left(\forall i, |l(h_i, T) - R(h_i)| \leq \frac{\varepsilon}{2}\right) \geq (1 - e^{-\epsilon^2/2m})^k \geq 1 - \frac{\varepsilon}{8}. \tag{11.11}$$

Therefore, by combining Eqs. (11.10) and (11.11), we obtain

$$\mathbb{P}\left(\exists i,\, \hat{R}_{S_i}(\mathcal{A}(S_i)) > e^{-\varepsilon}k\delta + \frac{15}{2}\epsilon + l(h_i, T)\right) > 1 - \frac{5\varepsilon}{8}.$$

Since q has sensitivity of 1, fix h_i, we have

$$\mathbb{P}\left(\mathcal{M}(h_i, \mathbf{S}, T) \leq \mathrm{OPT}(q(\mathbf{S}, T, i)) - m\varepsilon)\right) \geq 1 - \frac{\varepsilon}{4},$$

which leads to

$$\mathbb{P}\left(\hat{R}_{S_{i_{\mathcal{B}(\mathbf{S})}}}(h_{\mathcal{B}(\mathbf{S})}) > e^{-\varepsilon}k\delta + \frac{13}{2}\epsilon + \hat{R}_T(h_{\mathcal{B}(\mathbf{S})})\right) > 1 - \frac{7\varepsilon}{8}.$$

Then, using Eq. (11.11) again, we have

$$\mathbb{P}\left(\hat{R}_{S_{i_{\mathcal{B}(\mathbf{S})}}}(h_{\mathcal{B}(\mathbf{S})}) > e^{-\varepsilon}k\delta + 6\epsilon + R(h_{\mathcal{B}(\mathbf{S})})\right) > 1 - \varepsilon.$$

We complete the proof.

The Eq. (11.9) in Lemma 11.4 conflicts with Eq. (11.7). Thus, we proved Theorem 11.1.

We then compares our results with the existing works.

Comparison of Theorem 11.1.

There have been several high-probability generalization bounds for (ε, δ)-differentially private machine learning algorithms.

Dwork et al. (2015) proved that

$$\mathbb{P}\left[R(\mathcal{A}(S)) - \hat{R}_S(\mathcal{A}(S)) < 4\varepsilon\right] > 1 - 8\delta^{\varepsilon}.$$

Oneto et al. (2017) proved that

$$\mathbb{P}\left[\text{Diff } R < \sqrt{6\hat{R}_S(\mathcal{A}(S))\hat{\varepsilon}} + 6\left(\varepsilon^2 + 1/m\right)\right] > 1 - 3e^{-m\varepsilon^2},$$

and

$$\mathbb{P}\left[\text{Diff } R < \sqrt{4\hat{V}_S(\mathcal{A}(S))\hat{\varepsilon}} + \frac{5N}{N-1}\left(\varepsilon^2 + 1/m\right)\right] > 1 - 3e^{-m\varepsilon^2},$$

where

$$\text{Diff } R = R(\mathcal{A}(S)) - \hat{R}_S(\mathcal{A}(S)), \ \ \hat{\varepsilon} = \varepsilon + \sqrt{1/m},$$

and $\hat{V}_S(\mathcal{A}(S))$ is the empirical variance of $l(\mathcal{A}(S), \cdot)$:

$$\hat{V}_S(\mathcal{A}(S)) = \frac{1}{2m(m-1)} \sum_{i \neq j} \left[\ell\left(\mathcal{A}(S), z_i\right) - \ell\left(\mathcal{A}(S), z_j\right)\right]^2.$$

Nissim and Stemmer (2015) proved that

$$\mathbb{P}\left[R(\mathcal{A}(S)) - \hat{R}_S(\mathcal{A}(S)) < 13\varepsilon\right] > 1 - \frac{2\delta}{\varepsilon}\log\left(\frac{2}{\varepsilon}\right).$$

This is the existing tightest high-probability generalization bound in the literature. However, this bound only stands for binary classification problems. By contrast, our high-probability generalization bound holds for any machine learning algorithm.

Also, our bound is strictly tighter. All the bounds, including ours, are in the following form,

$$\mathbb{P}\left[R(\mathcal{A}(S)) - \hat{R}_S(\mathcal{A}(S)) < a\right] > 1 - b,$$

where a and b are two positive constant real numbers. Apparently, a smaller a and a smaller b imply a tighter generalization bound. Our bound improves the current tightest result from two aspects:

- Our bound tightens the term a from 13ε to 9ε.
- Our bound tightens the term b from $(2\delta/\varepsilon)\log(2/\varepsilon)$ to $(2e^{-\varepsilon}\delta/\varepsilon)\log(2/\varepsilon)$.

These improvements are significant. Abadi et al. (2016) conducted experiments on the differential privacy in deep learning. Their empirical results demonstrate that the factor ε can be as large as 10.

A comparison of Theorem 11.2.

There is only one related work in the literature that presents an on-average generalization bound for multi-database algorithms. Nissim and Stemmer (2015) proved that,

$$\left| \mathop{\mathbb{E}}_{\mathbf{S}\sim\mathcal{D}^{km}} \left[\mathop{\mathbb{E}}_{\mathcal{A}(\mathbf{S})} \left[\hat{R}_{S_{i_{\mathcal{A}(\mathbf{S})}}} \left(h_{\mathcal{A}(\mathbf{S})} \right) \right] - \mathop{\mathbb{E}}_{\mathcal{A}(\mathbf{S})} \left[R \left(h_{\mathcal{A}(\mathbf{S})} \right) \right] \right] \right| \leq k\delta + 2\varepsilon.$$

Our bound is tighter by a factor of e^{ε}. According to the empirical results by Abadi et al. (2016), this factor can be as large as $e^{10} \approx 20,000$. It is a significant multiplier for loss function. Furthermore, the result by Nissim and Stemmer stands only for binary classification, while our result applies to all differentially private learning algorithms.

11.2.2.2 How Does the Iterative Nature Contribute?

Most machine learning algorithms are iterative, which may degenerate the privacy-preserving ability as the number of iterations. This section studies the degenerative nature of the privacy preservation in iterative machine learning algorithms and its influence on the generalization.

We have the following composition theorem.

Theorem 11.4 (Composition Theorem I) *Suppose an iterative machine learning algorithm $\mathcal{A}$ has T steps: $\{Y_i(S)\}_{i=0}^T$, where Y_i is the learned hypothesis after the i-th iteration. Suppose the i-th iterator*

$$M_i : (Y_{i-1}, S) \mapsto Y_i$$

is (ε, δ)-differentially private. Then, the algorithm $\mathcal{A}$ is (ε', δ')-differentially private. The factor ε' is as follows,

$$\varepsilon' = \min \left\{ \varepsilon_1', \varepsilon_2', \varepsilon_3' \right\}, \tag{11.12}$$

where

$$\varepsilon_1' = \sum_{i=1}^{T} \varepsilon_i, \tag{11.13}$$

$$\varepsilon_2' = \sum_{i=1}^{T} \frac{(e^{\varepsilon_i} - 1)\,\varepsilon_i}{e^{\varepsilon_i} + 1} + \sqrt{2 \sum_{i=1}^{T} \varepsilon_i^2 \log \left(e + \frac{\sqrt{\sum_{i=1}^{T} \varepsilon_i^2}}{\tilde{\delta}} \right)},$$

$$\varepsilon_3' = \sum_{i=1}^{T} \frac{(e^{\varepsilon_i} - 1)\,\varepsilon_i}{e^{\varepsilon_i} + 1} + \sqrt{2 \log \left(\frac{1}{\tilde{\delta}} \right) \sum_{i=1}^{T} \varepsilon_i^2}, \tag{11.14}$$

and $\tilde{\delta}$ is an arbitrary positive real constant.

Correspondingly, the factor δ' is defined as the maximal value of the following equation with respect to $\{\alpha_i\}_{i=1}^T \in I$,

$$1 - \prod_{i=1}^{T} \left(1 - e^{\alpha_i} \frac{\delta_i}{1 + e^{\varepsilon_i}}\right) + 1 - \prod_{i=1}^{T} \left(1 - \frac{\delta_i}{1 + e^{\varepsilon_i}}\right) + \tilde{\delta}, \tag{11.15}$$

where

$$I = \left\{ \{\alpha_i\}_{i=1}^{T} : \sum_{i=1}^{T} \alpha_i = \varepsilon', \ |\{i : \alpha_i \neq \varepsilon_i, \ \alpha_i \neq 0\}| \leq 1 \right\},$$

and $\tilde{\delta}$ *is the same real constant mentioned above.*

Proof Skeleton

We proceed by providing a sketch of the proofs for Theorem 11.4, accompanied by the derivation of two additional composition theorems as incidental outcomes. While these two composition theorems are not as strong as Theorem 11.4, they play crucial roles in the overall proofs. The proof process unfolds across four distinct stages: (1) we approximate the KL-divergence between hypotheses learned from neighboring training sample sets; (2) we establish a composition bound for ε-differentially private learning algorithms; (3) we refine this composition theorem to derive a composition bound suitable for(ε, δ)-differentially private learning algorithms; and (4) we further tighten the results from the stage (2) to derive Theorem 11.4.

Stage 1: Approximate the KL-divergence between hypotheses acquired from adjacent training sample sets.

Addressing the differential privacy of an iterative learning algorithm by directly considering the differential privacy of each iteration would pose a significant technical challenge. To overcome this obstacle, we introduce KL divergence as an intermediary in this section. Specifically, for any ε-differentially private learning algorithm, we establish the following lemma to provide an approximation of the KL-divergence between hypotheses learned from neighboring training sample sets.

Lemma 11.5 *If* $\mathcal{A}$ *is an* ε*-differentially private algorithm , then for every neighboring database pair* S *and* S'*, the KL divergence between hypotheses* $\mathcal{A}(S)$ *and* $\mathcal{A}(S')$ *satisfies the following inequality,*

$$D_{KL}(\mathcal{A}(S) \| \mathcal{A}(S')) \leq \varepsilon \frac{e^{\varepsilon} - 1}{e^{\varepsilon} + 1}.$$

This lemma is new, and its proof involves non-trivial technical challenges. Lemma 11.1 is a crucial component in establishing the proof of Lemma 11.5. Notably, there are two related results in the literature, albeit considerably less stringent than ours. Dwork et al. (2010) demonstrated an inequality of the KL divergence as follows,

$$D_{KL}(\mathcal{A}(S) \| \mathcal{A}(S')) \leq \varepsilon(e^{\varepsilon} - 1).$$

Then, Dwork and Rothblum (2016) further improved it to

$$D_{KL}(\mathcal{A}(S)\|\mathcal{A}(S')) \le \frac{1}{2}\varepsilon(e^{\varepsilon}-1). \tag{11.16}$$

Compared with ours, Eq. (11.16) is larger by a factor of $(1+e^{\varepsilon})/2$, which can be very large in practice.

Proof By Lemma 11.1, we have a random variable $M(S)$ and $M(S')$, which satisfies

$$d_{\infty}(M(S)|M(S')) \le \varepsilon,\ d_{\infty}(M(s')|M(S)) \le \varepsilon,$$

and

$$d_{kl}(\mathcal{A}(S)|\mathcal{A}(S')) \le d_{kl}(M(S)|M(S')) = d_{kl}(M(S')|M(S)). \tag{11.17}$$

Therefore, we only need to derive a bound for $d_{kl}(M(S)|M(S'))$.
By direct calculation,

$$\begin{aligned}
&d_{kl}(M(S)|M(S'))\\
\overset{(*)}{=}&\frac{1}{2}\left[d_{kl}(M(S)|M(S')) + d_{kl}(M(S')|M(S))\right]\\
=&\frac{1}{2}\int \log\frac{\mathrm{d}\mathbb{P}(M(S))}{\mathrm{d}\mathbb{P}(M(S'))}\mathrm{d}\mathbb{P}(M(S)) + \frac{1}{2}\int \log\frac{\mathrm{d}\mathbb{P}(M(S'))}{\mathrm{d}\mathbb{P}(M(S))}\mathrm{d}\mathbb{P}(M(S'))\\
=&\frac{1}{2}\int \log\frac{\mathrm{d}\mathbb{P}(M(S))}{\mathrm{d}\mathbb{P}(M(S'))}\mathrm{d}\left[\mathbb{P}(M(S)) - \mathbb{P}(M(S'))\right]\\
&+\frac{1}{2}\int\left(\log\frac{\mathrm{d}\mathbb{P}(M(S'))}{\mathrm{d}\mathbb{P}(M(S))} + \log\frac{\mathrm{d}\mathbb{P}(M(S))}{\mathrm{d}\mathbb{P}(M(S'))}\right)\mathrm{d}\mathbb{P}(M(S'))\\
=&\frac{1}{2}\int \log\frac{\mathrm{d}\mathbb{P}(M(S))}{\mathrm{d}\mathbb{P}(M(S'))}\mathrm{d}\left[\mathbb{P}(M(S)) - \mathbb{P}(M(S'))\right] + \frac{1}{2}\int \log 1\mathrm{d}\mathbb{P}(M(S'))\\
=&\frac{1}{2}\int \log\frac{\mathrm{d}\mathbb{P}(M(S))}{\mathrm{d}\mathbb{P}(M(S'))}\mathrm{d}\left[\mathbb{P}(M(S)) - \mathbb{P}(M(S'))\right],
\end{aligned} \tag{11.18}$$

where Eq. $(*)$ comes from Eq. (11.17).
We now analyze the last integration in Eq. (11.18). We define

$$k(y) \triangleq \frac{\mathrm{d}\mathbb{P}(M(S)=y)}{\mathrm{d}\mathbb{P}(M(S')=y)} - 1. \tag{11.19}$$

Therefore,

$$k(y)\mathrm{d}\mathbb{P}(M(S')=y) = \mathrm{d}\mathbb{P}(M(S)=y) - \mathrm{d}\mathbb{P}(M(S')=y). \tag{11.20}$$

Additionally,

$$
\begin{aligned}
{}_{M(S')}k(M(S') &= \int_{y\in\langle} k(y)\mathrm{d}\mathbb{P}(M(S') = y) \\
&= \int_{y\in\langle} \mathrm{d}\left(\mathbb{P}(M(S) = y) - \mathrm{d}\mathbb{P}(M(S') = y)\right) \\
&= 0.
\end{aligned} \tag{11.21}
$$

By calculating the integration of both sides of Eq. (11.20), we have

$$
\int k(y)\mathrm{d}\mathbb{P}(M(S') = y) = 0.
$$

Also, combined with the definition of $k(y)$ (see Eq. 11.19), the right-hand side (rhs) of Eq. (11.18) becomes

$$
\text{rhs} = {}_{M(S')}k(M(S'))\log(k(M(S')) + 1). \tag{11.22}
$$

Since m is ε-differentially private, $k(y)$ is bounded from both sides as follows,

$$
e^{-\varepsilon} - 1 \le k(y) \le e^{\varepsilon} - 1. \tag{11.23}
$$

We now calculate the maximum of Eq. (11.22) subject to Eqs. (11.21) and (11.23).

First, we argue that the maximum is achieved when $k(M(S')) \in \{e^{-\varepsilon} - 1, e^{\varepsilon} - 1\}$ with a probability of 1 (almost surely). when $k(M(S')) \in \{e^{-\varepsilon} - 1, e^{\varepsilon} - 1\}$, almost surely, the distribution for $k(M(S'))$ is as follows,

$$
\begin{aligned}
\mathbb{P}^*(k(M(S')) = e^{\varepsilon} - 1) &= \frac{1}{1 + e^{\varepsilon}}, \\
\mathbb{P}^*(k(M(S')) = e^{-\varepsilon} - 1) &= \frac{e^{\varepsilon}}{1 + e^{\varepsilon}}.
\end{aligned}
$$

We argue that it is the distribution that maximizes $k(M(S'))$.

For brevity, we denote the probability measure for a given distribution q as $\mathbb{P}_q$. Similarly, $\mathbb{P}^*$ corresponds with the distribution q^*. We prove that q^* maximizes Eq. (11.22) in the following two cases: (1) $\mathbb{P}_q(k(M(S')) \ge 0) \le \mathbb{P}^*(k(M(S')) = e^{\varepsilon} - 1)$, and (2) $\mathbb{P}_q(k(M(S')) \ge 0) > \mathbb{P}^*(k(M(S')) = e^{\varepsilon} - 1)$

Case 1: $\mathbb{P}_q(k(M(S')) \ge 0) \le \mathbb{P}^*(k(M(S')) = e^{\varepsilon} - 1)$

We have

$$
\begin{aligned}
&\mathbb{e}_{M(S')\sim q^*}(k(M(S')\log(k(M(S')) + 1)) \\
=&\mathbb{P}^*(k(M(S')) = e^{\varepsilon} - 1) \cdot \varepsilon(e^{\varepsilon} - 1) - \mathbb{P}^*(k(M(S')) = e^{-\varepsilon} - 1) \cdot \varepsilon(e^{-\varepsilon} - 1) \\
=&(\mathbb{P}^*(k(M(S')) = e^{\varepsilon} - 1) - \mathbb{P}_q(k(M(S')) \ge 0)) \cdot \varepsilon(e^{\varepsilon} - 1)
\end{aligned}
$$

$$
\begin{aligned}
&+\mathbb{P}_q(k(M(S')) \geq 0) \cdot \varepsilon(e^{\varepsilon} - 1) - \mathbb{P}^*(k(M(S')) = e^{-\varepsilon} - 1) \cdot \varepsilon(e^{-\varepsilon} - 1) \\
\geq &\mathbb{P}_q(k(M(S')) \geq 0) \cdot \varepsilon(1 - e^{-\varepsilon}) - \mathbb{P}^*(k(M(S')) = e^{-\varepsilon} - 1) \cdot \varepsilon(1 - e^{-\varepsilon}) \\
&+\mathbb{P}_q(k(M(S')) \geq 0) \cdot \varepsilon(e^{\varepsilon} - 1) - \mathbb{P}^*(k(M(S')) = e^{-\varepsilon} - 1) \cdot \varepsilon(e^{-\varepsilon} - 1).
\end{aligned}
$$

Note that

$$
\begin{aligned}
\mathbb{P}_q(k(M(S')) < 0) = &\mathbb{P}^*(k(M(S')) = e^{\varepsilon} - 1) - \mathbb{P}_q(k(M(S')) \geq 0) \\
&+ \mathbb{P}^*(k(M(S')) = e^{-\varepsilon} - 1).
\end{aligned}
$$

Therefore, together with the condition in Eq. (11.23),

$$
\begin{aligned}
&\mathbb{e}_{M(S')\sim q}(k(M(S') \log(k(M(S')) + 1) i_{k(M(S') \leq 0)}) \\
\leq &(\mathbb{P}^*(k(M(S')) = e^{\varepsilon} - 1) - \mathbb{P}_q(k(M(S')) \geq 0)) \cdot \varepsilon(1 - e^{-\varepsilon}) \\
&+ \mathbb{P}^*(k(M(S')) = e^{-\varepsilon} - 1) \cdot \varepsilon(1 - e^{-\varepsilon}).
\end{aligned} \tag{11.24}
$$

Also,

$$
\mathbb{e}_{M(S')\sim q}(k(M(S') \log(k(M(S')) + 1) i_{k(M(S'))>0}) \leq \mathbb{P}_q(k(M(S')) \geq 0) \cdot \varepsilon(e^{\varepsilon} - 1). \tag{11.25}
$$

Therefore, by combining the inequalities in Eqs. (11.24) and (11.25), we have

$$
\mathbb{e}_{M(S')\sim q}(k(M(S') \log(k(M(S')) + 1)) \leq \mathbb{e}_{M(S')\sim q^*}(k(M(S') \log(k(M(S')) + 1)).
$$

Since the distribution q is arbitrary, the distribution q^* maximizes the $k(m(s') \log(k(m(s')) + 1)$.

Case 2: $\mathbb{P}_q(k(M(S')) \geq 0) > \mathbb{P}^*(k(M(S')) = e^{\varepsilon} - 1)$

We first prove that if $\mathbb{P}_q(1 - e^{-\varepsilon} < k(M(S')) < 0) \neq 0$, there exists a distribution q' such that

$$
\begin{aligned}
\mathbb{P}_{q'}(k(M(S')) \geq 0) &= \mathbb{P}_q(k(M(S')) \geq 0), \\
\mathbb{P}_{q'}(k(M(S')) < 0) &= \mathbb{P}_q(k(M(S')) < 0), \\
\mathbb{P}_{q'}(k(M(S')) < 0) &= \mathbb{P}_{q'}(k(M(S') = e^{-\varepsilon} - 1), \\
{}_{q'}(k(M(S') \log(k(M(S')) + 1)) &> {}_{q'}(k(M(S') \log(k(M(S')) + 1)),
\end{aligned}
$$

while the two conditions (Eqs. 11.21, 11.23) still hold.

Additionally, we have assumed that

$$
\mathbb{P}_q(k(M(S')) \geq 0) > \mathbb{P}^*(k(M(S')) = e^{\varepsilon} - 1).
$$

Therefore,

$$
\mathbb{P}_q(k(M(S')) \leq 0) < \mathbb{P}^*(k(M(S')) = e^{-\varepsilon} - 1).
$$

Also, since the distribution q' is arbitrary, let it satisfy

$$\mathbb{P}_{q'}(k(M(S')) < 0) = \mathbb{P}_q(k(M(S')) < 0) = \mathbb{P}_{q'}(k(M(S') = e^{-\varepsilon} - 1).$$

Then, in order to meet the condition Eq. (11.21), let

$$\mathbb{P}_{q'}(k(M(S') = e^{\varepsilon} - 1) > \mathbb{P}_q(k(M(S') = e^{\varepsilon} - 1),$$

and

$$\mathbb{P}_{q'}(0 < k(M(S')) < e^{\varepsilon} - 1) \leq \mathbb{P}_q(0 < k(M(S')) < e^{\varepsilon} - 1),$$

Since $x \log(x+1)$ increases when $x > 0$ and decreases when $x < 0$, we have

$$_{q'}(k(M(S') \log(k(M(S')) + 1)) > {}_q(k(M(S') \log(k(M(S')) + 1)).$$

Therefore, we have proved that the argument when $\mathbb{P}_q(k(M(S')) < 0) \neq \mathbb{P}_q(k(M(S')) = e^{-\varepsilon} - 1)$. We now prove the case that

$$\mathbb{P}_q(k(M(S')) < 0) = \mathbb{P}_q(k(M(S')) = e^{-\varepsilon} - 1),$$

where

$$_q(k(M(S') \log(k(M(S')) + 1) i_{k(M(S'))<0}) = \varepsilon(1 - e^{-\varepsilon})\mathbb{P}_q(k(M(S')) < 0).$$

By applying Jensen's inequality to bound the $_q(k(m(s') \log(k(m(s')) + 1) i_{k(m(s'))\geq 0})$, we have

$$\begin{aligned} &_q(k(M(S')) \log(k(M(S')) + 1) i_{k(M(S'))\geq 0}) \\ =&\mathbb{P}_q(M(S') \geq 0)_{q'}(k(M(S')) \log(k(M(S')) + 1) | k(M(S')) \geq 0) \\ \overset{(*)}{\leq}&\mathbb{P}_q(M(S') \geq 0)_q \left(k(M(S')) | k(M(S')) \geq 0\right) \cdot \log(_q \left(k(M(S')) | k(M(S')) \geq 0\right) + 1), \end{aligned} \tag{11.26}$$

where the inequality $(*)$ uses jensen's inequality ($x \log(1 + x)$ is convex with respect to x when $x > 0$). The upper bound in Eq. (11.26) is achieved as long as

$$\mathbb{P}_q(k(M(S')) \geq 0) = \mathbb{P}_q(k(M(S')) = {}_q(k(M(S')) | k(M(S')) \geq 0)).$$

Furthermore,

$$\mathbb{P}_q(k(M(S')) < 0) = \mathbb{P}_q(k(M(S')) = e^{-\varepsilon} - 1).$$

Therefore, the distribution q is determined by the cumulative density functions $\mathbb{P}_q(k(M(S')) < 0)$ and $\mathbb{P}_q(k(M(S')) \geq 0)$.

Hence, maximizing $_q(k(M(S')\log(k(M(S'))+1))$ is equivalent to maximizing the following object function,

$$g(q) = q(1-e^{-\varepsilon})\log e^{\varepsilon} + (1-q)\frac{q}{1-q}(1-e^{-\varepsilon})\log\left(\frac{q}{1-q}(1-e^{-\varepsilon})+1\right),$$

subject to

$$\frac{q}{1-q} \leq e^{\varepsilon}, \tag{11.27}$$

where $g(q)$ is the maximum of Eq. (11.22) subject to $\mathbb{P}_q(k(M(S')) < 0) = q$, and the condition Eq. (11.27) comes from the $\mathbb{P}_q(k(M(S')) \geq 0) > \mathbb{P}^*(k(M(S')) = e^{\varepsilon} - 1)$ (the assumption of case 2).

Additionally, $g(q)$ can be represented as follows,

$$q(1-e^{-\varepsilon})\log\left(\frac{q}{1-q}(e^{\varepsilon}-1)+\varepsilon\right).$$

Since both q and $q/(1-q)$ monotonously increase, $g(q)$ monotonously increases. Therefore, Q^* maximize Eq. (11.22), which completes the proof. □

Stage 2: Prove a weaker composition theorem where every iteration is ε-differential private.

Based on Lemma 11.5, we can prove the following composition theorem as a preparation theorem.

Theorem 11.5 (Composition Theorem IV) *Suppose an iterative machine learning algorithm $\mathcal{A}$ has T steps: $\{Y_i(S)\}_{i=1}^T$. Specifically, we define the i-th iterator as follows,*

$$M_i : (Y_{i-1}(S), S) \mapsto Y_i(S). \tag{11.28}$$

We assume that Y_0 is the initial hypothesis (which does not depend on S). If for any fixed observation y_{i-1} of the variable Y_{i-1}, $M_i(y_{i-1}, S)$ is ε_i-differentially private, then $\{Y_i(S)\}_{i=0}^T$ is (ε', δ')-differentially private that

$$\varepsilon' = \sqrt{2\log\left(\frac{1}{\delta'}\right)\left(\sum_{i=1}^T \varepsilon_i^2\right)} + \sum_{i=1}^T \varepsilon_i \frac{e^{\varepsilon_i}-1}{e^{\varepsilon_i}+1}.$$

Based on Lemma 11.5, we can prove the following composition theorem for ε-differential privacy as a preparation theorem of the general case.

Proof of Theorem 11.5 By calculating $\log \frac{\mathbb{P}(\{Y_i(S)=y_i\}_{i=0}^T)}{\mathbb{P}(\{Y_i(S')=y_i\}_{i=0}^T)}$, we have

$$\begin{aligned}
&\log \frac{\mathbb{P}\left(\{Y_i(S) = y_i\}_{i=0}^T\right)}{\mathbb{P}\left(\{Y_i(S') = y_i\}_{i=0}^T\right)} \\
=&\log \left(\prod_{i=0}^T \frac{\mathbb{P}\left(Y_i(S) = y_i | Y_{i-1}(S) = y_{i-1}, ..., Y_0(S) = y_0\right)}{\mathbb{P}\left(Y_i(S') = y_i | Y_{i-1}(S') = y_{i-1}, ..., Y_0(S') = y_0\right)}\right) \\
=&\sum_{i=0}^T \log \left(\frac{\mathbb{P}\left(Y_i(S) = y_i | Y_{i-1}(S) = y_{i-1}, ..., Y_0(S) = y_0\right)}{\mathbb{P}\left(Y_i(S') = y_i | Y_{i-1}(S') = y_{i-1}, ..., Y_0(S') = y_0\right)}\right) \\
\overset{(*)}{=}&\sum_{i=1}^T \log \left(\frac{\mathbb{P}\left(Y_i(S) = y_i | Y_{i-1}(S) = y_{i-1}, ..., Y_0(S) = y_0\right)}{\mathbb{P}\left(Y_i(S') = y_i | Y_{i-1}(S') = y_{i-1}, ..., Y_0(S') = y_0\right)}\right) \\
=&\sum_{i=1}^T \log \left(\frac{\mathbb{P}\left(M_i(y_{i-1}, S) = y_i | Y_{i-1}(S) = y_{i-1}, ..., Y_0(S) = y_0\right)}{\mathbb{P}\left(M_i(y_{i-1}, S') = y_i | Y_{i-1}(S') = y_{i-1}, ..., Y_0(S') = y_0\right)}\right) \\
\overset{(**)}{=}&\sum_{i=1}^T \log \left(\frac{\mathbb{P}\left(M_i(y_{i-1}, S) = y_i\right)}{\mathbb{P}\left(M_i(y_{i-1}, S') = y_i\right)}\right),
\end{aligned}$$

where Eq. $(*)$ comes from the independence of Y_0 with respect to S and Eq. $(**)$ is due the independence of M_i to Y_k $(k < i)$ when the Y_{i-1} is fixed.

By the definition of ε-differential privacy, one has for arbitrary y_{i-1} as the observation of Y_{i-1},

$$D_\infty\left(M_i(y_{i-1}, S) \| M_i(y_{i-1}, S')\right) < \varepsilon_i, \; D_\infty\left(M_i(y_{i-1}, S') \| M_i(y_{i-1}, S)\right) < \varepsilon_i.$$

Thus, by Lemma 11.5, we have

$$\begin{aligned}
&\mathbb{E}\left(\log \left(\frac{\mathbb{P}\left(M_i(Y_{i-1}, S) = Y_i\right)}{\mathbb{P}\left(M_i(Y_{i-1}, S') = Y_i\right)}\right) \middle| Y_{i-1}(S) = y_{i-1}, ..., Y_0(S) = y_0\right) \\
=&D_{KL}(M_i(y_{i-1}, S) \| M_i(y_{i-1}, S')) \\
\leq&\varepsilon_i \frac{e^{\varepsilon_i} - 1}{e^{\varepsilon_i} + 1}.
\end{aligned} \tag{11.29}$$

Combining Azuma Lemma (Lemma 11.3), Eq. (11.29) derives the following equation

$$\mathbb{P}\left(\{Y_i(S) = y_i\}_{i=0}^T : \frac{\mathbb{P}\left(\{Y_i(S) = y_i\}_{i=0}^T\right)}{\mathbb{P}\left(\{Y_i(S') = y_i\}_{i=0}^T\right)} > e^{\varepsilon'}\right) < \delta',$$

where S and S' are neighbouring sample sets.

Therefore, the algorithm $\mathcal{A}$ is ε'-differentially private.

The proof is completed.

Stage 3: Prove a weaker composition theorem where every iteration is $(\varepsilon_i, \delta_i)$-differentially private.

Based on Lemmas 11.1 and 11.3, we proved the following lemma that the maximum of the following function,

$$f\left(\{\alpha_i\}_{i=1}^T\right) = 1 - \prod_{i=1}^{T}(1 - \alpha_i A_i), \tag{11.30}$$

is achieved when $\{\alpha_i\}_{i=1}^T$ are at the boundary under some restrictions.

Lemma 11.6 *The maximum of function (11.30) when A_i is positive real such that,*

$$1 \leq \alpha_i \leq c_i, \text{ (here } c_i A_i \leq 1), \text{ and } \prod_{i=1}^{T} \alpha_i = c,$$

is achieved at the point when the cardinality follows the inequality:

$$|\{i : \alpha_i \neq c_i \text{ and } \alpha_i \neq 1\}| \leq 1. \tag{11.31}$$

Based on Lemmas 11.2 and 11.6, we can prove the following composition theorem whose estimate of ε' is somewhat looser than our main results.

Theorem 11.6 (Composition Theorem V) *Suppose an iterative machine learning algorithm $\mathcal{A}$ has T steps: $\{Y_i(S)\}_{i=1}^T$. Specifically, let the i-th iterator be as follows,*

$$M_i : (Y_{i-1}(S), S) \mapsto Y_i(S). \tag{11.32}$$

We assume that Y_0 is the initial hypothesis (which does not depend on S). If for any fixed observation y_{i-1} of the variable Y_{i-1}, $M_i(y_{i-1}, S)$ is $(\varepsilon_i, \delta_i)$-differentially private ($i \geq 1$), then $\{Y_i(S)\}_{i=0}^T$ is (ε', δ')-differentially private where

$$\varepsilon' = \sqrt{2\log\left(\frac{1}{\tilde{\delta}}\right)\left(\sum_{i=1}^{T}\varepsilon_i^2\right)} + \sum_{i=1}^{T}\varepsilon_i \frac{e^{\varepsilon_i} - 1}{e^{\varepsilon_i} + 1},$$

$$\delta' = \max_{\{\alpha_i\}_{i=1}^T \in I} 1 - \prod_{i=1}^{T}\left(1 - e^{\alpha_i}\frac{\delta_i}{1 + e^{\varepsilon_i}}\right) + 1 - \prod_{i=1}^{T}\left(1 - \frac{\delta_i}{1 + e^{\varepsilon_i}}\right) + \tilde{\delta},$$

and I is defined as the set of $\{\alpha_i\}_{i=1}^T$ such that

$$\sum_{i=1}^{T}\alpha_i = \varepsilon', \ |\{i : \alpha_i \neq \varepsilon_i \text{ and } \alpha_i \neq 0\}| \leq 1.$$

Now, we can prove our composition theorems for (ε, δ)-differential privacy. We first prove a composition algorithm of (ε, δ)-differential privacy whose estimate of ε' is somewhat looser than the existing results. Then, we tighten the results and obtain a composition theorem that is strictly tighter than the current estimate.

Proof of Theorem 11.6 It has been proved that the optimal privacy preservation can be achieved by a sequence of independent iterations (see Kairouz et al. (2017), Theorem 3.5). Therefore, without loss of generality, we assume that the iterations in our theorem are independent with each other.

Rewrite $Y_i(S)$ as Y_i^0, and $Y_i(S')$ as Y_i^1 ($i \geq 1$). Then, by Lemma 11.2, there exist random variables $\tilde{Y}_i^0$ and $\tilde{Y}_i^1$, such that

$$\Delta\left(Y_i^0 \| \tilde{Y}_i^0\right) \leq \frac{\delta_i}{1+e^{\varepsilon_i}}, \tag{11.33}$$

$$\Delta\left(Y_i^1 \| \tilde{Y}_i^1\right) \leq \frac{\delta_i}{1+e^{\varepsilon_i}}, \tag{11.34}$$

$$D_\infty\left(\tilde{Y}_i^0 \| \tilde{Y}_i^1\right) \leq \varepsilon_i, \tag{11.35}$$

$$D_\infty\left(\tilde{Y}_i^1 \| \tilde{Y}_i^0\right) \leq \varepsilon_i. \tag{11.36}$$

Applying Theorem 11.6 (here, $\delta = \tilde{\delta}$), we have

$$D_\infty^{\tilde{\delta}}\left(\{\tilde{Y}_i^0\}_{i=0}^T \| \{\tilde{Y}_i^1\}_{i=0}^T\right) \leq \varepsilon',$$

$$D_\infty^{\tilde{\delta}}\left(\{\tilde{Y}_i^1\}_{i=0}^T \| \{\tilde{Y}_i^0\}_{i=0}^T\right) \leq \varepsilon'.$$

Apparently,

$$\mathbb{P}(Y_i^0 \in B_i) - \min\left\{\frac{\delta_i}{1+e^{\varepsilon_i}}, \mathbb{P}(Y_i^0 \in B_i)\right\} \geq 0.$$

Therefore, for any sequence of hypothesis sets $B_0, \ldots, B_T$,

$$\begin{aligned}
&\mathbb{P}(Y_0^0 \in B_0)\left(\mathbb{P}(Y_1^0 \in B_1) - \min\left\{\frac{\delta_1}{1+e^{\varepsilon_1}}, \mathbb{P}(Y_1^0 \in B_1)\right\}\right) \\
&\cdots\left(\mathbb{P}(Y_T^0 \in B_T) - \min\left\{\frac{\delta_T}{1+e^{\varepsilon_T}}, \mathbb{P}(Y_T^0 \in B_1)\right\}\right) \\
\leq &\mathbb{P}(\tilde{Y}_0^0 \in B_0)\cdots\mathbb{P}(\tilde{Y}_T^0 \in B_T) \\
\leq &e^{\varepsilon'}\mathbb{P}(\tilde{Y}_0^1 \in B_0)\cdots\mathbb{P}(\tilde{Y}_T^1 \in B_T) + \tilde{\delta}.
\end{aligned} \tag{11.37}$$

Furthermore, by Eq. (11.36), we also have that

$$\mathbb{P}(\tilde{Y}_i^0 \in B_i) \le \min\left\{e^{\varepsilon_i}, \frac{1}{\mathbb{P}(\tilde{Y}_i^1 \in B_i)}\right\} \mathbb{P}(\tilde{Y}_i^1 \in B_i).$$

Therefore,

$$\mathbb{P}(\tilde{Y}_0^0 \in B_0) \cdots \mathbb{P}(\tilde{Y}_n^0 \in B_T) \le \prod_{i=1}^{T} \min\left\{e^{\varepsilon_i}, \frac{1}{\mathbb{P}(\tilde{Y}_i^1 \in B_i)}\right\} \mathbb{P}(\tilde{Y}_0^1 \in B_0) \cdots \mathbb{P}(\tilde{Y}_T^1 \in B_T) + \tilde{\delta}.$$

Then, we prove this theorem in two cases:
(1) $\prod_{i=1}^{T} \min\left\{e^{\varepsilon_i}, \frac{1}{\mathbb{P}(\tilde{Y}_i^1 \in B_i)}\right\} \le e^{\varepsilon'}$; and (2) $\prod_{i=1}^{T} \min\left\{e^{\varepsilon_i}, \frac{1}{\mathbb{P}(\tilde{Y}_i^1 \in B_i)}\right\} > e^{\varepsilon'}$.
Case 1- $\prod_{i=1}^{T} \min\left\{e^{\varepsilon_i}, \frac{1}{\mathbb{P}(\tilde{Y}_i^1 \in B_i)}\right\} \le e^{\varepsilon'}$.
We have

$$\begin{aligned}&\mathbb{P}(\tilde{Y}_0^1 \in B_0)\left(\mathbb{P}(\tilde{Y}_1^1 \in B_1) - \frac{\delta_1}{1+e^{\varepsilon_1}}\right) \cdots \left(\mathbb{P}(\tilde{Y}_T^1 \in B_T) - \frac{\delta_T}{1+e^{\varepsilon_T}}\right)\\ \le&\mathbb{P}(Y_0^1 \in B_0) \cdots \mathbb{P}(Y_T^1 \in B_T).\end{aligned}$$

By simple calculation, we have

$$\begin{aligned}&\prod_{i=1}^{T} \min\left\{e^{\varepsilon_i}, \frac{1}{\mathbb{P}(\tilde{Y}_i^1 \in B_i)}\right\} \mathbb{P}(\tilde{Y}_0^1 \in B_0) \cdots \mathbb{P}(\tilde{Y}_T^1 \in B_T)\\ \le&\prod_{i=1}^{T} \min\left\{e^{\varepsilon_i}, \frac{1}{\mathbb{P}(\tilde{Y}_i^1 \in B_i)}\right\} \mathbb{P}(Y_0^1 \in B_0) \cdots \mathbb{P}(Y_T^1 \in B_T)\\ &+\prod_{i=1}^{T} \min\left\{e^{\varepsilon_i}, \frac{1}{\mathbb{P}(\tilde{Y}_i^1 \in B_i)}\right\} \mathbb{P}(\tilde{Y}_0^1 \in B_0) \cdots \mathbb{P}(\tilde{Y}_T^1 \in B_T)\\ &-\prod_{i=1}^{n} \min\left\{e^{\varepsilon_i}, \frac{1}{\mathbb{P}(\tilde{Y}_i^1 \in B_i)}\right\} \mathbb{P}(\tilde{Y}_0^1 \in B_0)\\ &\quad\left(\mathbb{P}(\tilde{Y}_1^1 \in B_1) - \frac{\delta_1}{1+e^{\varepsilon_1}}\right) \cdots \left(\mathbb{P}(\tilde{Y}_T^1 \in B_T) - \frac{\delta_T}{1+e^{\varepsilon_T}}\right).\end{aligned}$$

Apparently,

$$\min\left\{e^{\varepsilon_i}, \frac{1}{\mathbb{P}(\tilde{Y}_i^1 \in B_i)}\right\} \mathbb{P}(\tilde{Y}_0^i \in B_i) \le 1,$$

and when $A > B$, $f(x) = Ax - (x-a)B$ increases when x increases.

Therefore, we have

$$\begin{aligned}
&\prod_{i=1}^{T}\min\left\{e^{\varepsilon_i},\frac{1}{\mathbb{P}(\tilde{Y}_i^1\in B_i)}\right\}\mathbb{P}(\tilde{Y}_0^1\in B_0)\cdots\mathbb{P}(\tilde{Y}_T^1\in B_T)\\
&-\prod_{i=1}^{T}\min\left\{e^{\varepsilon_i},\frac{1}{\mathbb{P}(\tilde{Y}_i^1\in B_i)}\right\}P(\tilde{Y}_0^1\in B_0)\\
&\qquad\left(\mathbb{P}(\tilde{Y}_1^1\in B_1)-\frac{\delta_1}{1+e^{\varepsilon_1}}\right)\cdots\left(\mathbb{P}(\tilde{Y}_T^1\in B_T)-\frac{\delta_T}{1+e^{\varepsilon_T}}\right)\\
&\leq 1-\prod_{i=1}^{T}\left(1-\min\left\{e^{\varepsilon_i},\frac{1}{\mathbb{P}(\tilde{Y}_i^1\in B_i)}\right\}\frac{\delta_i}{1+e^{\varepsilon_i}}\right).
\end{aligned}$$

Combining with Eq. (11.37), we have

$$\delta'\leq 1-\prod_{i=1}^{T}\left(1-\min\left\{e^{\varepsilon_i},\frac{1}{\mathbb{P}(\tilde{Y}_i^1\in B_i)}\right\}\frac{\delta_i}{1+e^{\varepsilon_i}}\right)+1-\prod_{i=1}^{T}\left(1-\frac{\delta_i}{1+e^{\varepsilon_i}}\right)+\tilde{\delta}.$$

Case 2- $\prod_{i=1}^{T}\min\left\{e^{\varepsilon_i},\frac{1}{\mathbb{P}(\tilde{Y}_i^1\in B_i)}\right\}>e^{\varepsilon'}$:
There exists a sequence of reals $\{\alpha_i\}_{i=1}^{T}$ such that

$$e^{\alpha_i}\leq\min\left\{e^{\varepsilon_i},\frac{1}{\mathbb{P}(\tilde{Y}_i^1\in B_i)}\right\},\quad\sum_{i=1}^{T}\alpha_i=\varepsilon'.$$

Therefore, similar to Case 1, we have that

$$\delta'\leq 1-\prod_{i=1}^{T}\left(1-e^{\alpha_i}\frac{\delta_i}{1+e^{\varepsilon_i}}\right)+1-\prod_{i=1}^{T}\left(1-\frac{\delta_i}{1+e^{\varepsilon_i}}\right).$$

Overall, we have proven that

$$\delta'\leq 1-\prod_{i=1}^{T}\left(1-e^{\alpha_i}\frac{\delta_i}{1+e^{\varepsilon_i}}\right)+1-\prod_{i=1}^{T}\left(1-\frac{\delta_i}{1+e^{\varepsilon_i}}\right),$$

where $\sum_{i=1}^{T}\alpha_i\leq\varepsilon'$ and $\alpha_i\leq\varepsilon_i$.

From Lemma 11.6, the minimum is realized on the boundary, which is exactly as this theorem claims.

The proof is completed.

Stage 4: Prove Theorem 11.4.

Applying Lemma 11.3 and Theorem 3.5 in Kairouz et al. (2015), we eventually extend the weaker versions to Theorem 11.4. Theorem 3.5 in Kairouz et al. (2017) relies on the term *privacy area*. A larger privacy area corresponds to a worse privacy preservation. In this section, we make a novel contribution that proves the moment generating function of the following random variable represents the worst case,

$$\log\left(\frac{\mathbb{P}(\cap_i Y_i(S) \in B_i)}{\mathbb{P}(\cap_i Y_i(S') \in B_i)}\right), \tag{11.38}$$

where $Y_i(S)$ and $Y_i(S')$ are the mechanisms achieving the largest privacy area. Thus, we can deliver an approximation of the differential privacy via this moment generating function.

Proof of Theorem 11.4 By applying Theorem 3.5 proposed in Kairouz et al. (2017) and replacing ε' in the proof of Theorem 11.6 as

$$\varepsilon' = \min\{I_1, I_2, I_3\},$$

where

$$I_1 = \sum_{i=1}^{T} \varepsilon_i,$$

$$I_2 = \sum_{i=1}^{T} \frac{(e^{\varepsilon_i} - 1)\,\varepsilon_i}{e^{\varepsilon_i} + 1} + \sqrt{\sum_{i=1}^{T} 2\varepsilon_i^2 \log\left(e + \frac{\sqrt{\sum_{i=1}^{T} \varepsilon_i^2}}{\tilde{\delta}}\right)},$$

$$I_3 = \sum_{i=1}^{T} \frac{(e^{\varepsilon_i} - 1)\,\varepsilon_i}{e^{\varepsilon_i} + 1} + \sqrt{\sum_{i=1}^{T} 2\varepsilon_i^2 \log\left(\frac{1}{\tilde{\delta}}\right)}.$$

The proof is completed.

Comparison with Existing Results

Our composition theorem is strictly tighter than the existing results.

A classic composition theorem is as follows (see Dwork and Roth (2014), Theorem 3.20 and Corollary 3.21, pp. 49–52),

$$\varepsilon' = \sum_{i=1}^{T} \varepsilon_i (e^{\varepsilon_i} - 1) + \sqrt{2\log\left(\frac{1}{\tilde{\delta}}\right) \sum_{i=1}^{T} \varepsilon_i^2}, \quad \delta' = \tilde{\delta} + \sum_{i=1}^{T} \delta_i,$$

where $\tilde{\delta}$ is an arbitrary positive real number, (ε', δ') is the differential privacy of the whole algorithm, and $(\varepsilon_i, \delta_i)$ is the differential privacy of the i-th iteration.

Currently, the tightest approximation is given by Kairouz et al. (2017) as follows,

$$\varepsilon' = \min\left\{\varepsilon_1', \varepsilon_2', \varepsilon_3'\right\}, \quad \delta' = 1 - (1-\delta)^T(1-\tilde{\delta}),$$

where

$$\varepsilon_1' = T\varepsilon,$$

$$\varepsilon_2' = \frac{(e^\varepsilon - 1)\,\varepsilon T}{e^\varepsilon + 1} + \varepsilon\sqrt{2T\log\left(e + \frac{\sqrt{T\varepsilon^2}}{\tilde{\delta}}\right)},$$

$$\varepsilon_3' = \frac{(e^\varepsilon - 1)\,\varepsilon T}{e^\varepsilon + 1} + \varepsilon\sqrt{2T\log\left(\frac{1}{\tilde{\delta}}\right)}.$$

The estimate of the ε' is the same as ours, while their δ' is also larger than ours approximately $\delta\frac{e^\varepsilon-1}{e^\varepsilon+1}\left(T - \left\lceil\frac{\varepsilon'}{\varepsilon}\right\rceil\right)$. In many situations, the number of iterations T is overwhelmingly large, which guarantees that our advantage is significant.

When all the iterations have the same privacy-preserving ability, we can tighten the approximation of the factor δ' as the following corollary.

Corollary 11.2 (Composition Theorem II) *When all the iterations are* (ε, δ)*-differential private,* δ' *is*

$$\begin{aligned}\delta' =& 1 - \left(1 - e^\varepsilon\frac{\delta}{1+e^\varepsilon}\right)^{\left\lceil\frac{\varepsilon'}{\varepsilon}\right\rceil}\left(1 - \frac{\delta}{1+e^\varepsilon}\right)^{T-\left\lceil\frac{\varepsilon'}{\varepsilon}\right\rceil} + 1 - \left(1 - \frac{\delta}{1+e^\varepsilon}\right)^T + \tilde{\delta}\\ =& \left(T - \left\lceil\frac{\varepsilon'}{\varepsilon}\right\rceil\right)\frac{2\delta}{1+e^\varepsilon} + \left\lceil\frac{\varepsilon'}{\varepsilon}\right\rceil\delta + \tilde{\delta} + O\left(\left(\frac{\delta}{1+e^\varepsilon}\right)^2\right).\end{aligned}$$

Proof The maximum of δ' is achieved when at most $T - \left\lceil\frac{\varepsilon'}{\varepsilon}\right\rceil$ elements $\alpha_i \neq 0$. We note that $(1-x)^n = 1 - nx + O(x^2)$. Then, the proof is completed by estimating the δ' in Theorem 11.4 as

$$\begin{aligned}\delta' =& 1 - \left(1 - e^\varepsilon\frac{\delta}{1+e^\varepsilon}\right)^{\left\lceil\frac{\varepsilon'}{\varepsilon}\right\rceil}\left(1 - \frac{\delta}{1+e^\varepsilon}\right)^{T-\left\lceil\frac{\varepsilon'}{\varepsilon}\right\rceil} + 1 - \left(1 - \frac{\delta}{1+e^\varepsilon}\right)^T + \tilde{\delta}\\ =& 1 + T\frac{\delta}{1+e^\varepsilon} + \tilde{\delta} + O\left(\left(\frac{\delta}{1+e^\varepsilon}\right)^2\right)\\ &- \left(1 - \frac{\left\lceil\frac{\varepsilon'}{\varepsilon}\right\rceil\delta}{1+e^\varepsilon} + O\left(\left(\frac{\delta}{1+e^\varepsilon}\right)^2\right)\right)\left(1 - \left(T - \left\lceil\frac{\varepsilon'}{\varepsilon}\right\rceil\right)\frac{\delta}{1+e^\varepsilon} + O\left(\left(\frac{\delta}{1+e^\varepsilon}\right)^2\right)\right)\end{aligned}$$

$$
\begin{aligned}
&= \left\lceil \frac{\varepsilon'}{\varepsilon} \right\rceil \frac{\delta}{1+e^{\varepsilon}} + \left(T - \left\lceil \frac{\varepsilon'}{\varepsilon} \right\rceil\right) \frac{\delta}{1+e^{\varepsilon}} + T\frac{\delta}{1+e^{\varepsilon}} + O\left(\left(\frac{\delta}{1+e^{\varepsilon}}\right)^2\right) + \tilde{\delta} \\
&\approx \left(T - \left\lceil \frac{\varepsilon'}{\varepsilon} \right\rceil\right) \frac{2\delta}{1+e^{\varepsilon}} + \left\lceil \frac{\varepsilon'}{\varepsilon} \right\rceil \delta + \tilde{\delta}.
\end{aligned}
$$

When all the iterators M_i are ε-differentially private, we can further tighten the third estimation of ε' in Theorem 11.4, Eq. (11.12) as the following theorem.

Corollary 11.3 (Composition Theorem III) *Suppose all the iterators M_i are ε-differentially private and all the other conditions in Theorem 11.4 hold. Then, the algorithm $\mathcal{A}$ is (ε', δ')-differentially private that*

$$
\begin{aligned}
\varepsilon' =& T\frac{(e^{\varepsilon}-1)\,\varepsilon}{e^{\varepsilon}+1} + \sqrt{2\log\left(\frac{1}{\tilde{\delta}}\right) T\varepsilon^2}, \\
\delta' =& 1 - \left(1 - e^{\varepsilon}\frac{\delta}{1+e^{\varepsilon}}\right)^{\left\lceil \frac{\varepsilon'}{\varepsilon} \right\rceil} \left(1 - \frac{\delta}{1+e^{\varepsilon}}\right)^{T-\left\lceil \frac{\varepsilon'}{\varepsilon} \right\rceil} + 1 - \left(1 - \frac{\delta}{1+e^{\varepsilon}}\right)^{T} + \delta'',
\end{aligned}
$$

where δ'' is defined as follows:

$$
\delta'' = e^{-\frac{\varepsilon'+T\varepsilon}{2}} \left(\frac{1}{1+e^{\varepsilon}}\left(\frac{2T\varepsilon}{T\varepsilon-\varepsilon'}\right)\right)^T \left(\frac{T\varepsilon+\varepsilon'}{T\varepsilon-\varepsilon'}\right)^{-\frac{\varepsilon'+T\varepsilon}{2\varepsilon}}.
$$

Proof of Corollary 11.3 Let $\mathcal{P}_0$ and $\mathcal{P}_1$ be two distributions whose cumulative distribution functions P_0 and P_1 are respectively defined as follows:

$$
P_0(x) = \begin{cases} \delta\,, & x=0 \\ \dfrac{(1-\delta)e^{\varepsilon}}{1+e^{\varepsilon}}\,, & x=1 \\ \dfrac{1-\delta}{1+e^{\varepsilon}}\,, & x=2 \\ 0\,, & x=3 \end{cases},
$$

and

$$
P_1(x) = \begin{cases} 0\,, & x=0 \\ \dfrac{(1-\delta)e^{\varepsilon}}{1+e^{\varepsilon}}\,, & x=1 \\ \dfrac{1-\delta}{1+e^{\varepsilon}}\,, & x=2 \\ \delta\,, & x=3 \end{cases}.
$$

According to Theorem 3.4 by Kairouz et al. (2017), the largest magnitude of the (ε', δ')-differential privacy can be calculated from the $\mathcal{P}_0^T$ and $\mathcal{P}_1^T$.

Construct $\tilde{\mathcal{P}}_0$ and $\tilde{\mathcal{P}}_1$, whose cumulative distribution functions are as follows,

$$\tilde{P}_0(x) = \begin{cases} \frac{e^{\varepsilon}\delta}{1+e^{\varepsilon}}, & x=0 \\ \frac{(1-\delta)e^{\varepsilon}}{1+e^{\varepsilon}}, & x=1 \\ \frac{1-\delta}{1+e^{\varepsilon}}, & x=2 \\ \frac{\delta}{1+e^{\varepsilon}}, & x=3 \end{cases},$$

and

$$\tilde{P}_1(x) \begin{cases} \frac{\delta}{1+e^{\varepsilon}}, & x=0 \\ \frac{(1-\delta)e^{\varepsilon}}{1+e^{\varepsilon}}, & x=1 \\ \frac{1-\delta}{1+e^{\varepsilon}}, & x=2 \\ \frac{e^{\varepsilon}\delta}{1+e^{\varepsilon}}, & x=3 \end{cases}.$$

One can easily verify that

$$\begin{aligned} \Delta(\mathcal{P}_0 \| \tilde{\mathcal{P}}_0) &\leq \frac{\delta}{1+e^{\varepsilon}}, \\ \Delta(\mathcal{P}_1 \| \tilde{\mathcal{P}}_1) &\leq \frac{\delta}{1+e^{\varepsilon}}, \\ D_{\infty}(\tilde{\mathcal{P}}_0 \| \tilde{\mathcal{P}}_1) &\leq \varepsilon, \\ D_{\infty}(\tilde{\mathcal{P}}_1 \| \tilde{\mathcal{P}}_0) &\leq \varepsilon_i. \end{aligned}$$

Let $V_i(x_i) = \log\left(\frac{\tilde{\mathcal{P}}_0(x_i)}{\tilde{\mathcal{P}}_1(x_i)}\right)$ and $S(x_1, \ldots, x_T) = \sum_{i=1}^{T} V_i(x_i)$. We have that for any $t > 0$,

$$\begin{aligned} \mathbb{P}_{\tilde{\mathcal{P}}_0^T}(\{x_i\} : S(\{x_i\}) > \varepsilon') &\leq e^{-\varepsilon' t} \mathbb{E}_{\tilde{\mathcal{P}}_0^T}(e^{tS}) \\ &= e^{-\varepsilon' t} \left(\frac{e^{t\varepsilon+\varepsilon}}{1+e^{\varepsilon}} + \frac{e^{-t\varepsilon}}{1+e^{\varepsilon}} \right)^T \\ &= e^{-\varepsilon' t - T t\varepsilon} \left(\frac{e^{2t\varepsilon+\varepsilon}}{1+e^{\varepsilon}} + \frac{1}{1+e^{\varepsilon}} \right)^T. \end{aligned} \tag{11.39}$$

By calculating the derivative, we can deduce that the minimum of the RHS of Eq. (11.39) is achieved at

$$e^{2\varepsilon t} = e^{-\varepsilon}\frac{T\varepsilon + \varepsilon'}{T\varepsilon - \varepsilon'}. \tag{11.40}$$

Since $\varepsilon' \geq T\frac{e^{\varepsilon}-1}{e^{\varepsilon}+1}$,

$$e^{-\varepsilon}\frac{T\varepsilon + \varepsilon'}{T\varepsilon - \varepsilon'} > 1.$$

Therefore, by applying Eq. (11.40) into the RHS of Eq. (11.39), we have

$$\mathbb{P}_{\tilde{\mathcal{P}}_0^T}(\{x_i\} : S(\{x_i\}) > \varepsilon') \leq e^{-\frac{\varepsilon'+T\varepsilon}{2}}\left(\frac{1}{1+e^{\varepsilon}}\left(\frac{2T\varepsilon}{T\varepsilon-\varepsilon'}\right)\right)^T\left(\frac{T\varepsilon+\varepsilon'}{T\varepsilon-\varepsilon'}\right)^{-\frac{\varepsilon'+T\varepsilon}{2\varepsilon}}. \tag{11.41}$$

We define the RHS of Eq. (11.41) as δ'. We have $(\tilde{\mathbb{P}}^b)^T$ $(b = 0, 1)$ is (ε', δ')-differentially private. Then, using similar analysis of the Proof of Theorem 11.6, we prove this theorem.

We now analyze the tightness of Corollary 11.3. Specifically, we compare it with our Theorem 11.4.

In the proof of Theorem 11.4, ε_3' is derived through Azuma Lemma (Lemma 11.3). Specifically, the δ' is derived by

$$\begin{aligned}
\mathbb{P}\left[S_T \geq \varepsilon' - T\frac{e^{\varepsilon}-1}{e^{\varepsilon}+1}\right] &\leq e^{-t(\varepsilon'-T\frac{e^{\varepsilon}-1}{e^{\varepsilon}+1})}\mathbb{E}\left[e^{tS_T}\right] \\
&= e^{-t(\epsilon'-T\frac{e^{\varepsilon}-1}{e^{\varepsilon}+1})}\mathbb{E}\left[e^{tS_{T-1}}\mathbb{E}\left[e^{t\mathbf{V}_T}|Y_1(S), \ldots, Y_{T-1}(S)\right]\right] \\
&\leq e^{-t(\epsilon'-T\frac{e^{\varepsilon}-1}{e^{\varepsilon}+1})}\mathbb{E}\left[e^{tS_{T-1}}\right]e^{4t^2\varepsilon^2/8} \\
&\leq e^{-t(\epsilon'-T\frac{e^{\varepsilon}-1}{e^{\varepsilon}+1})}e^{Tt^2\varepsilon^2/2},
\end{aligned}$$

where V_i is defined as $\log\frac{\mathbb{P}(Y_i(S))}{\mathbb{P}(Y_i(S'))} - \mathbb{E}\left[\log\frac{\mathbb{P}(Y_i(S))}{\mathbb{P}(Y_i(S'))}|Y_1(S), \ldots, Y_{i-1}(S)\right]$ and S_j is defined as $\sum_{i=1}^{j} V_i$.

Since $\mathbb{P}\left[S_T \geq \varepsilon' - T\frac{e^{\varepsilon}-1}{e^{\varepsilon}+1}\right]$ does not depend on t,

$$\mathbb{P}\left[S_T \geq \varepsilon' - T\frac{e^{\varepsilon}-1}{e^{\varepsilon}+1}\right] \leq \min_{t>0} e^{-\frac{(\epsilon'-T\frac{e^{\varepsilon}-1}{e^{\varepsilon}+1})^2}{2T\varepsilon^2}} = \delta',$$

By contrary, the approach here directly calculates $\mathbb{E}[e^{tS_T}]$, without the shrinkage in the proof of Theorem 11.4. Specifically,

$$e^{-\varepsilon' t - Tt\varepsilon}\left(\frac{e^{2t\varepsilon+\varepsilon}}{1+e^{\varepsilon}} + \frac{1}{1+e^{\varepsilon}}\right)^T = e^{-t\epsilon}\mathbb{E}\left[e^{tS_T}\right] \leq e^{-t(\epsilon'-T\frac{e^{\varepsilon}-1}{e^{\varepsilon}+1})}e^{Tt^2\varepsilon^2/2}.$$

Therefore,

$$\min_{t>0} e^{-\varepsilon' t - Tt\varepsilon}\left(\frac{e^{2t\varepsilon+\varepsilon}}{1+e^{\varepsilon}}+\frac{1}{1+e^{\varepsilon}}\right)^{T} \leq \min_{t>0} e^{-t(\epsilon' - T\frac{e^{\varepsilon}-1}{e^{\varepsilon}+1})}e^{Tt^{2}\varepsilon^{2}/2},$$

which leads to

$$e^{-\frac{\varepsilon'+T\varepsilon}{2}}\left(\frac{1}{1+e^{\varepsilon}}\left(\frac{2T\varepsilon}{T\varepsilon-\varepsilon'}\right)\right)^{T}\left(\frac{T\varepsilon+\varepsilon'}{T\varepsilon-\varepsilon'}\right)^{-\frac{\varepsilon'+T\varepsilon}{2\varepsilon}} \leq \delta'.$$

It ensures that this estimate further tightens δ' than Theorem 11.4 if the ε' is the same.

The trio of composition theorems expands upon the established connection between generalizability and privacy-preserving capabilities to encompass iterative machine learning algorithms. With these theorems, we establish the theoretical groundwork for understanding the generalizability of iterative differentially private machine learning algorithms.

11.2.3 Applications

Our theories apply to a wide spectrum of machine learning algorithms. This section implements them on two popular regimes as examples: (1) stochastic gradient Langevin dynamics (Welling and Teh 2011) as an example of the stochastic gradient Markov chain Monte Carlo scheme (Ma et al. 2015; Zhang et al. 2018); and (2) agnostic federated learning (Geyer et al. 2017; Mohri et al. 2019). Our results help deliver $O(\sqrt{\log m/m})$ high-probability generalization bounds and PAC-learnability guarantees for the two schemes.

11.2.3.1 Application in SGLD

Bayesian inference estimates the posterior distribution of model parameters in parametric machine learning models, enabling us to approach to the optimal parameters by iteratively increasing evidence. However, in practice, obtaining an analytical expression for the posterior distribution is challenging. Consequently, Markov Chain Monte Carlo (MCMC) methods are widely deployed to posterior inference (Hastings 1970; Duane et al. 1987). Unfortunately, traditional MCMC methods can be computationally demanding in general, especially when dealing with large-scale datasets. In response, stochastic gradient Markov chain Monte Carlo (SGMCMC, Ma et al. (2015)) methods have emerged by integrating stochastic gradient estimation techniques (Robbins and Monro 1951) into Bayesian inference. Stochastic Gradient Langevin Dynamics (SGLD) Welling and Teh (2011), serves as a representative example of SGMCMC methods. SGMCMC methods have found applications across a wide range of domains, including topic modeling (Larochelle and Lauly 2012; Zhang et al. 2020), Bayesian neural networks (Louizos and Welling 2017; Roth and Pernkopf 2018; Ban et al. 2019; Ye et al. 2020), and generative models (Wang et al. 2019; Kobyzev et al. 2020). In this section, we investigate the analysis of SGLD as an example of the SGMCMC framework. SGLD is illustrated as the following table.

Algorithm 1 Stochastic Gradient Langevin Dynamics

Require: Samples $S = \{z_1, ...z_m\}$, Gaussian noise variance σ, size of mini-batch τ, iteration steps T, learning rate $\{\eta_1, ...\eta_T\}$, Regularization function r, Lipschitz constant L of loss l.

Initialize θ_0 randomly.
For $t = 1$ to T do:
    Randomly sample a mini-batch $\mathcal{B}$ of size τ;
    Sample g_t from $N(0, \sigma^2 I)$;
    Update
$\theta_t \leftarrow \theta_{t-1} - \eta_t \left[\frac{1}{\tau}\nabla r(\theta_{t-1}) + \frac{1}{\tau}\sum_{z\in\mathcal{B}} \nabla l(z|\theta_{t-1}) + g_t\right]$.

The following theorem provides an estimation of the differential privacy and generalization bounds of SGLD.

Theorem 11.7 *SGLD is* (ε', δ')*-differentially private. The factor* ε' *is as follows,*

$$\varepsilon' = \sqrt{8\log\left(\frac{1}{\tilde{\delta}}\right)\left(\frac{\tau^2}{m^2}T\tilde{\varepsilon}^2\right)} + 2T\frac{\tau}{m}\tilde{\varepsilon}\frac{e^{2\frac{\tau}{m}\tilde{\varepsilon}} - 1}{e^{2\frac{\tau}{m}\tilde{\varepsilon}} + 1},$$

and the factor δ' *is as follows,*

$$1 - \left(1 - e^{2\frac{\tau}{m}\tilde{\varepsilon}}\frac{\frac{\tau}{m}\delta}{1 + e^{2\frac{\tau}{m}\tilde{\varepsilon}}}\right)^{\left\lceil\frac{m\varepsilon'}{2\tau\tilde{\varepsilon}}\right\rceil}\left(1 - \frac{\frac{\tau}{m}\delta}{1 + e^{2\frac{\tau}{m}\tilde{\varepsilon}}}\right)^{T - \left\lceil\frac{m\varepsilon'}{2\tau\tilde{\varepsilon}}\right\rceil} + 1 - \left(1 - \frac{\frac{\tau}{m}\delta}{1 + e^{2\frac{\tau}{m}\tilde{\varepsilon}}}\right)^T + \tilde{\delta},$$

where

$$\tilde{\varepsilon} = \frac{2\sqrt{2}L\sigma\frac{1}{\tau}\sqrt{\log\frac{1}{\delta} + \frac{4}{\tau^2}L^2}}{2\sigma^2},$$

and

$$\tilde{\delta} = e^{-\frac{\varepsilon' + \frac{\tau}{m}T\tilde{\varepsilon}}{2}}\left(\frac{1}{1 + e^{\frac{\tau}{m}\tilde{\varepsilon}}}\left(\frac{2\frac{\tau}{m}T\tilde{\varepsilon}}{\frac{\tau}{m}T\tilde{\varepsilon} - \varepsilon'}\right)\right)^T\left(\frac{\frac{\tau}{m}\tilde{\varepsilon}T + \varepsilon'}{\frac{\tau}{m}T\tilde{\varepsilon} - \varepsilon'}\right)^{-\frac{m\varepsilon' + \tau T\tilde{\varepsilon}}{2\tau\tilde{\varepsilon}}}.$$

Additionally, a generalization bound is delivered by combining with Theorem 11.1.

Some existing works have also studied the privacy-preservation and generalization of SGLD. Wang et al. (2015) proved that SGLD has "*privacy for free*" without injecting noise. Specifically, the authors proved that SGLD is (ε, δ)-differentially private if

$$T > \frac{\varepsilon^2 m}{32\tau \log(2/\delta)}.$$

Pensia et al. (2018) analyzed the generalizability of SGLD via information theory. Some works also deliver generalization bounds via algorithmic stability or the PAC-Bayesian framework (Hardt et al. 2016; Raginsky et al. 2017; Mou et al. 2018).

Our Theorem 11.7 also demonstrates that SGLD is PAC-learnable under the following assumption.

Assumption 11.8 There exist constants $K_1 > 0$, K_2, T_0, and m_0, such that, for $T > T_0$ and any $m > m_0$, we have

$$\hat{R}_S(\mathcal{A}(S)) \leq \exp(-K_1 T + K_2).$$

This assumption can be easily justified: the training error is possible to achieve a near-0 training error in machine learning. Then, we have the following remark.

Remark 11.1 Theorem 11.7 implies that

$$\mathbb{P}\left[\hat{R}_S(\mathcal{A}(S)) \leq O\left(\frac{T}{\sqrt{m}}\right) + R(\mathcal{A}(S))\right] \geq 1 - O\left(\frac{T}{\sqrt{m}}\right).$$

It leads to a PAC-learnable guarantee under Assumption 11.8.

Proof of Theorem 11.7 We first calculate the differential privacy of each step. We assume mini-batch $\mathcal{B}$ has been selected and define $\nabla \hat{R}_S^\tau(\theta)$ as following:

$$\nabla \hat{R}_S^\tau(\theta) = \nabla r(\theta) + \sum_{z \in \mathcal{B}} \nabla l(z|\theta).$$

For any two neighboring sample sets S and S' and fixed θ_{i-1}, we have

$$\begin{aligned}\max_{\theta_i} \frac{p(\theta_i^S = \theta_i | \theta_{i-1}^S = \theta_{i-1})}{p(\theta_i^{S'} = \theta_i | \theta_{i-1}^{S'} = \theta_{i-1})} &= \max_{\theta_i} \frac{p(\eta_i(-\frac{1}{\tau}\nabla \hat{R}_S^\tau(\theta_{i-1}) + N(0, \sigma^2\mathbf{I})) = \theta_i - \theta_{i-1})}{p(\eta_i(-\frac{1}{\tau}\nabla \hat{R}_{S'}^\tau(\theta_{i-1}) + N(0, \sigma^2\mathbf{I})) = \theta_i - \theta_{i-1})} \\ &= \max_{\theta_i'} \frac{p(\eta_i(-\frac{1}{\tau}\nabla \hat{R}_S^\tau(\theta_{i-1}) + N(0, \sigma^2\mathbf{I})) = \theta_i')}{p(\eta_i(-\frac{1}{\tau}\nabla \hat{R}_{S'}^\tau(\theta_{i-1}) + N(0, \sigma^2\mathbf{I})) = \theta_i')}.\end{aligned}$$

We define

$$D(\theta') = \log \frac{p(-\frac{1}{\tau}\nabla \hat{R}_S^\tau(\theta_{i-1}) + N(0, \sigma^2\mathbf{I}) = \theta')}{p(-\frac{1}{\tau}\nabla \hat{R}_{S'}^\tau(\theta_{i-1}) + N(0, \sigma^2\mathbf{I}) = \theta')},$$

where $\theta' = \frac{1}{\eta_i}\theta_i'$ obeys $-\frac{1}{\tau}\nabla \hat{R}_S^\tau(\theta_{i-1}) + N(0, \sigma^2\mathbf{I})$.

Let $\theta'' = \theta' + \frac{1}{\tau}\nabla \hat{R}_S^\tau(\theta_{i-1})$ and we rewrite $D(\theta')$ as:

$$
\begin{aligned}
D(\theta') &= \log \frac{e^{-\frac{\|\theta'+\frac{1}{\tau}\nabla \hat{R}_S^\tau(\theta_{i-1}))\|^2}{2\sigma^2}}}{e^{-\frac{\|\theta'+\frac{1}{\tau}\nabla \hat{R}_{S'}^\tau(\theta_{i-1})\|^2}{2\sigma^2}}} \\
&= -\frac{\|\theta'+\frac{1}{\tau}\nabla \hat{R}_S^\tau(\theta_{i-1}))\|^2}{2\sigma^2} + \frac{\|\theta'+\frac{1}{\tau}\nabla \hat{R}_{S'}^\tau(\theta_{i-1})\|^2}{2\sigma^2} \\
&= -\frac{\|\theta''\|^2}{2\sigma^2} + \frac{\|\theta''+\frac{1}{\tau}\nabla \hat{R}_{S'}^\tau(\theta_{i-1}) - \frac{1}{\tau}\nabla \hat{R}_S^\tau(\theta_{i-1}))\|^2}{2\sigma^2} \\
&= \frac{2\theta''^T \frac{1}{\tau}(\nabla \hat{R}_{S'}^\tau(\theta_{i-1}) - \nabla \hat{R}_S^\tau(\theta_{i-1})) + \frac{1}{\tau^2}(\|\nabla \hat{R}_{S'}^\tau(\theta_{i-1}) - \nabla \hat{R}_S^\tau(\theta_{i-1})\|^2)}{2\sigma^2}.
\end{aligned}
$$

We define $\nabla \hat{R}_{S'}^\tau(\theta_{i-1}) - \nabla \hat{R}_S^\tau(\theta_{i-1})$ as $\mathbf{v}$. By the definition of L, we have

$$\|\mathbf{v}\| < 2L.$$

Therefore, since $\theta''^T \mathbf{v} \sim N(0, \|\mathbf{v}\|^2\sigma^2)$, by the Chernoff Bound technique,

$$
\begin{aligned}
\mathbb{P}\left(\theta''^T \mathbf{v} \geq 2\sqrt{2}L\sigma\sqrt{\log\frac{1}{\delta}}\right) &\leq \mathbb{P}\left(\theta''^T \mathbf{v} \geq \sqrt{2}\|\mathbf{v}\|\sigma\sqrt{\log\frac{1}{\delta}}\right) \\
&\leq \min_t e^{-\sqrt{2}t\|\mathbf{v}\|\sigma\sqrt{\log\frac{1}{\delta}}}\mathbb{E}(e^{t\theta''^T\mathbf{v}}) \\
&= \delta.
\end{aligned}
$$

Therefore, with probability at least $1-\delta$ with respect to θ', we have

$$D(\theta') \leq \frac{2\sqrt{2}L\sigma\frac{1}{\tau}\sqrt{\log\frac{1}{\delta}} + \frac{4}{\tau^2}L^2}{2\sigma^2}.$$

We define $\varepsilon = \frac{1}{2}(2\sqrt{2}L\sigma\frac{1}{\tau}\sqrt{\log\frac{1}{\delta}} + \frac{4}{\tau^2}L^2)/\sigma^2$. By applying Lemma 4.4 in Beimel et al. (2010), we arrive at the conclusion iteration $-\frac{1}{\tau}\nabla \hat{R}_S^\tau(\theta_{i-1}) + N(0, \sigma^2\mathbf{I})$ is $(2\tau m^{-1}\varepsilon, \tau m^{-1}\delta)$-differentially private. By applying Theorem 11.3 and

$$\varepsilon' = \sqrt{8\log\left(\frac{1}{\tilde{\delta}}\right)\left(\frac{\tau^2}{m^2}T\varepsilon^2\right)} + 2T\frac{\tau}{m}\varepsilon\frac{e^{2\frac{\tau}{m}\varepsilon}-1}{e^{2\frac{\tau}{m}\varepsilon}+1},$$

we can prove the differential privacy.

By sampling $\mathcal{B}$ randomly and applying Theorem 11.1, we can prove the generalization bound.

The proof is completed.

11.2.3.2 Application in Agnostic Federated Learning

In today's data-driven world, enormous amounts of personal information, ranging from financial transactions to medical records, are routinely collected and used. While such data holds immense value for deriving insights on a population scale, it simultaneously poses significant concerns regarding individual privacy leaking. Federated learning (Shokri and Shmatikov 2015; Konečný et al. 2016; McMahan et al. 2017; Yang et al. 2019) aims to protect the privacy by adopting a decentralized approach. Specifically, instead of accessing raw data stored on personal devices, federated learning deploys learning models directly on these devices. The central server then aggregates gradients computed locally on each device and distributes weight updates without ever accessing the raw data. This innovative framework offers a promising solution to the privacy-preserving dilemma. Furthermore, algorithms developed by Geyer et al. (2017), Mohri et al. (2019) introduce additional layers of privacy protection by shielding client identities from potential differential attacks.

Algorithm 2 Differentially Private Federated Learning

Require: Clients $\{c_1, \ldots c_m\}$, Gaussian noise variance σ, size of mini-batch τ, iteration steps T, upper bound L of the step size.

Initialize θ_0 randomly.
For $t = 1$ to T do:
    Randomly sample a mini-batch of clients of size τ;
    Randomly sample b_t from $N(0, L^2\sigma^2 I)$;
    Central curator distributes θ_{t-1} to the clients in the mini-batch $\mathcal{B}$;
    Update $\theta_{t+1} \leftarrow \theta_t + \left(\frac{1}{B} \sum_{i \in \mathcal{B}} \frac{\text{ClientUpdate}(c_i, \theta_t)}{\max\left(1, \frac{\|h_i\|_2}{L}\right)} + b_t \right)$.

The following theorem provides an estimation of the differential privacy and a generalization bound for agnostic federated learning.

Theorem 11.9 (Differential Privacy and Generalization Bounds of Differentially Private Federated Learning) *Agnostic federated learning is* (ε', δ')*-differentially private. The factor* ε' *is as follows,*

$$\varepsilon' = \sqrt{8 \log \left(\frac{1}{\tilde{\delta}} \right) \left(\frac{\tau^2}{m^2} T \varepsilon^2 \right)} + 2T \frac{\tau}{m} \varepsilon \frac{e^{2 \frac{\tau}{m} \varepsilon} - 1}{e^{2 \frac{\tau}{m} \varepsilon} + 1}, \tag{11.42}$$

and the factor δ' *is defined as follows,*

$$1 - \left(1 - e^{2\frac{\tau}{m}\tilde{\varepsilon}} \frac{\frac{\tau}{m}\delta}{1 + e^{2\frac{\tau}{m}\tilde{\varepsilon}}}\right)^{\left\lceil \frac{m\varepsilon'}{2\tau\tilde{\varepsilon}} \right\rceil} \left(1 - \frac{\frac{\tau}{m}\delta}{1 + e^{2\frac{\tau}{m}\tilde{\varepsilon}}}\right)^{T - \left\lceil \frac{m\varepsilon'}{2\tau\tilde{\varepsilon}} \right\rceil} + 1 - \left(1 - \frac{\frac{\tau}{m}\delta}{1 + e^{2\frac{\tau}{m}\tilde{\varepsilon}}}\right)^T + \tilde{\delta},$$

where

$$\tilde{\varepsilon} = \frac{4\sigma \frac{1}{\tau} \sqrt{\log \frac{1}{\delta} + \frac{1}{\tau^2}}}{2\sigma^2},$$

and

$$\tilde{\delta} = e^{-\frac{\varepsilon' + \frac{\tau}{m}T\tilde{\varepsilon}}{2}} \left(\frac{1}{1 + e^{\frac{\tau}{m}\tilde{\varepsilon}}} \left(\frac{2\frac{\tau}{m}T\tilde{\varepsilon}}{\frac{\tau}{m}T\tilde{\varepsilon} - \varepsilon'} \right) \right)^T \left(\frac{\frac{\tau}{m}\tilde{\varepsilon}T + \varepsilon'}{\frac{\tau}{m}T\tilde{\varepsilon} - \varepsilon'} \right)^{-\frac{m\varepsilon' + \tau T\tilde{\varepsilon}}{2\tau\tilde{\varepsilon}}}.$$

Additionally, a generalization bound is delivered by combining with Theorem 11.1.

The following remark gives a PAC-learnable guarantee for agnostic federated learning.

Remark 11.2 Theorem 11.9 implies that

$$\mathbb{P}\left[\hat{R}_S(\mathcal{A}(S)) \leq O\left(\frac{T}{\sqrt{m}}\right) + R(\mathcal{A}(S)) \right] \geq 1 - O\left(\frac{T}{\sqrt{m}}\right).$$

It leads to a PAC-learnable guarantee under Assumption 11.8.

To prove Theorem 11.9, we only need to prove the differential privacy part of Theorem 11.9, while the rest of the proof is similar to that of Theorem 11.7.

Proof of Theorem 11.9 The proof bears resemblance to the proof of Theorem 11.7. One only has to notice that each update is still a Gauss mechanism, while

$$\left\| \frac{h_i^t}{\max(1, \frac{\|h_i^t\|_2}{L})} \right\| \leq L.$$

Then, in this situation, $D(\theta')$ is as follows:

$$D(\theta') = \log \frac{p\left(\frac{1}{\tau} \left(\sum_{c_k \in \mathcal{B}} \frac{h_i^t}{\max(1, \frac{\|h_i^t\|_2}{L})} \right) + N(0, L^2\sigma^2 \mathbf{I}) = \theta' \right)}{p\left(\frac{1}{\tau} \left(\sum_{c_k \in \mathcal{B}} \frac{h_i^t}{\max(1, \frac{\|h_i^t\|_2}{L})} \right) + N(0, L^2\sigma^2 \mathbf{I}) = \theta' \right)}.$$

All other reasoning is the same as the previous proof.

By Theorem 11.3 and

$$\varepsilon' = \sqrt{8 \log\left(\frac{1}{\tilde{\delta}}\right) \left(\frac{\tau^2}{m^2} T\varepsilon^2\right)} + 2T\frac{\tau}{m}\varepsilon \frac{e^{2\frac{\tau}{m}\varepsilon} - 1}{e^{2\frac{\tau}{m}\varepsilon} + 1},$$

we can calculate the differential privacy of federated learning.

The proof is completed.

References

Abadi, Martin, Andy Chu, Ian Goodfellow, H Brendan McMahan, Ilya Mironov, Kunal Talwar, and Li Zhang. 2016. Deep learning with differential privacy. In *ACM SIGSAC Conference on Computer and Communications Security*, 308–318.

Ban, Yutong, Xavier Alameda-Pineda, Laurent Girin, and Radu Horaud. Variational bayesian inference for audio-visual tracking of multiple speakers. *IEEE Transactions on Pattern Analysis and Machine Intelligence*, 2019.

Beimel, Amos, Shiva Prasad Kasiviswanathan, and Kobbi Nissim. 2010. Bounds on the sample complexity for private learning and private data release. In *Theory of Cryptography Conference*, 437–454. Springer.

Boucheron, Stéphane, Gábor Lugosi, and Pascal Massart. 2013. Concentration inequalities: a nonasymptotic theory of independence. Oxford University Press.

Bun, Mark, and Thomas Steinke. 2016. Concentrated differential privacy: simplifications, extensions, and lower bounds. In *Theory of Cryptography Conference*, 635–658.

Chaudhuri, Kamalika, Jacob Imola, and Ashwin Machanavajjhala. 2019. Capacity bounded differential privacy. *arXiv preprint* arXiv:1907.02159.

Cuff, Paul, and Lanqing Yu. 2016. Differential privacy as a mutual information constraint. In *ACM SIGSAC Conference on Computer and Communications Security*, 43–54.

Duane, Simon, Anthony D Kennedy, Brian J Pendleton, and Duncan Roweth. 1987. Hybrid Monte Carlo. *Physics Letters B* 195 (2): 216–222.

Dwork, Cynthia, and Aaron Roth. 2014. The algorithmic foundations of differential privacy. *Foundations and Trends® in Theoretical Computer Science* 9 (3–4): 211–407.

Dwork, Cynthia, and Deirdre K Mulligan. 2013. It's not privacy, and it's not fair. *Stanford Law Review Online* 66: 35.

Dwork, Cynthia, and Guy N Rothblum. 2016. Concentrated differential privacy. *arXiv preprint* arXiv:1603.01887.

Dwork, Cynthia, Guy N Rothblum, and Salil Vadhan. 2010. Boosting and differential privacy. In *IEEE Annual Symposium on Foundations of Computer Science*, 51–60.

Dwork, Cynthia, Vitaly Feldman, Moritz Hardt, Toniann Pitassi, Omer Reingold, and Aaron Leon Roth. 2015. Preserving statistical validity in adaptive data analysis. In *Annual ACM Symposium on Theory of Computing*, 117–126.

Geumlek, Joseph, Shuang Song, and Kamalika Chaudhuri. 2017. Renyi differential privacy mechanisms for posterior sampling. In *Advances in Neural Information Processing Systems*, 5289–5298.

Geyer, Robin C, Tassilo Klein, and Moin Nabi. 2017. Differentially private federated learning: a client level perspective. In *Advances in Neural Information Processing Systems*.

Hardt, Moritz, Ben Recht, and Yoram Singer. 2016. Train faster, generalize better: Stability of stochastic gradient descent. In *International Conference on Machine learning*, 1225–1234.

Hastings, W Keith. 1970. Monte Carlo sampling methods using Markov chains and their applications.

He, Fengxiang, Bohan Wang, and Dacheng Tao. 2020. Tighter generalization bounds for iterative differentially private learning algorithms. *arXiv preprint* arXiv:2007.09371.

Kairouz, Peter, Sewoong Oh, and Pramod Viswanath. 2015. The composition theorem for differential privacy. In *International Conference on Machine Learning*, 1376–1385.

Kairouz, Peter, Oh. Sewoong, and Pramod Viswanath. 2017. The composition theorem for differential privacy. *IEEE Transactions on Information Theory* 63 (6): 4037–4049.

Kobyzev, Ivan, Simon Prince, and Marcus Brubaker. 2020. Normalizing flows: an introduction and review of current methods. *IEEE Transactions on Pattern Analysis and Machine Intelligence*.

Konečný, Jakub, H Brendan McMahan, Felix X Yu, Peter Richtárik, Ananda Theertha Suresh, and Dave Bacon. 2016. Federated learning: Strategies for improving communication efficiency. In *Advances in Neural Information Processing Systems Workshop on Private Multi-Party Machine Learning*.

Larochelle, Hugo, and Stanislas Lauly. 2012. A neural autoregressive topic model. In *Advances in Neural Information Processing Systems*, 2708–2716.

Liao, Jiachun, Lalitha Sankar, Vincent YF Tan, and Flavio du Pin Calmon. 2017. Hypothesis testing under mutual information privacy constraints in the high privacy regime. *IEEE Transactions on Information Forensics and Security* 13 (4): 1058–1071.

Louizos, Christos, and Max Welling. 2017. Multiplicative normalizing flows for variational Bayesian neural networks. In *International Conference on Machine Learning*, 2218–2227.

Ma, Yi-An, Tianqi Chen, and Emily Fox. 2015. A complete recipe for stochastic gradient mcmc. In *Advances in Neural Information Processing Systems*, 2917–2925.

McMahan, H Brendan, Eider Moore, Daniel Ramage, Seth Hampson, and Blaise Aguera y Arcas. 2017. Communication-efficient learning of deep networks from decentralized data. In *International Conference on Artificial Intelligence and Statistics*.

McSherry, Frank, and Kunal Talwar. 2007. Mechanism design via differential privacy. In *48th Annual IEEE Symposium on Foundations of Computer Science (FOCS'07)*, 94–103. IEEE.

Mironov, Ilya. 2017. Rényi differential privacy. In *IEEE Computer Security Foundations Symposium*, 263–275.

Mohri, Mehryar, Afshin Rostamizadeh, and Ameet Talwalkar. 2018. *Foundations of Machine Learning*. MIT Press.

Mohri, Mehryar, Gary Sivek, and Ananda Theertha Suresh. 2019. Agnostic federated learning. In *International Conference on Machine Learning*, 4615–4625.

Mou, Wenlong, Liwei Wang, Xiyu Zhai, and Kai Zheng. 2018. Generalization bounds of sgld for non-convex learning: two theoretical viewpoints. In *Annual Conference on Learning Theory*, 605–638.

Nissim, Kobbi, and Uri Stemmer. 2015. On the generalization properties of differential privacy. *CoRR, abs/1504.05800.*

Oneto, Luca, Sandro Ridella, and Davide Anguita. 2017. Differential privacy and generalization: Sharper bounds with applications. *Pattern Recognition Letters* 89: 31–38.

Pensia, Ankit, Varun Jog, and Po-Ling Loh. 2018. Generalization error bounds for noisy, iterative algorithms. In *2018 IEEE International Symposium on Information Theory (ISIT)*, 546–550.

Pittaluga, Francesco, and Sanjeev Jagannatha Koppal. 2016. Pre-capture privacy for small vision sensors. *IEEE Transactions on Pattern Analysis and Machine Intelligence* 39 (11): 2215–2226.

Raginsky, Maxim, Alexander Rakhlin, and Matus Telgarsky. 2017. Non-convex learning via stochastic gradient Langevin dynamics: a nonasymptotic analysis. In *Conference on Learning Theory*, 1674–1703.

Robbins, Herbert, and Sutton Monro. 1951. A stochastic approximation method. *The Annals of Mathematical Statistics*, 400–407.

Roth, Wolfgang, and Franz Pernkopf. 2018. Bayesian neural networks with weight sharing using Dirichlet processes. *IEEE Transactions on Pattern Analysis and Machine Intelligence* 42 (1): 246–252.

Shokri, Reza, and Vitaly Shmatikov. 2015. Privacy-preserving deep learning. In *ACM SIGSAC Conference on Computer and Communications Security*, 1310–1321.

Vapnik, Vladimir. 2013. *The Nature of Statistical Learning Theory*. Springer Science & Business Media.

Wang, Yu-Xiang, Stephen Fienberg, and Alex Smola. 2015. Privacy for free: posterior sampling and stochastic gradient monte carlo. In *International Conference on Machine Learning*, 2493–2502.

Wang, Weina, Lei Ying, and Junshan Zhang. 2016. On the relation between identifiability, differential privacy, and mutual-information privacy. *IEEE Transactions on Information Theory* 62 (9): 5018–5029.

Wang, Hongwei, Jialin Wang, Jia Wang, Miao Zhao, Weinan Zhang, Fuzheng Zhang, Wenjie Li, Xing Xie, and Minyi Guo. 2019. Learning graph representation with generative adversarial nets. *IEEE Transactions on Knowledge and Data Engineering* 33 (8): 3090–3103.

Welling, Max, and Yee W Teh. 2011. Bayesian learning via stochastic gradient Langevin dynamics. In *International Conference on Machine Learning*, 681–688.

Yang, Qiang, Yang Liu, Tianjian Chen, and Yongxin Tong. 2019. Federated machine learning: concept and applications. *ACM Transactions on Intelligent Systems and Technology* 10 (2): 12:1–12:19. ISSN 2157-6904.

Ye, Qiaoling, Arash A Amini, and Qing Zhou. 2020. Optimizing regularized cholesky score for order-based learning of bayesian networks. *IEEE Transactions on Pattern Analysis and Machine Intelligence* 43 (10): 3555–3572.

Zhang, Hao, Bo Chen, Yulai Cong, Dandan Guo, Hongwei Liu, and Mingyuan Zhou. 2020. Deep autoencoding topic model with scalable hybrid Bayesian inference. *IEEE Transactions on Pattern Analysis and Machine Intelligence*.

Zhang, Cheng, Judith Bütepage, Hedvig Kjellström, and Stephan Mandt. 2018. Advances in variational inference. *IEEE Transactions on Pattern Analysis and Machine Intelligence* 41 (8): 2008–2026.

Chapter 12
Algorithmic Fairness

Deep learning technology has been widely deployed in increasingly critical decision-making tasks, such as mortgage approval, credit card assessment, college admissions, employee selection, and recidivism prediction. However, these areas are observed to be subject to long-standing systematic discrimination against certain people on the basis of diverse background traits, including race, gender, nationality, age, and religion. Unfortunately, the introduction of intelligent algorithms into these areas has failed to relieve the discrimination conundrum because people with minority backgrounds are institutionally underrepresented in the historical data that fuel the algorithmic systems. Thus, the unfairness residing in biased historical data is inherited and sometimes intensified by a machine learning (ML) model that is trained on such data. The consequent fairness concerns are particularly severe due to the black-box nature of ML algorithms. Therefore, mitigation of the fairness issues arising in ML applications is both urgent and important.

12.1 Definitions of Fairness

In general, there are two types of fairness related to ML: *individual fairness* and *group fairness*. The concept of individual fairness was first proposed by Dwork et al. (2012), based on the core principle that similar individuals should be treated similarly. Since then, many other individual fairness measures have been developed, including but not limited to *average individual fairness* (Kearns et al. 2019), *counterfactual fairness* (Kusner et al. 2017), *meritocratic fairness* (Kearns et al. 2017), and others (Yurochkin and Sun 2021). On the other hand, group fairness measures the level of bias across different groups of individuals. The most commonly used measures include *demographic parity* (Calders et al. 2009), *equalized odds* (Hardt et al. 2016),

F. He and D. Tao, *Foundations of Deep Learning*, Machine Learning: Foundations, Methodologies, and Applications, https://doi.org/10.1007/978-981-16-8233-9_12

and *equalized opportunity* (Hardt et al. 2016), among others (Zafar et al. 2017; Choi et al. 2020).

We consider a binary classification task. Each sample has the form $z = (x, a, y)$, where $x \in \mathcal{X}$ is the input feature, $a \in \mathcal{A} = \{0, 1\}$ represents one or more sensitive attributes (e.g., gender, race, and age), and $y \in \{0, 1\}$ is the prediction target. Let Z, X, A, and Y denote the random variables corresponding to z, x, a, and y, respectively. Then, the goal of fair ML is to learn a binary classifier $h : \mathcal{X} \times \mathcal{A} \to \{0, 1\}$ while ensuring a specific notion of fairness concerning the sensitive attributes A. For simplicity, we define $\hat{Y} := h(X, A)$ to denote the prediction of the classifier h for variable $Z = (X, A, Y)$.

Group fairness.

A major family of fairness concepts is *group fairness*, which aims to characterize discrimination across various groups of individuals. The most commonly used definitions of group fairness are *demographic parity* (Calders et al. 2009), *equalized odds* (Hardt et al. 2016), and *equal opportunity* (Hardt et al. 2016).

Given a data distribution $\mathcal{D}$, a classifier h satisfies

- demographic parity if the prediction $\hat{Y}$ is independent of the sensitive attributes A,
- equalized odds if the prediction $\hat{Y}$ and the sensitive attribute A are independent conditional on the target Y, and
- equal opportunity if $\mathbb{P}(\hat{Y} = 1|A = 0, Y = 1) = \mathbb{P}(\hat{Y} = 1|A = 1, Y = 1)$.

The following metrics are further designed to assess the degree to which a classifier satisfies the fairness constraints presented in Definition 12.1.

Given a data distribution $\mathcal{D}$, for a binary classifier h,

- the demographic parity Δ_{DP} is defined as

$$\Delta_{\text{DP}}(\hat{Y}, \mathcal{D}) = |\mathbb{P}(\hat{Y} = 1|A = 0) - \mathbb{P}(\hat{Y} = 1|A = 1)|, \tag{12.1}$$

- the equalized odds gap Δ_{EO} is defined as

$$\Delta_{\text{EO}}(\hat{Y}, \mathcal{D}) = \max_{y \in \{0,1\}} |\mathbb{P}(\hat{Y} = 1|A = 0, Y = y) - \mathbb{P}(\hat{Y} = 1|A = 1, Y = y)|, \tag{12.2}$$

  and
- the equal opportunity gap Δ_{EOP} is defined as

$$\Delta_{\text{EOP}}(\hat{Y}, \mathcal{D}) = |\mathbb{P}(\hat{Y} = 1|A = 0, Y = 1) - \mathbb{P}(\hat{Y} = 1|A = 1, Y = 1)|. \tag{12.3}$$

Combined with Definition 12.1, it is clear that a small value of any fairness measure in Definition 12.1 would indicate a strongly nondiscriminatory nature of the given classifier, and vice versa. When the metric Δ_{DP}, Δ_{EO}, or Δ_{EOP} is equal to zero, the classifier h perfectly satisfies demographic parity, equalized odds, or equal opportunity, respectively.

12.2 Fairness-Aware Algorithms

Based on the mathematical definition of fairness, many algorithms are believed to be able to mitigate existing biases. We argue that ML algorithms achieve bias mitigation via three dominant approaches and thus can be divided into three corresponding classes: *preprocessing methods*, *in-processing methods*, and *postprocessing methods*.

12.2.1 Preprocessing Methods

Preprocessing methods aim to eliminate the inherent bias in data before those data are fed into learning algorithms (Calders et al. 2009; Feldman et al. 2014; Calmon et al. 2017; Louizos et al. 2015; Choi et al. 2020; Kamiran and Calders 2012; Zemel et al. 2013; Zhao et al. 2020). For example, Kamiran and Calders (2012) reviewed several data preprocessing techniques, including (1) removing attributes that are most relevant to the target sensitive attribute, (2) changing data labels, and (3) reweighting or resampling the data. Zemel et al. (2013) proposed learning fair representations by learning a linear transform to encode all information in the input features except information that could lead to biased decision-making. This method can enhance both individual fairness and group fairness. Zhao et al. (2020) focused on group fairness and further proposed a novel fair representation learning technique that can mitigate unfairness in terms of both accuracy parity and equalized odds simultaneously while achieving a better utility-fairness trade-off.

12.2.2 In-processing Methods

In-processing methods enforce fairness by imposing constraints or regularizers during the training procedure (Mozannar et al. 2020; Baharlouei et al. 2020; Yurochkin and Sun 2021; Zafar et al. 2017; Martinez et al. 2020; Agarwal et al. 2018; Kamishima et al. 2012; Menon and Williamson 2018; Kearns et al. 2018). We present an example to illustrate how these methods work (Mozannar et al. 2020).

Let $\Delta_{\mathcal{H}}$ be the set of all possible distributions over the hypothesis space $\mathcal{H}$, where each hypothesis depends only on X. Let $\gamma_{y,a}(\tilde{Y})$ be $\mathbb{P}(\tilde{Y} = 1|Y = y, A = a)$. Then, the goal is to learn a predictor that approximates the optimal nondiscriminatory distribution:

$$Y^* = \arg\min_{Q \in \Delta_{\mathcal{H}}} \mathbb{P}(Q(X) \neq Y)$$
$$s.t.\ \gamma_{y,a}(Q) = \gamma_{y,0}(Q),\ \forall y \in \{0, 1\}, a \in A.$$

Several works have explored methods of finding a near-optimal nondiscriminatory solution. Mozannar et al. (2020) proposed a two-step method for learning a near-optimal predictor. They divided the whole dataset into two equal parts, $\mathcal{S}_1$ and $\mathcal{S}_2$. First, they trained a predictor $\hat{Y}$, defined as follows:

$$\hat{Y} = \arg\min_{Q \in \Delta_{\mathcal{H}}} \mathbb{P}_{\mathcal{S}_1}(Q(X) \neq Y)$$
$$s.t.\ |\gamma_{y,a}^{S_1}(Q) - \gamma_{y,0}^{S_1}(Q)| \leq \alpha_n,\ \forall y \in \{0, 1\}, a \in A.$$

To solve this optimization problem, they used the Lagrange method to minimize the following loss function with Lagrange multipliers $\lambda \in \mathbb{R}_+^K$:

$$L(Q, \lambda) = \mathbb{P}_{\mathcal{S}_1}(Q(X) \neq Y) + \sum \lambda_k(|\gamma_{y,a}^{S_1}(Q) - \gamma_{y,0}^{S_1}(Q)| - \alpha_n).$$

Thus, they obtained a predictor $\hat{Y}$ that depended only on X. To extend this to a predictor $\tilde{Y}$ that would also depend on A, based on the given predictor $\hat{Y}$, they found $\tilde{Y}$ by punishing the gap Δ_{EOP}. Let $p_{y,a}$ be $\mathbb{P}(\tilde{Y} = 1|\hat{Y} = y, A = a)$. Then, they found the transition probability p by solving the following problem:

$$p^* = \arg\min_p \left(\mathbb{P}_{\mathcal{S}_2}(\hat{Y} = y, A = a, Y = 0) - \mathbb{P}_{\mathcal{S}_2}(\hat{Y} = y, A = a, Y = 1)\right) p_{y,a}$$
$$s.t.\ |p_{0,a}\mathbb{P}_{\mathcal{S}_2}(\hat{Y} = 0|A = a, Y = y) + p_{1,a}\mathbb{P}_{\mathcal{S}_2}(\hat{Y} = 1|A = a, Y = y)$$
$$- p_{0,0}\mathbb{P}_{\mathcal{S}_2}(\hat{Y} = 0|A = 0, Y = y) - p_{1,0}\mathbb{P}_{\mathcal{S}_2}(\hat{Y} = 1|A = 0, Y = y)| \leq \tilde{\alpha}_n, \forall y, a.$$

The transition probability p learned in this way is the optimal solution to make $\tilde{Y}$ closer to being nondiscrinminatory. Based on these two steps, the authors obtained a both near-optimal and near-nondiscrinminatory predictor $\tilde{Y}$.

Moreover, Zafar et al. (2017) studied binary classification and proposed a fair learning algorithm by restricting the covariance among the target sensitive attributes and restricting the distance between the data features and the decision boundary. Baharlouei et al. (2020) employed Rényi correlation as a regularizer and proposed a min-max optimization algorithm that can reduce arbitrary dependence between the predictions of a model and the sensitive attributes of the input data. Yurochkin and Sun (2021) further defined the concept of *distributional individual fairness* and then designed an approximation algorithm to enforce such a fairness restriction by means of regularization.

12.2.3 Postprocessing Methods

Postprocessing methods aim to refine the outputs of learning systems to remove potential discrimination (Woodworth et al. 2017; Kim et al. 2019; Hardt et al. 2016; Fish et al. 2016; Lahoti et al. 2020; Dwork et al. 2018). Fish et al. (2016) proposed the *shifted decision boundary* algorithm based on predicted confidences; this algorithm finds the shift for the data that best balances fairness and decision accuracy and shifts the data accordingly. Woodworth et al. (2017) argued that the *post hoc correction* method proposed by Hardt et al. (2016) to achieve fair learning through postprocessing is not optimal and may fail to reduce bias in some situations. To address this shortcoming, the authors further suggested a two-step learning framework that employs a second-moment fairness constraint. Based on the concept of *multiaccuracy*, which is defined as a reweighting of the prediction accuracy of a model, Kim et al. (2019) proposed the *Multiaccuracy Boost* algorithm to mitigate the decision bias of any black-box learning system.

References

Agarwal, Alekh, Alina Beygelzimer, Miroslav Dudík, John Langford, and Hanna Wallach. 2018. A reductions approach to fair classification. *arXiv preprint* arXiv:1803.02453.

Baharlouei, Sina, Maher Nouiehed, Ahmad Beirami, and Meisam Razaviyayn. 2020. Rényi fair inference. In *International Conference on Learning Representations.*

Calders, Toon, Faisal Kamiran, and Mykola Pechenizkiy. 2009. Building classifiers with independency constraints. In *IEEE International Conference on Data Mining Workshops*, 13–18.

Calmon, Flavio, Dennis Wei, Bhanukiran Vinzamuri, Karthikeyan Natesan Ramamurthy, and Kush R Varshney. 2017. Optimized pre-processing for discrimination prevention. *Advances in Neural Information Processing Systems* 30: 3992–4001.

Choi, Kristy, Aditya Grover, Trisha Singh, Rui Shu, and Stefano Ermon. 2020. Fair generative modeling via weak supervision. In *International Conference on Machine Learning*, 1887–1898.

Dwork, Cynthia, Moritz Hardt, Toniann Pitassi, Omer Reingold, and Richard Zemel. 2012. Fairness through awareness. In *Innovations in Theoretical Computer Science Conference*, 214–226.

Dwork, Cynthia, Nicole Immorlica, Adam Tauman Kalai, and Max Leiserson. 2018. Decoupled classifiers for group-fair and efficient machine learning. In *Conference on Fairness, Accountability and Transparency*, 119–133.

Feldman, Michael, Sorelle Friedler, John Moeller, Carlos Scheidegger, and Suresh Venkatasubramanian. 2014. Certifying and removing disparate impact.

Fish, Benjamin, Jeremy Kun, and Ádám D Lelkes. 2016. A confidence-based approach for balancing fairness and accuracy. In *2016 SIAM International Conference on Data Mining*, 144–152. SIAM.

Hardt, Moritz, Eric Price, and Nati Srebro. 2016. Equality of opportunity in supervised learning. In *Advances in Neural Information Processing Systems*, 3315–3323.

Kamiran, Faisal, and Toon Calders. 2012. Data preprocessing techniques for classification without discrimination. *Knowledge and Information Systems* 33 (1): 1–33.

Kamishima, Toshihiro, Shotaro Akaho, Hideki Asoh, and Jun Sakuma. 2012. Fairness-aware classifier with prejudice remover regularizer. In *Joint European Conference on Machine Learning and Knowledge Discovery in Databases*, 35–50.

Kearns, Michael, Aaron Roth, and Saeed Sharifi-Malvajerdi. 2019. Average individual fairness: Algorithms, generalization and experiments. In *Advances in Neural Information Processing Systems*, vol. 32.

Kearns, Michael, Aaron Roth, and Zhiwei Steven Wu. 2017. Meritocratic fairness for cross-population selection. In *International Conference on Machine Learning*, 1828–1836.

Kearns, Michael, Seth Neel, Aaron Roth, and Zhiwei Steven Wu. 2018. Preventing fairness gerrymandering: Auditing and learning for subgroup fairness. In *International Conference on Machine learning*, 2564–2572.

Kim, Michael P, Amirata Ghorbani, and James Zou. 2019. Multiaccuracy: Black-box post-processing for fairness in classification. In *2019 AAAI/ACM Conference on AI, Ethics, and Society*, 247–254.

Kusner, Matt J, Joshua Loftus, Chris Russell, and Ricardo Silva. 2017. Counterfactual fairness. In *Advances in Neural Information Processing Systems*, vol. 30.

Lahoti, Preethi, Alex Beutel, Jilin Chen, Kang Lee, Flavien Prost, Nithum Thain, Xuezhi Wang, and Ed Chi. 2020. Fairness without demographics through adversarially reweighted learning. In *Advances in Neural Information Processing Systems*, vol. 33, 728–740.

Louizos, Christos, Kevin Swersky, Yujia Li, Max Welling, and Richard Zemel. 2015. The variational fair autoencoder. *arXiv preprint* arXiv:1511.00830.

Martinez, Natalia, Martin Bertran, and Guillermo Sapiro. 2020. Minimax Pareto fairness: A multi objective perspective. In *International Conference on Machine Learning*, 6755–6764.

Menon, Aditya Krishna, and Robert C Williamson. 2018. The cost of fairness in binary classification. In *Conference on Fairness, Accountability and Transparency*, 107–118.

Mozannar, Hussein, Mesrob Ohannessian, and Nathan Srebro. 2020. Fair learning with private demographic data. In *International Conference on Machine Learning*, 7066–7075.

Nesterov, Yurii E. 1983. A method for solving the convex programming problem with convergence rate o (1/k^ 2). In *Dokl. Akad. Nauk Sssr*, vol. 269, 543–547.

Woodworth, Blake, Suriya Gunasekar, Mesrob I Ohannessian, and Nathan Srebro. 2017. Learning non-discriminatory predictors. *arXiv preprint* arXiv:1702.06081.

Yurochkin, Mikhail, and Yuekai Sun. 2021. SenSeI: Sensitive set invariance for enforcing individual fairness. In *International Conference on Learning Representations*.

Zafar, Muhammad Bilal, Isabel Valera, Manuel Gomez Rogriguez, and Krishna P Gummadi. 2017. Fairness constraints: Mechanisms for fair classification. In *Artificial Intelligence and Statistics*, 962–970.

Zemel, Rich, Yu Wu, Kevin Swersky, Toni Pitassi, and Cynthia Dwork. 2013. Learning fair representations. In *International Conference on Machine Learning*, 325–333.

Zhao, Han, Amanda Coston, Tameem Adel, and Geoffrey J. Gordon. 2020. Conditional learning of fair representations. In *International Conference on Learning Representations*.

Chapter 13
Adversarial Robustness

Szegedy et al. (2014) discovered that neural networks are vulnerable to adversarial examples, which are true examples perturbed with small artificial noise to fake the classifiers. Since that finding was reported, many methods have been proposed to study attacks on neural network using adversarial examples and defences against such adversarial attacks. This chapter discusses the theory of adversarial robustness and its relation to generalizability and privacy preservation.

13.1 Adversarial Attacks and Defences

Generalizability in the presence of adversarial examples. Schmidt et al. (2018) first studied the generalizability in the problems with adversarial examples. They propose new notion of *adversarial robustness generalization* to measure the generalizability of neural networks trained on data including adversarial examples. In contrast, the generalizability of neural networks trained without adversarial examples can be evaluated with the *standard generalization* measure. They proved that to ensure the same generalizability, adversarial robustness generalization requires a larger sample size when the data are drawn from certain specific distributions and the adversarial examples are generated under the l_∞ norm.

Cullina et al. (2018) extended the Probably Approximately Correct (PAC) learning framework to that with adversarial examples. The authors proposed *corrupted hypothesis classes*, which are defined as the hypothesis classes for a classification problem with adversaries. The Vapnik-Chervonenkis (VC) dimension of a corrupted hypothesis class is then defined as the *adversarial VC dimension*, while the *standard VC dimension* is defined for clean datasets. They showed that the adversarial VC dimension can be either larger or smaller than the standard VC dimension. Montasser et al. (2019) further defined *agnostic robust PAC learnability* and *realizable*

F. He and D. Tao, *Foundations of Deep Learning*, Machine Learning: Foundations, Methodologies, and Applications, https://doi.org/10.1007/978-981-16-8233-9_13

robust PAC learnability. When a learning algorithm meets the requirements for both agnostic robust PAC learnability and realizable robust PAC learnability, the learning rule is defined as improper. They then showed that under an improper learning rule, any hypothesis space $\mathcal{H}$ is robustly PAC learnable if its VC dimension is finite.

Generalizability of adversarial training. Among the proposed defence approaches, adversarial training (Dai et al. 2018; Li et al. 2018a; Baluja and Fischer 2018; Zheng et al. 2019) have promising performance on improving the adversarial robustness of deep neural networks against adversarial examples. Mathematically, adversarial training can be formulated as solving the following minimax loss problem (Tu et al. 2019; Yin et al. 2019; Khim and Loh 2018):

$$\min_{\theta} \frac{1}{m} \sum_{i=1}^{m} \max_{\|x_i' - x_i\| \leq \rho} l(h_\theta(x_i'), y_i),$$

where h_θ is the hypothesis parameterized by θ, m is the training sample size, x_i is a feature, y_i is the corresponding label, and l is the loss function. Intuitively, adversarial training optimizes a neural network on the data with adversarial examples.

As for such minimax loss problem, several generalization bounds have been proposed (Yin et al. 2019; Tu et al. 2019; Khim and Loh 2018). All these results suggest that adversarial training will comprise generalizability while improving adversarial robustness.

Yin et al. (2019) studied how adversarial training influences generalizability by leveraging the techniques developed by Bartlett et al. (2017) and Raghunathan et al. (2018). They derived a tight upper bound on the Rademacher complexity and obtained a generalization bound. However, this generalization bound is only applicable to neural networks with one hidden layer and rectified linear unit (ReLU) activation functions.

To address this drawback, Khim and Loh (2018) proved a generalization bound via function transformation. An upper bound for the Rademacher complexity is then proved based on the techniques developed by Golowich et al. (2018).

Tu et al. (2019) proposed an innovative pathway for assessing the generalizability of machine learning models, comprising the following three steps. (1) We demonstrated the feasibility of eliminating the maximization operation from the objective function within the minimax framework. This was achieved through a reparameterization mapping that minimally altered the parameter distribution under the Wasserstein distribution. (2) We established a local worst-case risk bound for the resulting reparameterized minimization problem, providing valuable insights into the model's performance. (3) We derived a comprehensive generalization bound based on the local worst-case risk, offering a broader understanding of the model's predictive ability. Notably, this generalization bound extends its applicability in two perspectives: (1) it covers adversarial examples generated by bounded adversaries under the l_q norm, accommodating various threat levels; and (2) the method is highly versatile, applicable to multiclass classification tasks and compatible with a wide range of practical loss functions, including the hinge loss and ramp loss.

In contrast, Bhagoji et al. (2019) provided valuable insights by establishing a lower bound for the generalization error of classifiers when considering adversarial examples. Meanwhile, Min et al. (2020) delved deeper into the dynamics between adversarial training performance and training sample size. Their study categorized all learning problems into three distinct regimes, shedding light on various scenarios: (1) In the weak adversary regime, enlarging the training sample size corresponds to an improvement in generalizability. (2) Within the medium adversary regime, an intriguing phenomenon emerged, where the generalization error followed a double-descent curve with increasing training sample size, suggesting a complex interplay between model complexity and adversarial robustness. (3) In contrast, within the strong adversary regime, increasing the training sample size actually led to a rise in the generalization error, revealing the limitations imposed by formidable adversaries. By and large, these regimes, categorized based on the adversaries' strength, provide a nuanced understanding of model behavior. Additionally, Chen et al. (2020) corroborated similar observations akin to those found in the medium adversary regime, further reinforcing the significance of these findings.

Several recent theoretical analyses have significantly enriched our understanding of machine learning models' generalization and robustness. Attias et al. (2019) advanced the field by deriving an improved generalization bound through Rademacher complexity. Meanwhile, Zhang et al. (2019) conducted a thorough exploration into the delicate trade-offs between accuracy and robustness, shedding light on the challenges posed by adversarial settings. Additionally, Pydi and Jog (2020) introduced a generalization bound framework based on optimal transport and optimal couplings, offering a new perspective on the model's generalization abilities under different learning scenarios. These theoretical analyses collectively contribute to a more comprehensive understanding of the generalization, robustness, and convergence underlying machine learning algorithms.

13.2 Interplay Between Adversarial Robustness, Privacy Preservation, and Generalizability

Adversarial training (Dai et al. 2018; Li et al. 2018a; Baluja and Fischer 2018; Zheng et al. 2019) has promising performance on improving the adversarial robustness of deep neural networks against adversarial examples (Biggio et al. 2013; Szegedy et al. 2013; Goodfellow et al. 2014; Papernot et al. 2016). Specifically, adversarial training can be formulated as solving the following minimax loss problem:

$$\min_{\theta} \frac{1}{m} \sum_{i=1}^{m} \max_{\|x_i' - x_i\| \leq \rho} l(h_\theta(x_i'), y_i),$$

where h_θ is the hypothesis parameterized by θ, m is the number of training samples, x_i is a feature, y_i is the corresponding label, and l is the loss function. Intuitively, adversarial training optimizes a neural network in accordance with its performance on worst-case examples, which are most likely to be adversarial examples.

This section studies how adversarial training influences the privacy-preserving ability (Dwork and Mulligan 2013; Dwork and Roth 2014) and generalizability (Vapnik 2013; Mohri et al. 2018) of neural networks, both of which are considerably important in machine learning. Based on both theoretical and empirical evidence, we prove the following:

> *The minimax-based approach can degrade a neural network's privacy preservation and generalization abilities while enhancing its adversarial robustness.*

The first question to be addressed is *how to measure adversarial robustness?* Adversarial accuracy, which is defined as the accuracy on adversarial examples, is the most straightforward measure for this purpose. However, it might be difficult to develop theoretical foundations based on this measure. The radius ρ is also straightforward; however, it has been observed that a larger ρ does not necessarily imply a higher adversarial accuracy, and a single adversarial accuracy value might correspond to multiple ρ values.

Addressing the pressing challenge of safe machine learning algorithms against adversarial attacks, we introduce a new metric called the "robustified intensity". This metric serves as a yardstick for assessing how effectively a learning algorithm can withstand such attacks. It measures the disparity in gradient norms that arise from adversarial training, providing a tangible measure of robustness.

Moreover, to bridge theory with practical application, we develop an empirical estimator termed the "empirical robustified intensity". This estimator offers a pragmatic approach to quantifying robustness in real-world scenarios. Through rigorous analysis, we demonstrate that the empirical robustified intensity asymptotically converges to its theoretical counterpart as the sample size grows, ensuring its consistency and applicability.

Moving beyond theoretical foundations, we conduct a comprehensive empirical study to validate the utility of the robustified intensity. Our findings reveal a direct and positive correlation between the robustified intensity and adversarial accuracy. This empirical evidence validates the robustified intensity as not just a theoretical construct but a practical and informative metric for evaluating the resilience of learning algorithms against adversarial attacks.

Exploring the intricate dynamics between privacy and robustness in neural networks, we shift our focus to adversarial training methodologies. Unlike conventional approaches that aim for the optimal performance across all training examples, adversarial training takes a more cautious route. Specifically, neural networks are trained to withstand worst-case scenarios, where adversaries attempt to exploit vulnerabilities by perturbing input data. This strategic adaptation forces the model to rely more heavily on a small subset of training examples, heightening susceptibility to differential attacks. These attacks involve replacing a training example with a perturbed version and then using the model's output to infer other training examples.

Our investigation extends beyond practical implications to theoretical underpinnings. We establish the privacy-preserving nature of adversarial training, demonstrating that it achieves (ε, δ)-differential privacy when employing stochastic gradient descent (SGD) to minimize the minimax loss function. Furthermore, we uncover an intriguing positive correlation between the magnitudes of ε and δ and the robustified intensity. This relationship marks a significant step forward in understanding the delicate balance between privacy guarantees and model robustness.

Building upon the concept of privacy preservation, we unveil insights into the generalization performance of adversarial training methods. Our analysis yields two key generalization bounds tailored for adversarial scenarios: an $O(\sqrt{\log m}/m)$ on-average generalization bound and an $O(1/\sqrt{m})$ high-probability bound, where m denotes the size of the training sample. These two bounds are derived from a new theorem that establishes a profound link between algorithmic stability and differential privacy.

Importantly, our generalization bounds offer a distinct advantage—they are not contingent on the number of model parameters. This is particularly valuable in the context of deep learning, where models can exhibit substantial complexity. Remarkably, the terms governing these bounds, except for the gradient norm, exhibit no explicit dependence on model size. Through empirical studies, we confirm that the gradient norm remains consistently small.

Our experimental study involves a meticulous examination conducted across three widely-used datasets: CIFAR-10, CIFAR-100 (Krizhevsky and Hinton 2009), and Tiny ImageNet (Le and Yang 2015). We carefully controlled unrelated variables, ensuring the reproducibility of our results. Throughout our experiments, we systematically vary training settings to analyze their impact on model performance.

We focus on several key metrics, including generalization gaps, membership inference attack accuracies, and empirical robustified intensities, to assess the efficacy of different training configurations. Remarkably, our empirical findings consistently support our hypotheses.

13.2.1 Measurement of Robustness

The most straightforward measure of adversarial robustness is the adversarial accuracy, which is the accuracy on adversarial examples. However, it might be difficult to establish theoretical foundations concerning the relationships between adversarial accuracy, privacy preservation, and generalization based on this measure. Another natural choice is the radius ρ. However, our experiments show that there is no one-to-one mapping from ρ to the adversarial accuracy (see Fig. 13.1a).

This section proposes a new metric, the *robustified intensity*, and its asymptotically consistent empirical estimator, the *empirical robustified intensity*, to measure adversarial robustness.

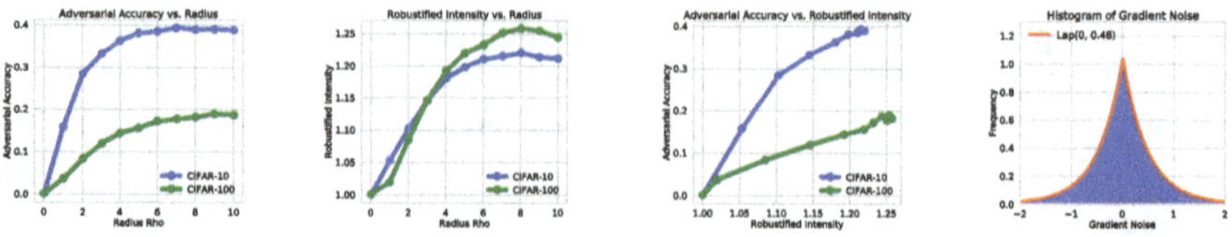

(a) (b) (c) (d)

Fig. 13.1 The first three subfigures are plots of **a** adversarial accuracy versus radius, **b** robustified intensity versus radius, and **c** adversarial accuracy versus robustified intensity. The green and blue curves correspond to CIFAR-10 and CIFAR-100, respectively. Subfigure **d** shows a histogram of 8, 000, 000 gradient noise instances on Tiny ImageNet versus the probability density function of Lap(0, 0.48)

13.2.1.1 Robustified Intensity

In conventional methods, the empirical risk minimization (ERM) problem is solved to approach the optimal hypothesis, as follows:

$$\min_{\theta} \hat{R}_S(\theta) = \min_{\theta} \frac{1}{m} \sum_{i=1}^{m} l(h_\theta(x_i), y_i).$$

Stochastic-gradient-based optimizers are then employed for ERM in deep learning, including SGD (Robbins and Monro 1951), momentum (Nesterov 1983; Tseng 1998), and Adam (Kingma and Ba 2015). For brevity, we consider SGD in this analysis. The analysis for other stochastic-gradient-based optimizers is similar.

Suppose that $\mathcal{B}$ is a minibatch with a batch size of τ. Then, the stochastic gradient on $\mathcal{B}$ is as follows:

$$\hat{g}^{\mathrm{ERM}}(\theta) = \frac{1}{\tau} \sum_{(x_i, y_i) \in \mathcal{B}} \nabla_\theta l(h_\theta(x_i), y_i).$$

In iteration t, the weights are updated as follows:

$$\theta_{t+1} = \theta_t - \eta_t \hat{g}^{\mathrm{ERM}}(\theta_t), \tag{13.1}$$

where θ_t is the weight vector in the t-th iteration and η_t is the corresponding learning rate.

In adversarial training, SGD is employed to solve the following minimax loss problem:

$$\min_{\theta} \hat{R}_S^A(\theta) = \min_{\theta} \frac{1}{m} \sum_{i=1}^{m} \max_{\|x_i' - x_i\| \le \rho} l(h_\theta(x_i'), y_i), \tag{13.2}$$

where ρ is the radius of a ball centred at example (x_i, y_i). Here, we call $\hat{\mathcal{R}}_S^A(\theta)$ the adversarial empirical risk. Correspondingly, the stochastic gradient on a minibatch $\mathcal{B}$ and the weight update are calculated as follows:

$$\hat{g}^A(\theta) = \frac{1}{|\mathcal{B}|} \sum_{(x_i, y_i) \in \mathcal{B}} \nabla_\theta \max_{\|x_i' - x_i\| \leq \rho} l(h_\theta(x_i'), y_i),$$
$$\theta_{t+1} = \theta_t - \eta_t \hat{g}^A(\theta_t). \tag{13.3}$$

We then define the robustified intensity based on the gradient norms as follows.

Definition 13.1 (Robustified intensity) For adversarial training (Eq. 13.2), the robustified intensity is defined as

$$I = \frac{\max_{\theta,x,y} \left\| \nabla_\theta \max_{\|x' - x\| \leq \rho} l(h_\theta(x'), y) \right\|}{\max_{\theta,x,y} \|\nabla_\theta l(h_\theta(x), y)\|}, \tag{13.4}$$

where $\|\cdot\|$ is a norm defined in the space of the gradient.

For brevity, we denote the numerator and denominator by L_A and L_{ERM}, respectively; i.e., $I = L_A / L_{ERM}$. Specifically, when $\rho = 0$, we have $L_A = L_{ERM}$, and thus, $I = 1$. Usually, $I > 1$ in our experiments (see Fig. 13.1b).

13.2.1.2 How can the Robustified Intensity be Estimated?

Conducting a rigorous search for the maximal r value for either adversarial training or ERM within every ball of radius ρ in the Euclidean space is technically impossible. Therefore, we define an *empirical robustified intensity* for practical utilization to empirically estimate the robustified intensity, as follows.

Definition 13.2 *(Empirical robustified intensity)* For adversarial training (Eq. 13.2), the empirical robustified intensity is defined as

$$\hat{I} = \frac{\max_{(x_i,y_i) \in \mathcal{B},\theta} \left\| \nabla_\theta \max_{\|x_i' - x_i\| \leq \rho} l(h_\theta(x_i'), y_i) \right\|}{\max_{(x_i,y_i) \in \mathcal{B},\theta} \|\nabla_\theta l(h_\theta(x_i), y_i)\|}, \tag{13.5}$$

where $\|\cdot\|$ is a norm defined in the space of the gradient.

To calculate the empirical robustified intensity, the maximal values of the gradient norm for either adversarial training or ERM in the robustified intensity are replaced by the maximal gradient norm in all minibatches.

We can now prove the following theorem.

Theorem 13.1 (Asymptotic consistency of the empirical robustified intensity) *Suppose that the empirical robustified intensity is $\hat{I}_m$ when the number of training samples is m. The empirical robustified intensity is an unbiased estimator of the robustified intensity, i.e.,* $\lim_{m \to \infty} \hat{I}_m = I$.

This theorem ensures that when the number of training sample is sufficiently large, the empirical robustified intensity is rigorously equal to the robustified intensity. Theorem 13.1 requires only one mild assumption, as shown below.

Assumption 13.2 The gradient of the loss function satisfies $\nabla_\theta l(h_\theta(x), y) \in C^0(\mathcal{Z})$; *i.e.*, for any hypothesis $h_\theta \in \mathcal{H}$, $\nabla_\theta l(h_\theta(x), y)$ is continuous w.r.t. z. □

To prove Theorem 13.1, we first recall additional preliminaries that are necessary for the proofs.

Suppose that every example z is sampled in an independent and identically distributed (i.i.d.) manner from the data distribution $\mathcal{D}$, i.e., $z \sim \mathcal{D}$. Thus, the training sample set satisfies $S \sim \mathcal{D}^m$, where m is the number of training samples.

To prove our theorem, we establish the following definitions.

Definition 13.3 (*Ball and sphere)* The ball of radius $r > 0$ in terms of norm $\|\cdot\|$ in space $\mathcal{H}$ centred at point $x \in \mathcal{H}$ is defined as

$$B_r(h) = \{x : \|x - h\| \leq r\}.$$

The sphere $\partial B_r(h)$ corresponding to ball $B_r(h)$ is defined as

$$\partial B_r(h) = \{x : \|x - h\| = r\}.$$

Definition 13.4 (*Complementary set)* For a subset $A \subset \mathcal{H}$ of a space $\mathcal{H}$, its complementary set A^c is defined as follows:

$$A' = \{h : h \in \mathcal{H}, h \notin A\}.$$

Proof of Theorem 13.1 We need only to prove that

$$\lim_{m\to\infty} \max_{\theta, x_i, y_i} \left\| \nabla_\theta \max_{\|x_i' - x_i\| \leq \rho} l\left(h_\theta\left(x_i'\right), y_i\right) \right\| = \max_{\theta, x, y} \left\| \nabla_\theta \max_{\|x' - x\| \leq \rho} l\left(h_\theta\left(x'\right), y\right) \right\| \tag{13.6}$$

almost surely and

$$\lim_{m\to\infty} \max_{\theta, x_i, y_i} \|\nabla_\theta l\left(h_\theta\left(x_i\right), y_i\right)\| = \max_{\theta, x, y} \|\nabla_\theta l\left(h_\theta\left(x\right), y\right)\| \tag{13.7}$$

almost surely.

We first prove that for any positive real

$$g(\theta, z) = \nabla_\theta \max_{x' \in B_x(\rho)} l\left(h_\theta\left(x'\right), y\right), \ \rho > 0$$

is a continuous function with respect to $z = (x, y)$, where $B_x(\rho) = \{x' : \|x - x'\| \leq \rho\}$ is a ball centred at x with a radius of ρ.

By fixing $y \in \mathcal{Y}$, we define $T_y(x) = \arg\max_{x' \in B_x(\rho)} l(h_\theta(x'), y)$ as a mapping from $\mathcal{X}$ to $\mathcal{X}$. We will prove by reduction to absurdity that $T_y(x)$ is continuous with respect to (x, y). Suppose that there exist a sequence

$$\{z_i = (x_i, y_i)\}_{i=1}^{\infty}, \lim_{i \to \infty} z_i = z_0$$

and a constant $\varepsilon_0 > 0$ such that

$$\|T_{y_i}(x_i) - T_{y_0}(x_0)\| \geq \varepsilon_0.$$

Since $\{T_{y_i}(x_i)\}_{i=1}^{\infty}$ is a bounded set, there exists an increasing subsequence $\{k_i\}_{i=1}^{\infty} \subseteq Z^+$ such that $\{T_{y_{k_i}}(x_{k_i})\}_{i=1}^{\infty}$ converges to some point T_∞. Then, we have

$$T_\infty \in \cap_{i=1}^{\infty} B_{x_{k_i}}(\rho) \subset B_{x_0}(\rho).$$

Furthermore, for any $\varepsilon \geq 0$, there exists a $\delta > 0$, such that for any $x \in B_{T_{y_0}(x_0)}(\delta)$, $l(h_\theta(x), y_0)) \geq l(h_\theta(T_{y_0}(x_0)), y_0) - \varepsilon$. In the case of $T_{y_0}(x_0) \in \partial B_{x_0}(\rho)$ such that $T_{y_0}(x_0) \notin \cap_{i=1}^{\infty} B_{x_{k_i}}(\rho)$, let $x' \in B_{T_{y_0}(x_0)}(\delta)$ be an inner point of $B_{x_0}(\rho)$. When i is sufficiently large, we have $x' \in B_{x_{k_i}}(\rho)$, which yields

$$l(h_\theta(x'), y_{k_i}) \leq l(h_\theta(T_{y_{k_i}}(x_{k_i})), y_{k_i}).$$

Let i approach ∞; then, we have

$$l(h_\theta(T_{y_0}(x_0), y_0) - \varepsilon \leq l(h_\theta(x'), y_0) \leq l(h_\theta(T_\infty), y_0).$$

Since ε is arbitrarily selected, we then have

$$l(h_\theta(T_{y_0}(x_0), y_0)) \leq l(h_\theta(T_\infty), y_0) \leq l(h_\theta(T_{y_0}(x_0), y_0)).$$

Therefore, $T_\infty = T_{y_0}(x_0)$, which leads to a contradiction since $\|T_{y_i}(x_i) - T_{y_0}(x_0)\| \geq \varepsilon_0$.

Since $g(\theta, z)$ can be rewritten as

$$g(\theta, z) = \nabla_\theta \max_{x' \in B_x(\rho)} l\left(h_\theta\left(x'\right), y\right) = \nabla_\theta l(h_\theta(T_y(x)), y),$$

from Assumption 13.2, we know that $g(\theta, z)$ is continuous with respect to z.

We can now prove Eqs. (13.6) and (13.7). For Eq. (13.6), there exist a θ_0 and a $z_0 = (x_0, y_0)$ such that

$$\left\|\nabla_{\theta_0} \max_{\|x'-x_0\| \leq \rho} l\left(h_{\theta_0}\left(x'\right), y_0\right)\right\| = \max_{\theta,x,y} \left\|\nabla_\theta \max_{\|x'-x\| \leq \rho} l\left(h_\theta\left(x'\right), y\right)\right\|.$$

For any $\varepsilon > 0$, since $g(\theta_0, z)$ is continuous with respect to z, there exists a $\delta > 0$ such that for any $(x', y') \in B_{(x_0, y_0)}(\delta)$,

$$g(\theta_0, (x', y')) \geq g(\theta_0, (x_0, y_0)) - \varepsilon.$$

Therefore,

$$\{(x, y) : g(\theta_0, (x, y)) < g(\theta_0, (x_0, y_0)) - \varepsilon\} \subset \left(B_{(x_0, y_0)}(\delta)\right)^c,$$

and we have

$$\begin{aligned}\mathbb{P}_{S\sim\mathcal{D}^m}\left(\max_{\theta,z\in S} g(\theta, z) \le \max_{\theta,z} g(\theta, z) - \varepsilon\right) &\le \mathbb{P}_{S\sim\mathcal{D}^m}\left(\max_{z\in S} g(\theta_0, z) \le g(\theta_0, z_0) - \varepsilon\right) \\ &\le \mathbb{P}_{S\sim\mathcal{D}^m}\left(S \cap B_{(x_0,y_0)}(\delta) = \emptyset\right) \\ &= \left(1 - \mathbb{P}_{z\sim\mathcal{D}}\left(z \in B_{(x_0,y_0)}(\delta)\right)\right)^m.\end{aligned}$$

As $m \to \infty$, we have

$$\lim_{m\to\infty} \mathbb{P}_{S\sim\mathcal{D}^m}\left(\max_{\theta,z\in S} g(\theta, z) \le \max_{\theta,z} g(\theta, z) - \varepsilon\right) = 0.$$

Since ε is arbitrarily selected, we have

$$\lim_{m\to\infty} \mathbb{P}_{S\sim\mathcal{D}^m}\left(\max_{\theta,z\in S} g(\theta, z) \le \max_{\theta,z} g(\theta, z)\right) = 0,$$

which proves Eq. (13.6).

By replacing $g(\theta, z) = \nabla_\theta l(h_\theta(x), y)$, we can prove Eq. (13.7) via the same procedure.

The proof is complete.

13.2.1.3 Is the Robustified Intensity Informative?

We present a comprehensive empirical study conducted to compare the robustified intensity, radius ρ, and adversarial accuracy on CIFAR-10 and CIFAR-100.

In Fig. 13.1, we present three plots, namely, the adversarial accuracy versus the radius ρ, the robustified intensity versus the radius ρ, and the adversarial accuracy versus the robustified intensity. From these plots, we can draw three major observations: (1) The adversarial test accuracy has a clear positive correlation with the radius ρ when ρ is smaller than 5 on CIFAR-10 or smaller than 9 on CIFAR-100; after those points, no significant correlation between the adversarial test accuracy and radius is observed. This observation suggests that the radius ρ is not always an informative choice for measuring the robustness. (2) A bell curve shape is observed in the plot of the robustified intensity I versus the radius ρ, which suggests that both maximal and minimal values of the robustified intensity I are achieved, i.e., the full domain of the potential robustified intensity I is discovered. (3) A clear positive correlation is observed between the adversarial test accuracy and the robustified intensity I throughout the full interval of I, which suggests that there is a one-to-one mapping from robustified intensity to adversarial accuracy. These three observations verify that the robustified intensity I is an informative measure of robustness.

13.2.2 *Privacy–Robustness Trade-Off*

This section studies the relationship between privacy preservation and robustness in adversarial training. We prove that adversarial training is (ε, δ)-differentially private.

13.2.2.1 What Is the Distribution of the Gradient Noise?

In stochastic-gradient-based optimization, noise is an inherent factor. This leads us to ponder: what exactly characterizes the distribution of this noise? Past research efforts by Kushner and Yin (2003), Ljung et al. (2012), Mandt et al. (2017) have leaned towards assuming a Gaussian distribution for the gradient noise. However, in our quest for deeper understanding, we embark on a large-scale experiment to scrutinize this assumption more closely.

Our investigation takes us to the Tiny ImageNet dataset, where we conduct extensive experiments. The results, as depicted in Fig. 13.1d, reveal an intriguing insight: the gradient noise seems to align more closely with a Laplacian distribution. This pivotal finding prompts us to adopt a new foundational assumption moving forward, one that acknowledges the Laplacian nature of the gradient noise distribution.

Assumption 13.3 The gradient calculated from a minibatch is drawn from a Laplacian distribution centred on the empirical risk,

$$\frac{1}{\tau}\sum_{(x,y)\in\mathcal{B}}\nabla_\theta \max_{\|x'-x\|\le\rho} l(h_\theta(x'), y) \sim \text{Lap}\left(\nabla_\theta \hat{R}_S^A(\theta), b\right).$$

13.2.2.2 Theoretical Evidence

Then, following the Laplacian mechanism, we approximate the differential privacy of adversarial training as follows.

Theorem 13.4 *Suppose that SGD is employed for adversarial training. L_{ERM} is the maximal gradient norm in ERM. Additionally, suppose that the whole training procedure consists of T iterations. Then, the adversarial training is (ε, δ)-differentially private, where*

$$\varepsilon = \varepsilon_0\sqrt{2T\log\frac{m}{\delta'}} + T\varepsilon_0(e^{\varepsilon_0} - 1),$$

$$\delta = \frac{\delta'}{m},$$

with

$$\varepsilon_0 = \frac{2L_{ERM}}{mb} I.$$

Here, δ' is a positive real number, τ is the batch size, I is the robustified intensity, and b is the Laplace parameter.

Remark 13.1 The approximation of the differential privacy given by Theorem 13.4 is $(\mathcal{O}(\sqrt{\log m}/m), \mathcal{O}(1/m))$.

To prove Theorem 13.4, we will first prove two lemmas (Lemma 13.1 and Lemma 13.2 below).

In practice, it is easier to obtain high-probability approximations of ε-differential privacy from concentration inequalities. Lemma 13.1 presents a relationship between high-probability approximations of ε-differential privacy and approximations of (ε, δ)-differential privacy. Similar arguments have been used in some related works; see, for example, the proof of Theorem 3.20 in (Dwork and Roth 2014). Here, we present a detailed proof to conclude our work in this study.

Lemma 13.1 *Suppose that $\mathcal{A} : \mathcal{Z}^m \to \mathcal{H}$ is a stochastic algorithm, whose output hypothesis learned on training sample set S is $\mathcal{A}(S)$. For any hypothesis $h \in \mathcal{H}$, if the condition*

$$\log \left[\frac{\mathbb{P}\left[\mathcal{A}(S) = h\right]}{\mathbb{P}\left[\mathcal{A}(S') = h\right]} \right] \leq \varepsilon \tag{13.8}$$

is satisfied with probability at least $1 - \delta$ over the randomness of $\mathcal{A}(S)$, then algorithm $\mathcal{A}$ is (ε, δ)-differentially private.

Proof After rearranging Eq. (13.8), we have that, with probability at least $1 - \delta$,

$$\mathbb{P}\left[\mathcal{A}(S) = h\right] \leq \mathbb{P}\left[\mathcal{A}(S') = h\right] e^{\varepsilon}. \tag{13.9}$$

For brevity, we define an event as follows:

$$B_0 = \left\{ h : \log \left[\frac{\mathbb{P}\left[\mathcal{A}(S) = h\right]}{\mathbb{P}\left[\mathcal{A}(S') = h\right]} \right] \leq \varepsilon \right\}.$$

Additionally, we define

$$B_0^c = \mathcal{H} - B_0.$$

Apparently, for any subset $B \in \mathcal{H}$,

$$\begin{aligned} \mathbb{P}\left[\mathcal{A}(S) \in B_0 \cap B\right] \leq& \mathbb{P}\left[\mathcal{A}(S') \in B_0 \cap B\right] e^{\varepsilon}, & (13.10) \\ \mathbb{P}\left[\mathcal{A}(S) \in B_0\right] \geq& 1 - \delta, \\ \mathbb{P}\left[\mathcal{A}(S) \in B_0^c\right] \leq& \delta. & (13.11) \end{aligned}$$

Then, for any subset $B \in \Theta$, we have

$$\begin{aligned}\mathbb{P}[\mathcal{A}(S) \in B] =& \mathbb{P}\left[\mathcal{A}(S) \in B \cap \left(B_0 \cup B_0^c\right)\right] \\ =& \mathbb{P}[\mathcal{A}(S) \in B \cap B_0] + \mathbb{P}\left[\mathcal{A}(S) \in B \cap B_0^c\right] \\ \leq& \mathbb{P}[\mathcal{A}(S) \in B \cap B_0] + \mathbb{P}\left[\mathcal{A}(S) \in B_0^c\right].\end{aligned}$$

Combining Eqs. (13.9), (13.10), and (13.11), we obtain

$$\mathbb{P}[\mathcal{A}(S) \in B] \leq e^{\varepsilon}\mathbb{P}\left[\mathcal{A}(S') \in B \cap B_0\right] + \delta \leq e^{\varepsilon}\mathbb{P}\left[\mathcal{A}(S') \in B\right] + \delta.$$

Therefore, the stochastic algorithm $\mathcal{A}$ is (ε, δ)-differentially private.
The proof is complete. □

Quantifying the level of differential privacy for an iterative algorithm can be a daunting task. Yet, by dissecting the algorithm into its individual steps, we find that assessing the differential privacy of each step is more manageable. The following advanced composition lemma enables us to estimate the overall level of differential privacy for the entire iterative algorithm. It achieves this by aggregating the differential privacy of each step.

Lemma 13.2 (Advanced composition; cf. (Dwork and Roth 2014), Theorem 3.20) *Suppose that an $(\varepsilon_0, \delta_0)$-differentially private process is run repeatedly T times. Then, the whole algorithm is (ϵ, δ)-differentially private, where*

$$\begin{aligned}\varepsilon =& \sqrt{2T \log \frac{1}{\delta'}}\varepsilon_0 + T\varepsilon_0(e^{\varepsilon_0} - 1), \\ \delta =& T\delta_0 + \delta',\end{aligned}$$

and δ' is a positive real number.

We now prove Theorem 13.4.

Proof of Theorem 13.4 We assume that the gradients calculated from a randomly sampled minibatch $\mathcal{B}$ with a size of τ are random variables drawn from a Laplace distribution, as justified previously:

$$\frac{1}{\tau}\sum_{z\in\mathcal{B}} \nabla_\theta \max_{\|x'-x\|\leq\rho} l(\theta, x, y) \sim \text{Lap}\left(\nabla_\theta \hat{R}_S^A(\theta), b\right).$$

Correspondingly, the counterpart of this distribution on the training sample set S' is as follows:

$$\frac{1}{\tau}\sum_{z\in\mathcal{B}'} \nabla_\theta \max_{\|x'-x\|\leq\rho} l(\theta, x, y) \sim \text{Lap}\left(\nabla_\theta \hat{R}_{S'}^A(\theta), b\right),$$

where $\mathcal{B}'$ is uniformly sampled from S', which is also of size τ.

The output hypothesis is uniquely indexed by the weights. Specifically, we denote the weights after the t-th iteration by θ_{t+1}. Furthermore, the weight updates $\Delta\theta_t = \theta_{t+1} - \theta_t$ are uniquely determined by the gradients. Therefore, we can calculate the probability of the gradients to approximate the differential privacy. For any $\hat{g}_t^A$,

$$\log \left[\frac{p\left[\mathrm{Lap}(\nabla_\theta \hat{R}_S^A(\theta_t), b) = \hat{g}_t^A\right]}{p\left[\mathrm{Lap}(\nabla_\theta \hat{R}_{S'}^A(\theta_t), b) = \hat{g}_t^A\right]} \right] = \log \left[\frac{\exp\left\{-\left\|\nabla_\theta \hat{R}_S^A(\theta_t) - \hat{g}_t^A\right\| / b\right\}}{\exp\left\{-\left\|\nabla_\theta \hat{R}_{S'}^A(\theta_t) - \hat{g}_t^A\right\| / b\right\}} \right]$$
$$= \frac{1}{b}\left[-\left\|\nabla_\theta \hat{R}_S^A(\theta_t) - \hat{g}_t^A\right\| + \left\|\nabla_\theta \hat{R}_{S'}^A(\theta_t) - \hat{g}_t^A\right\|\right]. \tag{13.12}$$

We define

$$L_A = \max_{\theta_t, x, y} \left\| \nabla_\theta \max_{\|x'-x\| \le \rho} l(\theta_t, x, y) \right\|$$

and

$$v = \nabla_\theta \hat{R}_{S'}^A(\theta_t) - \nabla_\theta \hat{R}_S^A(\theta_t).$$

Since the training sample sets S and S' differ by only one pair of examples, we have

$$\|v\| \le \frac{2L_A}{m}. \tag{13.13}$$

Combining Eqs. (13.12) and (13.13), we obtain

$$\log \left[\frac{p\left[\mathrm{Lap}(\nabla_\theta \hat{R}_S^A(\theta_t), b) = \hat{g}_t^A\right]}{p\left[\mathrm{Lap}(\nabla_\theta \hat{R}_{S'}^A(\theta_t), b) = \hat{g}_t^A\right]} \right]$$
$$= \frac{1}{b}\left[-\left\|\nabla_\theta \hat{R}_S^A(\theta_t) - \hat{g}_t^A\right\| + \left\|\nabla_\theta \hat{R}_{S'}^A(\theta_t) - \hat{g}_t^A\right\|\right]$$
$$= \frac{2L_A}{mb}. \tag{13.14}$$

Since $L_A = I L_{ERM}$, we have

$$\log \left[\frac{p\left[\mathrm{Lap}(\nabla_\theta \hat{R}_S^A(\theta_t), b) = \hat{g}_t^A\right]}{p\left[\mathrm{Lap}(\nabla_\theta \hat{R}_{S'}^A(\theta_t), b) = \hat{g}_t^A\right]} \right] \le \frac{2L_{ERM}}{mb} I.$$

We define

$$\varepsilon_0 = \frac{2L_{ERM}}{mb} I.$$

By applying Lemma 13.2 with $\varepsilon_0 = \varepsilon_0$, $\delta' = \delta'/m$, and $\delta_0 = 0$, the proof is completed.

Similarly, we can calculate the differential privacy of ERM.

Corollary 13.1 *Suppose that SGD is employed for ERM and that the whole training procedure consists of T iterations. Then, the ERM process is (ε, δ)-differentially private, where*

$$\varepsilon = \varepsilon_0^{ERM} \sqrt{2T \log \frac{m}{\delta'}} + T \varepsilon_0^{ERM} (e^{\varepsilon_0^{ERM}} - 1),$$
$$\delta = \frac{\delta'}{m},$$

with

$$\varepsilon_0^{ERM} = \frac{2L_{ERM}}{mb}.$$

By comparing the results between adversarial training and ERM, we can conclude that $\varepsilon_0 = I \cdot \varepsilon_0^{\text{ERM}}$.

Theorem 13.4 and Corollary 13.1 show that the factors ε and δ are both positively correlated with the robustified intensity, which suggests a trade-off between privacy preservation and adversarial robustness.

13.2.2.3 Empirical Evidence

We conducted an extensive empirical study on the privacy–robustness trade-off based on the ResNet network architecture and the CIFAR-10, CIFAR-100, and Tiny ImageNet datasets. We employed membership inference attack (Shokri et al. 2017), a standard privacy attack tool, to measure the privacy preservation ability. A membership inference attack attempts to infer whether a data point is present in the training sample from the output of the model. Therefore, the membership inference attack accuracy intuitively expresses the algorithm's privacy preservation ability. Specifically, a higher membership inference attack accuracy suggests worse privacy preservation.

We collected data of both the membership inference attack accuracy and the empirical robustified intensity across all models. This comprehensive dataset enabled us to construct the four plots showcased in Fig. 13.2. Understanding these plots is crucial: a higher membership inference attack accuracy implies an increased susceptibility to privacy breaches, indicating weaker privacy preservation measures.

By investigating the figure, we can observe a pattern: there's a noticeable positive correlation between the membership inference attack accuracy and the robustified intensity I. Essentially, this correlation highlights a critical trade-off between privacy preservation and model robustness. The stronger the model's defense against

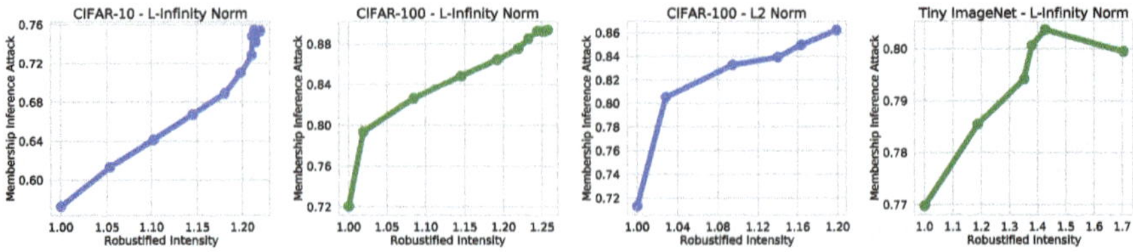

Fig. 13.2 Plots of membership inference attack accuracy versus empirical robustified intensity. For the four plots, the datasets and norms used in projected gradient descent (PGD) are (1) CIFAR-10 and the L_∞ norm, (2) CIFAR-100 and the L_∞ norm, (3) CIFAR-100 and the L_2 norm, and (4) Tiny ImageNet and the L_∞ norm

adversarial attacks (reflected in its robustness), the more vulnerable it becomes to privacy breaches. This observation suggests the intricate balance required in developing learning models that prioritize both privacy and robustness.

13.2.3 Generalization–Robustness Trade-Off

This section studies the relationship between generalization and robustness.

We first prove a high-probability generalization bound for an (ε, δ)-differentially private machine learning algorithm. Since we have already approximated the (ε, δ)-differential privacy of adversarial training, this bound can help us study the generalizability of adversarial training.

Theorem 13.5 (High-probability generalization bound in terms of differential privacy) *Suppose that all conditions of Theorem 13.4 hold. Then, algorithm $\mathcal{A}$ has a high-probability generalization bound as follows. Specifically, the following inequality holds with probability at least $1-\gamma$:*

$$\mathbb{E}_{\mathcal{A}} R(\mathcal{A}(S)) - \mathbb{E}_{\mathcal{A}} \hat{R}_S(\mathcal{A}(S)) \leq c \left(M(1 - e^{-\varepsilon} + e^{-\varepsilon}\delta) \log m \log \frac{m}{\gamma} + \sqrt{\frac{\log 1/\gamma}{m}} \right), \tag{13.15}$$

where γ is an arbitrary probability mass, M is the bound on the loss l, m is the number of training samples, c is a universal constant for any sample distribution, and the probability is defined over the sample set S.

The proof of Theorem 13.5 relies on the following lemma from Oneto et al. (2017).

Lemma 13.3 (cf. (Feldman and Vondrak 2019), Theorem 1) *Suppose that a deterministic machine learning algorithm $\mathcal{A}$ is stable with uniform stability β. Suppose that $l \leq 1$. Then, for any sample distribution and any $\gamma \in (0, 1)$, there exists a universal constant c such that, with probability at least $1-\gamma$ over the draw of the sample, the generalization error can be upper bounded as follows:*

$$\mathbb{E}_z l(\mathcal{A}(S), z) - \frac{1}{m}\sum_{z \in S} l(\mathcal{A}(S), z) \leq c\left(\beta \log m log\frac{m}{\gamma} + \sqrt{\frac{\log 1/\gamma}{m}}\right).$$

Proof of Theorem 13.5 By combining Lemmas 13.5 and 13.3, we can directly prove Theorem 13.5.

Remark 13.2 By the postprocessing property of differential privacy, since $\mathcal{B}: h \to \max_{x' \in \mathbb{B}_*(\rho)} l(h, (x', *))$ is a one-to-one mapping, $\max_{x' \in \mathbb{B}_*(\rho)} l(\mathcal{A}, (x', *))$ is (ε, δ)-differentially private. Therefore, Theorems 13.5 and 13.6 hold for adversarial learning algorithms.

In our endeavor to quantify the generalizability of a learning model, we not only establish a high-probability generalization bound but also derive an on-average generalization bound. This on-average bound offers a glimpse into the "expected" performance of the learning model. It's essential to understand the theoretical possibility of deriving on-average bounds from high-probability bounds through integration. However, in practice, executing such calculations becomes dauntingly challenging. As a result, we opt for a more feasible and independent approach to achieve the on-average bound.

Theorem 13.6 (On-average generalization bound in terms of differential privacy) *Suppose that all conditions of Theorem 13.4 hold. Then, the on-average generalization error of algorithm $\mathcal{A}$ is upper bounded by*

$$\mathbb{E}_{S,\mathcal{A}}\left[R(\mathcal{A}(S)) - \hat{R}_S(\mathcal{A}(S))\right] \leq M\delta e^{-\varepsilon} + M(1 - e^{-\varepsilon}).$$

The proof of Theorem 13.6 relies on the following lemma from Dwork et al. (2015a).

Lemma 13.4 (Lemma 11, cf. (Shalev-Shwartz et al. 2010)) *Suppose that the loss function is upper bounded. For any machine learning algorithm with β replace-one stability, its generalization error is upper bound as follows:*

$$R(\mathcal{A}(S)) - \hat{R}_S(\mathcal{A}(S)) \leq \beta.$$

Proof of Theorem 13.6 By combining Lemma 13.5 and Lemma 13.4, we can directly prove Theorem 13.6.

By combining Theorem 13.4, Corollary 13.1, Theorem 13.5, and Theorem 13.6, we can obtain generalization bounds for both adversarial training and conventional ERM.

Both generalization bounds are positively correlated with the magnitude of the differential privacy, which, in turn, is positively correlated with the adversarial robustness. This leads to the following corollary.

Corollary 13.2 *A trade-off exists between generalizability and adversarial robustness (as measured by the robustified intensity) in adversarial training.*

13.2.3.1 Establishing Generalization Bounds Based on Algorithmic Stability

Theorems 13.5 and 13.6 are established via algorithmic stability, which measures how stable an algorithm is when the training sample is exposed to disturbance (Rogers and Wagner 1978; Kearns and Ron 1999; Bousquet and Elisseeff 2002). While algorithmic stability has many different definitions, this work mainly discusses uniform stability.

Definition 13.5 (*Uniform stability; cf.* (Bousquet and Elisseeff 2002)) A machine learning algorithm $\mathcal{A}$ is uniformly stable if, for any neighbouring sample pair S and S' that differ by only one example, the following inequality holds:

$$\left|\mathbb{E}_{\mathcal{A}} l(\mathcal{A}(S), z) - \mathbb{E}_{\mathcal{A}} l(\mathcal{A}(S'), z)\right| \leq \beta,$$

where z is an arbitrary example; $\mathcal{A}(S)$ and $\mathcal{A}(S')$ are the output hypotheses learned on training sets S and S', respectively; and β is a positive real constant. The constant β is called the uniform stability of $\mathcal{A}$.

In this section, we prove that (ε, δ)-differentially private machine learning algorithms are algorithmically stable, as shown by the following lemma.

Lemma 13.5 (Stability–privacy relationship) *Suppose that a machine learning algorithm $\mathcal{A}$ is (ε, δ)-differentially private. Additionally, suppose that the loss function l is upper bounded by a positive real constant $M > 0$. Then, the algorithm $\mathcal{A}$ is uniformly stable:*

$$\left|\mathbb{E}_{\mathcal{A}} l(\mathcal{A}(S), z) - \mathbb{E}_{\mathcal{A}} l(\mathcal{A}(S'), z)\right| \leq M\delta e^{-\varepsilon} + M(1 - e^{-\varepsilon}).$$

To prove Lemma 13.5, we first prove a weaker version of it that holds when algorithm $\mathcal{A}$ has ε-pure differential privacy.

Lemma 13.6 *Suppose that a machine learning algorithm $\mathcal{A}$ is ε-differentially private. Additionally, suppose that the loss function l is upper bounded by a positive real constant $M > 0$. Then, the algorithm $\mathcal{A}$ is uniformly stable:*

$$\left|\mathbb{E}_{\mathcal{A}(S)} l(\mathcal{A}(S), z) - \mathbb{E}_{\mathcal{A}(S')} l(\mathcal{A}(S'), z)\right| \leq M(1 - e^{-\varepsilon}).$$

Proof We define a set B as $B = \{h \in H : l(h, z) > t\}$, where t is an arbitrary real number. Then, for any $t \in \mathbb{R}$,

$$\mathbb{P}_{\mathcal{A}(S)}(\mathcal{A}(S) \in B) \leq e^{\varepsilon} \mathbb{P}_{\mathcal{A}(S')}(\mathcal{A}(S') \in B). \tag{13.16}$$

By rearranging Eq. (13.16), we obtain

$$e^{-\varepsilon} \mathbb{P}_{\mathcal{A}(S)}(\mathcal{A}(S) \in B) \leq \mathbb{P}_{\mathcal{A}(S')}(\mathcal{A}(S') \in B),$$

$$(e^{-\varepsilon} - 1)\mathbb{P}_{\mathcal{A}(S)}(\mathcal{A}(S) \in B) \leq \mathbb{P}_{\mathcal{A}(S')}(\mathcal{A}(S') \in B) - \mathbb{P}_{\mathcal{A}(S)}(\mathcal{A}(S) \in B).$$

Since $\varepsilon > 0$, we have $e^{-\varepsilon} < 1$. Therefore,

$$(e^{-\varepsilon} - 1) \leq \mathbb{P}_{\mathcal{A}(S')}(\mathcal{A}(S') \in B) - \mathbb{P}_{\mathcal{A}(S)}(\mathcal{A}(S) \in B). \tag{13.17}$$

Equation (13.17) holds for every pair of neighbouring sample sets S and S'. Thus,

$$e^{-\varepsilon} - 1 \leq \min_{S \text{ and } S' \text{ neighbour}} (\mathbb{P}_{\mathcal{A}(S')}(\mathcal{A}(S') \in B) - \mathbb{P}_{\mathcal{A}(S)}(\mathcal{A}(S) \in B)).$$

Therefore,

$$\max_{S \text{ and } S' \text{ neighbour}} \left|\left[\mathbb{P}_{\mathcal{A}(S')}(\mathcal{A}(S') \in B) - \mathbb{P}_{\mathcal{A}(S)}(\mathcal{A}(S) \in B)\right]\right| \leq 1 - e^{-\varepsilon}.$$

Thus,

$$\begin{aligned}
&|\mathbb{E}_{\mathcal{A}(S')} l(\mathcal{A}(S'), z) - \mathbb{E}_{\mathcal{A}(S)} l(\mathcal{A}(S), z)| \\
=&\left|\int l(\mathcal{A}(S), z)\, d\mathbb{P}_{\mathcal{A}(S)} - \int l(\mathcal{A}(S'), z) \mathrm{d}\, \mathbb{P}_{\mathcal{A}(S')}\right| \\
\overset{(*)}{\leq}& \max\{I_1, I_2\} \leq M(1 - e^{-\varepsilon}),
\end{aligned}$$

where I_1 and I_2 in inequality $(*)$ are defined as

$$\begin{aligned}
I_1 &= \int_{\mathbb{P}_{\mathcal{A}(S)} > \mathbb{P}_{\mathcal{A}(S')}} l(\mathcal{A}(S), z) \left(\mathrm{d}\mathbb{P}_{\mathcal{A}(S)} - \mathrm{d}\mathbb{P}_{\mathcal{A}(S')}\right), \\
I_2 &= \int_{\mathbb{P}_{\mathcal{A}(S)} \leq \mathbb{P}_{\mathcal{A}(S')}} l(\mathcal{A}(S), z) \left(\mathrm{d}\mathbb{P}_{\mathcal{A}(S')} - \mathrm{d}\mathbb{P}_{\mathcal{A}(S)}\right).
\end{aligned}$$

The proof is complete. □

Now, we prove Lemma 13.5 using a different method.

Proof of Lemma 13.5 Since algorithm $\mathcal{A}$ is (ε, δ)-differentially private, we have

$$\mathbb{P}_{\mathcal{A}(S)}(\mathcal{A}(S) \in B) \leq e^{\varepsilon} \mathbb{P}_{\mathcal{A}(S')}(\mathcal{A}(S') \in B) + \delta,$$

where B is an arbitrary subset drawn from the hypothesis space $\mathcal{H}$. Let $B = \{h \in H : l(h, z) > t\}$. Then, we have the following inequality:

$$\mathbb{P}_{\mathcal{A}(S)}(l(\mathcal{A}(S), z) > t) \leq e^{\varepsilon} \mathbb{P}_{\mathcal{A}(S')}(l(\mathcal{A}(S'), z) > t) + \delta. \tag{13.18}$$

Additionally, $\mathbb{E}_{\mathcal{A}(S)} l(\mathcal{A}(S), z)$ is calculated as follows:

$$\mathbb{E}_{\mathcal{A}(S)} l(\mathcal{A}(S), z) = \int_0^M \mathbb{P}_{\mathcal{A}(S)}(l(\mathcal{A}(S), z) > t) dt.$$

By applying Eq. (13.18), we obtain

$$\begin{aligned}\mathbb{E}_{\mathcal{A}(S)} l(\mathcal{A}(S), z) &= \int_0^M \mathbb{P}_{\mathcal{A}(S)}(l(\mathcal{A}(S), z) > t) dt \\ &\leq e^{\varepsilon} \int_0^M \mathbb{P}_{\mathcal{A}(S')}(l(\mathcal{A}(S'), z) > t) dt + M\delta \\ &= e^{\varepsilon} \mathbb{E}_{\mathcal{A}(S')} l(\mathcal{A}(S'), z) + M\delta. \end{aligned} \tag{13.19}$$

By rearranging Eq. (13.19), we obtain

$$\begin{aligned} e^{-\varepsilon} \mathbb{E}_{\mathcal{A}(S)} l(\mathcal{A}(S), z) &\leq \mathbb{E}_{\mathcal{A}(S')} l(\mathcal{A}(S'), z) + e^{-\varepsilon} M\delta, \\ (e^{-\varepsilon} - 1) \mathbb{E}_{\mathcal{A}(S)} l(\mathcal{A}(S), z) &\leq \mathbb{E}_{\mathcal{A}(S')} l(\mathcal{A}(S'), z) - \mathbb{E}_{\mathcal{A}(S)} l(\mathcal{A}(S), z) + e^{-\varepsilon} M\delta. \end{aligned}$$

Therefore,

$$\mathbb{E}_{\mathcal{A}(S')} l(\mathcal{A}(S'), z) - \mathbb{E}_{\mathcal{A}(S)} l(\mathcal{A}(S), z) \leq e^{-\varepsilon} M\delta + (1 - e^{-\varepsilon}) \mathbb{E}_{\mathcal{A}(S)} l(\mathcal{A}(S), z).$$

Similarly, we can obtain the following inequality:

$$-\mathbb{E}_{\mathcal{A}(S')} l(\mathcal{A}(S'), z) + \mathbb{E}_{\mathcal{A}(S)} l(\mathcal{A}(S), z) \leq e^{-\varepsilon} M\delta + (1 - e^{-\varepsilon}) \mathbb{E}_{\mathcal{A}(S)} l(\mathcal{A}(S), z).$$

Thus,

$$\left| \mathbb{E}_{\mathcal{A}(S)} l(\mathcal{A}(S), Z) - \mathbb{E}_{\mathcal{A}(S')} l(\mathcal{A}(S'), Z) \right| \leq M\delta e^{-\varepsilon} + M(1 - e^{-\varepsilon}).$$

The proof is complete.

Theorems 13.5 and 13.6 are further established based on Lemma 13.5 with Feldman and Vondrak (2019) and Bousquet and Elisseeff (2002), respectively.

13.2.3.2 Tightness of Generalization Bounds

Dependency on the number of training samples m.

In Sect. 13.2.2, we have approximated the differential privacy rate of adversarial training with respect to the number of training samples m; see Remark 13.1. By combining Theorems 13.5 and 13.6, we can approximate the tightness of the high-probability generalization bound and the on-average generalization bound as shown in the following remarks.

Remark 13.3 The high-probability generalization bound given by Theorem 13.5 is $O(1/\sqrt{m})$.

Remark 13.4 The on-average generalization bound given by Theorem 13.6 is $O(\sqrt{\log m}/m)$.

Dependency on the model size.

Our generalization bounds do not explicitly depend on the number of parameters, which typically is prohibitively large in deep learning. The only term that could implicitly depend on the model size is the gradient norm. We trained two different networks, namely, ResNet and Tiny ImageNet, within the frameworks of ERM and adversarial training with different values of the radius ρ and two different norms in projected gradient descent (PGD), namely, the L_2 norm and the L_∞ norm. Some of the collected data are shown in Fig. 13.3. These box plots clearly illustrate that the gradient norm is very small in comparison to the number of parameters (which is on the order of tens of millions): the mean of the gradient norm is approximately 1, and the largest value is approximately 5.

Existing works have proven generalization bounds based on hypothesis complexity. Yin et al. (2019) proved an $O(1/\sqrt{m})$ generalization bound for models learned via adversarial training based on the Rademacher complexity of the deep neural networks. Khim and Loh (2018) also proved an $O(1/\sqrt{m})$ generalization bound based on the Rademacher complexity of the hypothesis space. Tu et al. (2019) proved an $O(1/\sqrt{m})$ generalization bound based on the covering number of the hypothesis space. However, the hypothesis complexity will be prohibitively large in deep learning.

13.2.3.3 Experimental Verification

Our investigation into the trade-off between generalization and robustness is further grounded in empirical analysis, leveraging the ResNet architecture and datasets

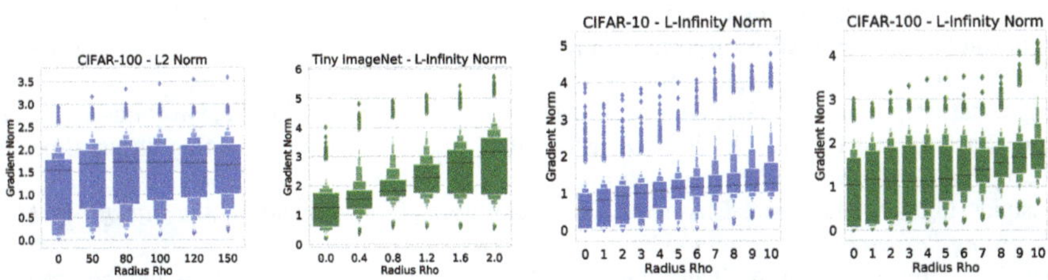

Fig. 13.3 Box plots of the gradient norms in ERM and adversarial training on CIFAR-100 and Tiny ImageNet with six different ρ values and two different norms in PGD. The sample sizes are 55, 000 and 120, 000, respectively.The last two plots correspond to experimental settings A and B, respectively

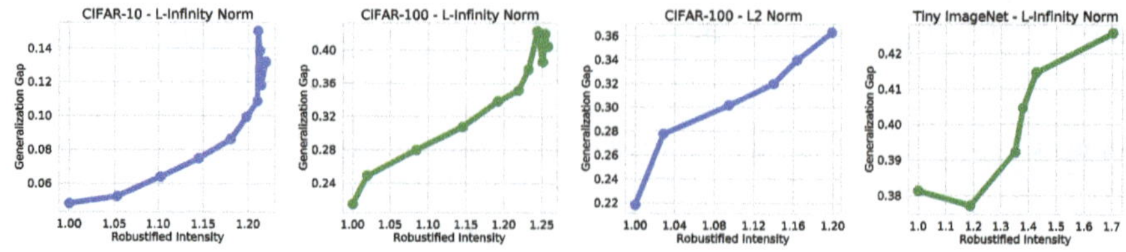

Fig. 13.4 Plots of the generalization gap versus the empirical robustified intensity. For the four plots, the datasets and norms used in PGD are (1) CIFAR-10 and the L_∞ norm, (2) CIFAR-100 and the L_∞ norm, (3) CIFAR-100 and the L_2 norm, and (4) Tiny ImageNet and the L_∞ norm

including CIFAR-10, CIFAR-100, and Tiny ImageNet. To gauge the generalizability of our models, we employ a key metric known as the generalization gap. This metric quantifies the disparity between training and test accuracies, offering valuable insights into the model's ability to perform well on unseen data.

We collected data on both the generalization gaps and empirical robustified intensities across all models, using this information to construct four informative plots presented in Fig. 13.4. A larger generalization gap signifies poorer generalizability of the models to unseen data. Upon scrutinizing these plots, a notable trend emerges: there's a distinct positive correlation between the generalization gap and the empirical robustified intensity. This observation serves as compelling evidence confirming the existence of a trade-off between generalization and robustness in the examined deep learning models.

References

Attias, Idan, Aryeh Kontorovich, and Yishay Mansour. 2019. Improved generalization bounds for robust learning. In *Algorithmic Learning Theory*, 162–183.

Baluja, Shumeet, and Ian Fischer. 2018. Learning to attack: adversarial transformation networks. In *AAAI Conference on Artificial Intelligence*, vol. 1, 3.

Bartlett, Peter L, Dylan J Foster, and Matus J Telgarsky. 2017. Spectrally-normalized margin bounds for neural networks. In *Advances in Neural Information Processing Systems*, 6240–6249.

Bhagoji, Arjun Nitin, Daniel Cullina, and Prateek Mittal. 2019. Lower bounds on adversarial robustness from optimal transport. In *Advances in Neural Information Processing Systems*, 7498–7510.

Biggio, Battista, Igino Corona, Davide Maiorca, Blaine Nelson, Nedim Šrndić, Pavel Laskov, Giorgio Giacinto, and Fabio Roli. 2013. Evasion attacks against machine learning at test time. In *European Conference on Machine Learning*.

Bousquet, Olivier, and André Elisseeff. 2002. Stability and generalization. *Journal of Machine Learning Research* 2 (Mar): 499–526.

Chen, Lin, Yifei Min, Mingrui Zhang, and Amin Karbasi. 2020. More data can expand the generalization gap between adversarially robust and standard models. *arXiv preprint* arXiv:2002.04725.

Cullina, Daniel, Arjun Nitin Bhagoji, and Prateek Mittal. 2018. PAC-learning in the presence of adversaries. In *Advances in Neural Information Processing Systems*, 230–241.

Dai, Hanjun, Hui Li, Tian Tian, Xin Huang, Lin Wang, Jun Zhu, and Le Song. 2018. Adversarial attack on graph structured data. *arXiv preprint* arXiv:1806.02371.

Dwork, Cynthia, and Aaron Roth. 2014. The algorithmic foundations of differential privacy. *Foundations and Trends® in Theoretical Computer Science* 9 (3–4): 211–407.

Dwork, Cynthia, and Deirdre K Mulligan. 2013. It's not privacy, and it's not fair. *Stanford Law Review Online* 66: 35.

Dwork, Cynthia, Vitaly Feldman, Moritz Hardt, Toni Pitassi, Omer Reingold, and Aaron Roth. 2015. Generalization in adaptive data analysis and holdout reuse. In *Advances in Neural Information Processing Systems*, 2350–2358.

Feldman, Vitaly, and Jan Vondrak. 2019. High probability generalization bounds for uniformly stable algorithms with nearly optimal rate. *arXiv preprint* arXiv:1902.10710.

Golowich, Noah, Alexander Rakhlin, and Ohad Shamir. 2018. Size-independent sample complexity of neural networks. In *Annual Conference on Learning Theory*, 297–299.

Goodfellow, Ian J, Jonathon Shlens, and Christian Szegedy. 2014. Explaining and harnessing adversarial examples. *arXiv preprint* arXiv:1412.6572.

Kearns, Michael, and Dana Ron. 1999. Algorithmic stability and sanity-check bounds for leave-one-out cross-validation. *Neural Computation* 11 (6): 1427–1453.

Khim, Justin, and Po-Ling Loh. 2018. Adversarial risk bounds via function transformation. *arXiv preprint* arXiv:1810.09519.

Kingma, Diederik P, and Jimmy Ba. 2015. Adam: a method for stochastic optimization. In *International Conference on Learning Systems*.

Krizhevsky, Alex, and Geoffrey Hinton. 2009. *Learning multiple layers of features from tiny images*. Technical report, Citeseer.

Kushner, Harold, and G George Yin. 2003. *Stochastic Approximation and Recursive Algorithms and Applications*, vol. 35. Springer Science & Business Media.

Le, Ya, and Xuan Yang. 2015. Tiny imagenet visual recognition challenge. *CS 231N* 7 (7): 3.

Li, Bai, Changyou Chen, Wenlin Wang, and Lawrence Carin. 2018. Second-order adversarial attack and certifiable robustness.

Ljung, Lennart, Georg Pflug, and Harro Walk. 2012. *Stochastic Approximation and Optimization of Random Systems*, vol. 17. Birkhäuser.

Mandt, Stephan, Matthew D Hoffman, and David M Blei. 2017. Stochastic gradient descent as approximate Bayesian inference. *Journal of Machine Learning Research* 18 (1): 4873–4907.

Min, Yifei, Lin Chen, and Amin Karbasi. 2020. The curious case of adversarially robust models: More data can help, double descend, or hurt generalization. *arXiv preprint* arXiv:2002.11080.

Mohri, Mehryar, Afshin Rostamizadeh, and Ameet Talwalkar. 2018. *Foundations of Machine Learning*. MIT Press.

Montasser, Omar, Steve Hanneke, and Nathan Srebro. 2019. VC classes are adversarially robustly learnable, but only improperly. In *Conference on Learning Theory*, 2512–2530.

Nesterov, Yurii E. 1983. A method for solving the convex programming problem with convergence rate o (1/k^ 2). In *Dokl. Akad. Nauk Sssr*, vol. 269, 543–547.

Oneto, Luca, Sandro Ridella, and Davide Anguita. 2017. Differential privacy and generalization: Sharper bounds with applications. *Pattern Recognition Letters* 89: 31–38.

Papernot, Nicolas, Patrick McDaniel, Somesh Jha, Matt Fredrikson, Z Berkay Celik, and Ananthram Swami. 2016. The limitations of deep learning in adversarial settings. In *IEEE European Symposium on Security and Privacy*, 372–387.

Pydi, Muni Sreenivas, and Varun Jog. 2020. Adversarial risk via optimal transport and optimal couplings. In *International Conference on Machine Learning*, 7814–7823.

Raghunathan, Aditi, Jacob Steinhardt, and Percy Liang. 2018. Certified defenses against adversarial examples. In *International Conference on Learning Representations*.

Robbins, Herbert, and Sutton Monro. 1951. A stochastic approximation method. *The Annals of Mathematical Statistics*, 400–407.

Rogers, William H, and Terry J Wagner. 1978. A finite sample distribution-free performance bound for local discrimination rules. *The Annals of Statistics*, 506–514.

Schmidt, Ludwig, Shibani Santurkar, Dimitris Tsipras, Kunal Talwar, and Aleksander Madry. 2018. Adversarially robust generalization requires more data. In *Advances in Neural Information Processing Systems*, 5014–5026.

Shalev-Shwartz, Shai, Ohad Shamir, Nathan Srebro, and Karthik Sridharan. 2010. Learnability, stability and uniform convergence. *Journal of Machine Learning Research* 11 (Oct): 2635–2670.

Shokri, Reza, Marco Stronati, Congzheng Song, and Vitaly Shmatikov. 2017. Membership inference attacks against machine learning models. In *IEEE Symposium on Security and Privacy (SP)*, 3–18.

Szegedy, Christian, Wojciech Zaremba, Ilya Sutskever, Joan Bruna, Dumitru Erhan, Ian Goodfellow, and Rob Fergus. 2013. Intriguing properties of neural networks. *arXiv preprint* arXiv:1312.6199.

Szegedy, Christian, Wojciech Zaremba, Ilya Sutskever, Joan Bruna, Dumitru Erhan, Ian Goodfellow, and Rob Fergus. 2014. Intriguing properties of neural networks. In *International Conference on Learning Representations*.

Tseng, Paul. 1998. An incremental gradient (-projection) method with momentum term and adaptive stepsize rule. *SIAM Journal on Optimization* 8 (2): 506–531.

Tu, Zhuozhuo, Jingwei Zhang, and Dacheng Tao. 2019. Theoretical analysis of adversarial learning: a minimax approach. In *Advances in Neural Information Processing Systems*, 12280–12290.

Vapnik, Vladimir. 2013. *The Nature of Statistical Learning Theory*. Springer Science & Business Media.

Yin, Dong, Ramchandran Kannan, and Peter Bartlett. 2019. Rademacher complexity for adversarially robust generalization. In *International Conference on Machine Learning*, 7085–7094.

Zhang, Hongyang, Yaodong Yu, Jiantao Jiao, Eric P Xing, Laurent El Ghaoui, and Michael I Jordan. 2019. Theoretically principled trade-off between robustness and accuracy. *arXiv preprint* arXiv:1901.08573.

Zheng, Tianhang, Changyou Chen, and Kui Ren. 2019. Distributionally adversarial attack. In *AAAI Conference on Artificial Intelligence*, vol. 33, 2253–2260.

www.ingramcontent.com/pod-product-compliance
Lightning Source LLC
Chambersburg PA
CBHW071511160225
22041CB00001B/15